Differential Equations
with Maxima

PURE AND APPLIED MATHEMATICS

A Program of Monographs, Textbooks, and Lecture Notes

MONOGRAPHS AND TEXTBOOKS IN PURE AND APPLIED MATHEMATICS

Recent Titles

Vincenzo Ancona and Bernard Gaveau, Differential Forms on Singular Varieties: De Rham and Hodge Theory Simplified (2005)

Santiago Alves Tavares, Generation of Multivariate Hermite Interpolating Polynomials (2005)

Sergio Macías, Topics on Continua (2005)

Mircea Sofonea, Weimin Han, and Meir Shillor, Analysis and Approximation of Contact Problems with Adhesion or Damage (2006)

Marwan Moubachir and Jean-Paul Zolésio, Moving Shape Analysis and Control: Applications to Fluid Structure Interactions (2006)

Alfred Geroldinger and Franz Halter-Koch, Non-Unique Factorizations: Algebraic, Combinatorial and Analytic Theory (2006)

Kevin J. Hastings, Introduction to the Mathematics of Operations Research with *Mathematica*®, Second Edition (2006)

Robert Carlson, A Concrete Introduction to Real Analysis (2006)

John Dauns and Yiqiang Zhou, Classes of Modules (2006)

N. K. Govil, H. N. Mhaskar, Ram N. Mohapatra, Zuhair Nashed, and J. Szabados, Frontiers in Interpolation and Approximation (2006)

Luca Lorenzi and Marcello Bertoldi, Analytical Methods for Markov Semigroups (2006)

M. A. Al-Gwaiz and S. A. Elsanousi, Elements of Real Analysis (2006)

Theodore G. Faticoni, Direct Sum Decompositions of Torsion-Free Finite Rank Groups (2007)

R. Sivaramakrishnan, Certain Number-Theoretic Episodes in Algebra (2006)

Aderemi Kuku, Representation Theory and Higher Algebraic K-Theory (2006)

Robert Piziak and P. L. Odell, Matrix Theory: From Generalized Inverses to Jordan Form (2007)

Norman L. Johnson, Vikram Jha, and Mauro Biliotti, Handbook of Finite Translation Planes (2007)

Lieven Le Bruyn, Noncommutative Geometry and Cayley-smooth Orders (2008)

Fritz Schwarz, Algorithmic Lie Theory for Solving Ordinary Differential Equations (2008)

Jane Cronin, Ordinary Differential Equations: Introduction and Qualitative Theory, Third Edition (2008)

Su Gao, Invariant Descriptive Set Theory (2009)

Christopher Apelian and Steve Surace, Real and Complex Analysis (2010)

Norman L. Johnson, Combinatorics of Spreads and Parallelisms (2010)

Lawrence Narici and Edward Beckenstein, Topological Vector Spaces, Second Edition (2010)

Moshe Sniedovich, Dynamic Programming: Foundations and Principles, Second Edition (2010)

Drumi D. Bainov and Snezhana G. Hristova, Differential Equations with Maxima (2011)

Differential Equations with Maxima

Drumi D. Bainov

Medical University

Sofia, Bulgaria

Snezhana G. Hristova

Plovdiv University

Plovdiv, Bulgaria

CRC Press

Taylor & Francis Group

Boca Raton London New York

CRC Press is an imprint of the

Taylor & Francis Group, an **informa** business

A CHAPMAN & HALL BOOK

CRC Press
Taylor & Francis Group
6000 Broken Sound Parkway NW, Suite 300
Boca Raton, FL 33487-2742

First issued in paperback 2019

© 2011 by Taylor and Francis Group, LLC
Chapman & Hall/CRC is an imprint of Taylor & Francis Group, an Informa business

No claim to original U.S. Government works

ISBN-13: 978-1-4398-6757-0 (hbk)
ISBN-13: 978-0-367-38282-7 (pbk)

Visit the Taylor & Francis Web site at
http://www.taylorandfrancis.com

and the CRC Press Web site at
http://www.crcpress.com

Contents

Preface

The current book is an introduction to the theory of differential equations with "maxima." Differential equations with "maxima" are a special type of differential equations that contain the maximum of the unknown function over a previous interval(s). Such equations adequately model real world processes whose present state significantly depends on the maximum value of the state on a past time interval. For example, in the theory of automatic control in various technical systems, often the law of regulation depends on the maximum values of some regulated state parameters over certain time intervals and their behavior is modeled by differential equations with "maxima." Recently, the interest in differential equations with "maxima" has increased exponentially. The theoretical results and investigations of differential equations with "maxima" opens the door to enormous possibilities for their applications to real world processes and phenomena.

This book presents the qualitative theory and develops some approximate methods for differential equations with "maxima."

Chapter 2 gives an introduction to the mathematical apparatus of integral inequalities, involving maxima of unknown functions. Different types of linear and nonlinear integral inequalities with "maxima" are solved. Both cases of single integral inequalities and double integral inequalities are studied. Several direct applications of the solved inequalities are illustrated on various types of differential equations with "maxima."

In Chapter 3 are studied some general properties of the solutions of differential equations with "maxima." Several existence results for initial value problems and boundary value problems are presented.

In Chapter 4 several stability results for differential equations with "maxima" are given. The investigations are based on appropriate modifications of the Razumikhin technique by applying Lyapunov functions.

Appropriate definitions about different types of stability are given and sufficient conditions are obtained.

Chapter 5 deals with the theory of oscillation for differential equations with "maxima." The asymptotic and oscillatory behavior of solutions of n-th order differential equations with "maxima" is studied. Several sufficient conditions for oscillation as well as almost oscillation are obtained. Several differential equations with "maxima" and their corresponding delay differential equations are examined and the oscillatory properties of their solutions are studied. The influence of the presence of maxima function on the behavior of the solutions is demonstrated.

In Chapter 6 two approximate methods for solving differential equations with "maxima" are applied to initial value problems as well as boundary value problems for differential equations with "maxima." The considered methods combine the method of lower and upper solutions with appropriate monotone methods. Algorithms for constructing sequences of successive approximation to the solutions are introduced. Each term of the constructed sequences is a solution of an appropriately chosen linear equation.

In Chapter 7 a systematic development of the average method for differential equations with "maxima" is given. This method is applied to first-order differential equations with "maxima" and neutral differential equations with "maxima." Different schemes for averaging, such as ordinary averaging, partial averaging, partially additive averaging, and partially multiplicative averaging are suggested.

This book, being the first one in the field, gives a good overview of the entire field of differential equations with "maxima" and serves as a stimulating guide for the theoretical and applied researchers in mathematics. It is a helpful tool for further investigations and applications of these equations for better and more adequate studying of real world problems.

The current book is intended for a wide audience, including mathematicians, applied researchers and practitioners, whose interest extends beyond the boundaries of qualitative analysis of well-known differential equations.

Sofia, Bulgaria   *Drumi D. Bainov*
July, 2010   *Snezhana G. Hristova*

Chapter 1

Introduction

Differential equations are a basic yet powerful mathematical apparatus for studying real world objects and phenomena. These equations, when used as models and combined with information technology tools, allow us to conduct theoretical investigations and to predict the behavior of real systems. However, since real life processes are quite intricate, complex mathematical equations are required to study them. One natural setting is the case of evolutionary equations using past history. These equations are now referred to as functional differential equations. Picard (1908), at the international congress of mathematicians, stressed the importance of functional differential equations in physical problems. Interest in this topic increased after 1940, but there was little progress on qualitative theory until the mid-1950s. Perhaps the reason was that even though the evolution of the system at present time was determined by some of the past history, the primary object of study was where the system was at the present time. In 1956 Krasovskii made the important observation that the state of a system at any time described by a functional differential equations should be the system at that time together with the past history that is required to determine the future evolution of the system. With his contribution, he made a huge impact on the qualitative investigations of functional differential equations. By the early 1970s, a framework for the qualitative theory of functional differential equations had been outlined. In the last decades, different qualitative properties of the solutions of functional differential equations have been obtained (see, for example, monographs [El'sgol'ts and Norkin 1973], [Hale 1977], [Hale and Lune 1993], [Kolmanovski and Nosov 1986] and references cited therein).

One special type of functional differential equations is the case when the evolutionary equations use the maximum of the studied function on a past time interval. Since the maximum function has very specific properties, it makes the equations strongly nonlinear. As a result these equations gain an important place in the theory of differential equations and are called differential equations with "maxima." Differential equations with "maxima" first appeared as an object of investigation about thirty years ago in connection with modeling of some applied problems.

For example, in the theory of automatic control of various technical systems, it often occurs that the law of regulation depends on the maximum values of some regulated state parameters over certain time intervals. E. P. Popov (see [Popov 1966]) in 1966 considered the system for regulating the voltage of a generator of constant current. The object of the experiment was a generator of constant current with parallel simulation and the regulated quantity was the voltage at the source electric current. The equation describing the work of the regulator involves the maximum of the unknown function and it has the form (see [Popov 1966])

$$T_0 u'(t) + u(t) + q \max_{s \in [t-h,t]} u(s) = f(t),$$

where T_0 and q are constants characterizing the object, $u(t)$ is the regulated voltage and $f(t)$ is the perturbed effect.

Note that some modifications of the above differential equation are used to model the vision process in the compound eye ([Hadeler 1979]), the Hausrath equation (see [Hale 1977]). Also differential equations are used in optimal control theory in [Kichmarenko 2006] and [Plotnikov and Kichmarenko 2006].

On the other hand, it is relevant to mention here the opinion of A. D. Myshkis, who in his survey ([Myshkis 1977]) also distinguished the equations with "maxima" as differential equations with deviating argument of complex structure.

At the same time differential equations with "maxima" have very different properties than the well-known in the literature differential equations with delay. For example, let us consider the following two scalar equations:

(A) differential equation with delay

$$x' = \Big(x\big(t - \tau(t)\big) \Big)^2,$$

where the function $\tau \in C(\mathbb{R}_+, \mathbb{R}_+)$, $\tau(t) \not\equiv 0$;

(B) differential equation with "maxima"

$$x' = \left(\max_{s \in [t-h,t]} x(s) \right)^2,$$

where h is a positive constant.

Note that equation (A) seems to be very similar to (B), especially in the case of $\tau(t) \equiv h$. In both equations the right part is nonnegative and the solution $x(t)$ is nondecreasing. However, the equation (B) reduces to the ordinary differential equation $x' = x^2$ and the initial condition is required only at one single point. On the other hand, the equation (A) could not be reduced to an ordinary differential equation for any nontrivial function $\tau(t)$.

Generally, differential equations with "maxima" are characterized by two main parts:

1. differential equations;

2. maximum of the unknown function over a past time interval.

The first part, differential equations, could be ordinary differential equations of any order, linear or nonlinear, partial differential equations, etc.

The second part makes the set of differential equations with "maxima" too wide since the maximum of the unknown function $x(t)$ could be given

- on an interval with fixed length, i.e., $\max_{s \in [t-r,t]} x(s)$, $r = const > 0$;

- on a retarded interval with variable length, i.e., $\max_{s \in [\sigma(t), \tau(t)]} x(s)$, where $\sigma(t) \leq \tau(t) \leq t$;

- on several different intervals with fixed lengths or variable lengths; etc.

Remark 1.0.1. *If $\sigma(t) = \tau(t)$ for some value(s) of t from the domain of the functions, then we will assume that $\max_{s \in [\sigma(t), \tau(t)]} x(s) = x(\tau(t))$.*

We will give a brief description of the main types of differential equations with "maxima."

Let first order ordinary differential equations be used to describe the differential equations with "maxima:"

$$x' = f\left(t, x(t), \max_{s \in S(t)} x(s)\right), \quad t \in [a, b), \tag{1.1}$$

where $x \in \mathbb{R}^n$, $S(t) = [\sigma(t), \tau(t)]$, σ, $\tau : \mathbb{R} \to \mathbb{R}$, $a =$const., $b \leq \infty$.

Note that in the multidimensional case $x \in \mathbb{R}^n$, $x = (x_1, x_2, \ldots, x_n)$, the following notation

$$\max_{s \in S(t)} x(s) = \left(\max_{s \in S(t)} x_1(s), \max_{s \in S(t)} x_2(s), \ldots, \max_{s \in S(t)} x_n(s)\right)$$

is used.

The right side of the equation (1.1) could be very complex and the initial conditions for the differential equations with "maxima" depend significantly on the type of the interval $S(t)$.

We emphasize that if $S(t) \subset [a, b)$ for all $t \in [a, b)$, then the initial condition to the differential equation with "maxima" (1.1) will be $x(a) = x_0$. In the opposite case, however, one can expect different types of additional initial conditions depending on applications. For example, it can be required that $x(t) = \varphi(t)$ outside the interval $[a, b)$, where $\varphi \in \mathbb{R}^n$ is a given function, or $x(t) = x(a)$ for $t \leq a$ and $x(t) = x(b)$ for $t \geq b$ (if $b < \infty$). Furthermore, the right side of the differential equation in (1.1) can also be more complex, e.g., there can occur the dependence on maximum values of different components of the state vector x on different time intervals. It is also worth noting that in some real systems the law of regulation depends only on the past and present state, and thus $S(t) \subset (-\infty, t]$.

In this book we will consider only the case of retarded intervals, i.e., $S(t) \subset (-\infty, t]$, i.e., $\sigma(t) \leq \tau(t) \leq t$ for all values of t from the domain of the functions $\sigma(t)$, $\tau(t)$.

Now we will give some particular cases of differential equations with "maxima" (1.1).

Let $h > 0$ be a constant, $\sigma(t) \equiv t - h$, $\tau(t) \equiv t$. Then $S(t) = [t - h, t]$ and the differential equation with "maxima" is written in the form

$$x' = f\left(t, x(t), \max_{s \in [t-h, t]} x(s)\right), \quad t \geq t_0. \tag{1.2}$$

The initial condition for the equation (1.2) is in the form

$$x(t) = \varphi(t), \quad t \in [t_0 - h, t_0], \tag{1.3}$$

where $\varphi : [t_0 - h, t_0] \to \mathbb{R}$.

Let $t_0 \geq 0$ be a given point, functions $\sigma(t)$, $\tau(t) \in C(\mathbb{R}_+, \mathbb{R}_+)$ be such that $\sigma(t) \leq \tau(t) \leq t$ for $t \geq 0$. Then the differential equation with "maxima" (1.1) is written in the form

$$x' = f\left(t, x(t), \max_{s \in [\sigma(t), \tau(t)]} x(s)\right), \quad t \geq t_0. \tag{1.4}$$

The initial condition for the equation (1.4) is in the form

$$x(t) = \varphi(t), \quad t \in I_{t_0}, \tag{1.5}$$

where $\varphi : I_{t_0} \to \mathbb{R}$, I_{t_0} is an initial interval, which depends significantly on the functions $\sigma(t)$ and $\tau(t)$.

We will consider some particular examples of functions $\sigma(t)$ and $\tau(t)$ to illustrate the complexity of the initial interval I_{t_0} in condition (1.5).

Let $\sigma(t) = \sin(t)$, $\tau(t) = |\sin(t)|$, $t_0 = \frac{\pi}{2}$.

On the interval $[\frac{\pi}{2}, \pi]$ the equation (1.4) reduces to the following functional differential equation

$$x' = f(t, x(t), x(\sin(t))).$$

On the interval $[\pi, 2\pi]$ the equation (1.4) has the form

$$x' = f\left(t, x(t), \max_{s \in [\sin(t), -\sin(t)]} x(s)\right).$$

Since the interval $[\sin(t), -\sin(t)] \subseteq [-1, 1]$ for $t \in [\pi, 2\pi]$ and $\sin(t) \in [0, 1]$ for $t \in [\frac{\pi}{2}, \pi]$, in this case the initial interval for the equation (1.4) is $I_{t_0} \equiv [-1, 1] \cup \{\frac{\pi}{2}\}$, i.e., the initial condition (1.5) is $x(t) = \varphi(t)$ for $t \in [-1, 1]$, $x(\frac{\pi}{2}) = x_0$.

Note that if the function $x(t)$ is an increasing/decreasing function, then its maximum is at the right/left end of the interval and the differential equation with "maxima" (1.1) reduces to the well-known in the literature differential equation.

Note that if first order differential equations are used and the maximum of the first derivative of the unknown function is included in the right-hand side of the equations, then the differential equations with "maxima" could be written in the form:

$$x' = f\left(t, x(t), \max_{s \in S(t)} x(s), \max_{s \in S(t)} x'(s)\right), \quad t \geq t_0, \tag{1.6}$$

where $x \in \mathbb{R}^n$, $x = (x_1, x_2, \ldots, x_n)$ and the notation

$$\max_{s \in S(t)} x'(s) = \left(\max_{s \in S(t)} x_1'(s), \max_{s \in S(t)} x_2'(s), \ldots, \max_{s \in S(t)} x_n'(s)\right)$$

is used.

The initial condition for the equation (1.6) could be

$$x(t) = \varphi(t), \quad x'(t) = \varphi'(t), \qquad t \in I_{t_0}, \tag{1.7}$$

where I_{t_0} is the initial interval.

We will illustrate the influence of the maximum of the function on the behavior of the solution.

Example 1. Consider the scalar differential equation with "maxima"

$$x'(t) = \max_{s \in [t-\frac{\pi}{2}, t]} x(s) \quad \text{for } t \geq 0, \tag{1.8}$$

with the initial condition

$$x(t) = -a\sin(t) \quad \text{for } t \in \left[-\frac{\pi}{2}, 0\right], \tag{1.9}$$

where $a > 0$ is a constant.

The solution of the initial value problem (1.8), (1.9) is given by

$$x(t) = \begin{cases} -a\sin(t) & \text{for } t \in \left[-\frac{\pi}{2}, 0\right], \\ a\sin(t) & \text{for } t \in \left[0, \frac{\pi}{4}\right], \\ \frac{a\sqrt{2}}{2} e^{t-\frac{\pi}{4}} & \text{for } t \geq \frac{\pi}{4}. \end{cases} \tag{1.10}$$

The solution of the initial value problem (1.8), (1.9) is unbounded.

Now let us consider the differential equation with "maxima" (1.8) with initial condition

$$x(t) = a\sin(t) \quad \text{for } t \in \left[-\frac{\pi}{2}, 0\right], \tag{1.11}$$

where $a > 0$ is a constant.

The solution of the initial value problems (1.8) and (1.11) is given by

$$x(t) = \begin{cases} -a\sin(t) & \text{for } t \in \left[-\frac{\pi}{2}, 0\right], \\ 0 & \text{for } t \geq 0. \end{cases} \tag{1.12}$$

Now let us consider the differential equation with delay

$$x'(t) = x\left(t - \frac{\pi}{2}\right) \quad \text{for } t \geq 0, \tag{1.13}$$

with initial condition (1.9). The solution of the initial value problems (1.9) and (1.13) is $x(t) = a\sin(t)$ for $t \geq -\frac{\pi}{2}$ and it is bounded. $\square$

The above example illustrates the differences between the behavior of the solutions of the differential equations with "maxima" and differential equations with delay. It proves the necessity of separately deeply studying the properties of solutions of differential equations with "maxima."

We emphasize that several types of differential equations, known in the literature, could be obtained as partial cases of differential equations with "maxima:"

1. if $\sigma(t) \equiv \tau(t) \equiv t$, then the differential equations with "maxima" (1.6) reduce to ordinary differential equations $x' = f(t, x(t), x(t))$;

2. if $\sigma(t) \equiv \tau(t) \equiv t - h$, $h = const > 0$, then the differential equations with "maxima" (1.6) reduce to differential equations with a constant delay $x' = f(t, x(t), x(t - h))$;

3. if $\sigma(t) \equiv \tau(t) \leq t$ for all $t \geq t_0$, then the differential equations with "maxima" (1.6) reduce to differential equations with variable delay $x' = f(t, x(t), x(\tau(t)))$.

In the last few decades the mathematical importance of various types of differential equations with "maxima" has grown exponentially. It is mostly due to their applications in investigations of multiple problems in optimal control theory. The study of equations which includes a maximum of the unknown function is spread to various types of differential and difference equations. We will mention only some of them:

- *differential equations with "maxima:"* Some properties of the solutions are studied by A. R. Magomedov et al. ([Magomedov 1993], [Magomedov and Ryabov 1975], [Magomedov and Nabiev 1986], [Rjabov and Magomedov 1978]). Differential equations with "maxima" are also studied by D. D. Bainov and his scholars, and several properties are investigated such as:

  – oscillatory properties (see [Bainov et al. 1997], [Bainov et al. 1995e], [Bainov and Zahariev 1984], [Dontchev et al. 2010a], [Kolev et al. 2010a], [Kolev et al. 2010b], [Kolev and Markova 2010], [Markova and Nenov 2010], [Markova and Simeonov 2010]);

  – stability (see [Bainov and Hristova 2010], [Henderson and Hristova 2010], [Hristova 2009b], [Hristova 2010d], [Hristova and Gluhcheva 2010], [Voulov 1995], [Voulov and Bainov 1991]);

- periodicity (see [Arolska and Bainov 1980], [Teryokhin and Kiryushkin 2010]);

- approximate solutions (see [Bainov and Hristova 1995], [Bainov and Kazakova 1992], [Bainov and Sarafova 1981], [Golev et al. 2010], [Sarafova and Bainov 1981], [Simeonov and Bainov 1985]); averaging (see [Bainov and Milusheva 1983], [Milusheva and Bainov 1986a], [Milusheva and Bainov 1986b], [Plotnokov and Kichmarenko 2009], [Plotnikov and Kichmarenko 2006], [Plotnikov and Kichmarenko 2002]).

Qualitative properties of the solutions of differential equations with "maxima" are also studied by many other authors (see [Bantsur and Trofimchuk 1998], [Gonzalez and Pinto 2007], [Gonzalez and Pinto 2002], [Hristova 1982], [Hristova and Roberts 2000a], [Jankowski 2002], [Jankowski 1997], [Kichmarenko 2009], [Muntyan and Shpakovich 1987], [Nabiev 1985], [Nabiev 1984b], [Otrocol and Ioan 2008a], [Otrocol and Ioan 2008b], [Petrov 1998], [Plotnikov and Kichmarenko 2002], [Ronto 1999], [Samoilenko et al. 1998], [Sarafova 1984], [Shabadikov and Yuldashev 1989], [Shpakovich and Muntyan 1987], [Shpakovich and Muntyan 1986], [Sobeih and Aly 1991], [Stepanov 1997], [Voulov 1995], [Voulov 1991], [Xu and Liz 1996], [Yuldashev 1995], [Zhang and Petrov 2000], [Zhang and Zhang 2000], [Zhang and Zhang 1999]);

- *integro-differential equations with "maxima:"* Some results are published in [Bainov et al. 1993], [Muntyan 1987];

- *partial differential equations with "maxima:"* Qualitative investigations of solutions are obtained in [Bainov and Minchev 1999], [Bainov and Minchev 1998], [Bainov and Mishev 1991], [Donchev et al. 2010b], [Mishev 1989], [Mishev 1990], [Mishev 1986], [Mishev and Musa 2007];

- *impulsive differential equations with supremum*: At the beginning of the 1990s, D.D. Bainov and his collaborators S. Hristova and S. Milusheva combined the ideas of impulsive differential equations with differential equations with "maxima" and initiated the investigations of these types of equations. These equations are adequate models of real processes whose present state has instantaneous changes at certain moments and it depends significantly on the maximal value of the state on a past time interval. Several results for impulsive differential equations with supremum are obtained in [Bainov et al. 1996], [Bainov et al. 1994a], [Bainov et al. 1994b], [Bainov et al. 1993], [He et al. 2003], [Hristova 2010b], [Hristova 2010c], [Hristova 2010d], [Hristova 2009c], [Hristova 2009d], [Hristova 2000], [Hristova and Bainov 1993], [Hristova and Bainov 1991], [Hristova and Markova 2010], [Hristova and Roberts 2001], [Milusheva and Bainov 1991], [Oyelami and Ale 2010], [Qi 2004], [Qi and Chen 2008], [Shi and Wang 2010];

- *integral equations with supremum*: Some properties of the solutions are studied in [Caballero et al. 2005], [Darwish 2008];

- *difference equations with "maxima:"* Some investigations of properties of the solutions are done in [Atici et al. 2006], [Atici et al. 2002], [Berenhaut et al. 2006], [Cinar et al. 2005], [Fan 2004], [Fan et al. 2005], [Gelisken and Cinar 2009], [Gelisken et al. 2010], [Gelisken et al. 2008] , [Iricanin and Elsayed 2010], [Li et al. 2008], [Li and Zhou 2007], [Liu et al. 2006], [Liz et al. 2003], [Luo 2000], [Luo and Bainov 2001], [Migda and Zhang 2006], [Mishev et al. 2002], [Stevic 2010], [Stevic 2009], [Sun 2008], [Voulov 2008], [Voulov 2003], [Yalcinkaya et al. 2007], [Yang et al. 2006], [Wu 2003], [Zhang and Liu 2007];

- *integral equations with "maxima" and fractional derivatives*: For some results see [El-Borai at al. 2006];

- *stochatic differential equations with "maxima."* Some theoretical results and applications to the financial market are done in [Appleby and Wu 2008].

The purpose of this monograph is to present a general method of analysis for nonlinear differential equations with "maxima." This monograph is the first in which basic qualitative results for differential equations with "maxima" are given as well as some approximate methods for their solution are presented.

In Chapter 2, we develop the method of integral inequalities as a mathematical tool in the investigations of qualitative properties of the solutions of differential equations with "maxima." In connections with the main goal of the book, different types of integral inequalities involving the maximum of the unknown functions are solved. These results generalize the classical integral inequalities of Gronwall–Bellman [Gronwall 1919], [Bellman 1943] and Bihari types [Bihari 1956], which are well studied for ordinary differential equations and delay differential equations (see, for example, [Bainov and Simeonov 1989], [Lakshmikantham and Leela 1969], [Walter 1970] and references cited therein). At the same time, the investigation of differential equations with "maxima" requires a new type of inequalities, since the known ones are not applicable. Several results for integral inequalities containing a maximum value of the unknown function are solved in (see [Hristova and Stefanova 2010a], [Hristova and Stefanova 2010b]).

Section 2.1 deals with linear integral inequalities that involve the maximum of the unknown function. In Section 2.2 several nonlinear integral inequalities for scalar functions are studied. Linear and nonlinear integral inequalities involving a maximum of the unknown scalar function of two variables are solved in Section 2.3. Note that the integral inequalities are successfully employed for studying existence, uniqueness, continuous dependence, comparison, perturbation, boundedness, and stability of the solutions of differential and integral equations. Section 2.4 demonstrates the direct applications of the integral inequalities solved in previous sections for investigation of some properties of the solutions of differential equations with "maxima." Nonlinear differential equations with "maxima" and partial differential equations with "maxima" are studied.

Chapter 3 introduces some fundamental concepts and theorems in the qualitative analysis of the solutions of several types of differential equations with "maxima." Some qualitative properties of the solutions of different types of differential equations with "maxima," such as existence and continuous dependence, are studied in [Angelov and Bainov 1983], [Bantsur and Trofimchuk 1998], [Georgiev and Angelov 2002], [Gonzalez and Pinto 2007], [Gonzalez and Pinto 2002], [Muntyan and Shpakovich 1987], [Sobeih and Aly 1991], [Voulov and Bainov 1995], [Zhang and Zhang 1999].

In Section 3.1 some existence results for initial value problems for differential equations with "maxima" are given. Both scalar case and multidimensional case are studied.

In Section 3.2 the existence of solutions of a boundary value problem for first order nonlinear differential equations with "maxima" is proven. The method of a priori estimate, based on the Leray-Schader topological degree theory, is developed and applied to the main proofs.

In Section 3.3 the existence, uniqueness, and data dependence results for the initial value problem for a nonlinear scalar differential equation with "maxima" are studied. The main proofs are based on the weakly Picard operator theory.

In Chapter 4, we establish notations, definitions, and develop stability theory for nonlinear differential equations with "maxima." Our focus is on Lyapunov functions used as means to investigate stability properties of solutions of differential equations with "maxima." A. M. Lyapunov introduced and began systematically to apply this special type of functions in his famous studies [Lyapunov 1956]. Later, B. S. Razumikhin proposed in 1956 (see [Razumikhin 1956]) a method to investigate the stability of solutions of systems with delays. This method is based on the application of Lyapunov functions in combination with the general concept of "impossibility of the first breakdown." The method allowed to "rehabilitate" an application of Lyapunov functions to functional differential equations to a considerable extent. Such application was found in some cases to be simpler and more visual than an application of general functional. As a result the method was later developed both by Razumikhin himself (in the most explained form in monograph [Razumikhin 1988]) and by other authors (see monographs [El'sgol'ts and Norkin 1973], [Hale 1977], [Hale and Lune 1993], papers [Myshkis 1977], [Myshkis 1995]). In Chapter 4, the extension of the Razumikhin method to differential equations with "maxima" is studied. In order to generalize considerations, two different measures for the initial data and for the solutions are used and fundamental results for different types of stability are obtained.

Section 4.1 deals with the stability and uniform stability of differential equations with "maxima." Several sufficient conditions, based on Lyapunov's functions and comparison results, are obtained. Both cases of regular norm and two different measures are studied.

In Section 4.2 the definition of integral stability in terms of two measures is modified to differential equations with "maxima" and several sufficient conditions are given. Integral stability for ordinary differential equations was introduced by I. Vrkoc (see [Vrkoc 1959]) and later studied for various types of differential equations by many authors (see, for example, [Hristova 2010a], [Hristova 2009a], [Soliman and Abdalla 2008]). The concept of integral stability occurs in connection with the stability under persistent perturbations when the perturbations are small enough everywhere except on a small interval. The presence of maximum in the equation requires initially a well-defined and proven comparison result. In this section scalar comparison differential equations and perturbed Lyapunov functions are used.

In Section 4.3 the case of cone valued Lyapunov functions is studied. A new type of stability is defined. The introduced stability is based on the application of two different measures and a dot product on a cone. The fixed vector included in the definitions serves as a proxy for the weight of the components of the solution. Several sufficient conditions are obtained. Some examples illustrate the advantages of the introduced type of stability.

In Section 4.4 the practical stability and the eventual practical stability for differential equations with "maxima" is defined. The definitions are based on the application of two different measures as well as on the application of a scalar product on a cone. This allows us to use cone valued Lyapunov functions for investigation of stability properties of the solutions. Some sufficient conditions for d-practical stability and for d-eventual practical stability in terms of two measures of nonlinear differential equations with "maxima" are obtained. An example illustrates the application of some of the proven results.

The main purpose of Chapter 5 is to give a brief overview of the oscillation theory of differential equations with "maxima." In recent years the literature on oscillation theory of differential equations has started growing very fast. This applies even more so for neutral delay differential equations which is a relatively new field with intriguing applications in real world problems. In fact, the neutral delay differential equations appear in modeling of the networks containing lossless transmission lines (as in high-speed computers where the lossless transmission lines are used to interconnect switching circuits), in the study of vibrating masses attached to an elastic bar, as the Euler equation in some variation problems, theory of automatic control and in neuromechanical systems in which inertia plays an important role (see [Hale 1977], [Popov 1966] and reference cited therein). The systematic investigation of the oscillatory properties and asymptotic behavior of the solutions of functional-differential equations has been published in some literature already (see [Bainov and Mishev 1991], [Gopalsamy 1992], [Ladde et al. 1987]). Also several works about oscillatory properties of the solutions of neutral differential equations with "maxima" have been conducted by Bainov et al. (see [Bainov et al. 1997], [Bainov et al. 1995e]).

In Section 5.1 both linear delay differential equations with "maxima" and corresponding linear delay-differential equation are studied. The behavior of their solutions is investigated. The influence of the presence of maximum of the solution in the right side of the equation on the oscillatory behavior of the solutions is demonstrated.

The main goal of Section 5.2 is to comprehensively discuss the oscillation and nonoscillation of differential equations with "maxima."

Sections 5.3 and 5.4 are concerned with n-th order differential equations with "maxima." The asymptotic and oscillatory behavior of their solutions is studied. Several sufficient conditions for oscillation as well as almost oscillation are obtained.

In Section 5.5 sufficient conditions for oscillation of all bounded solutions of differential inequalities with "maxima" are obtained.

Chapter 6 deals with some approximate methods for solving various types of problems for differential equations with "maxima." Note that the finite difference method for differential equations with "maxima" is considered in [Kazakova 1990a], [Kazakova 1990b]. In the book two different methods, based on an application of the method of lower and upper solutions, are analyzed.

In Sections 6.1 through 6.3 a monotone iterative technique is applied to initial value problems and periodic boundary value problems for first and second order differential equations with "maxima." The method of upper and lower solutions together with a monotone iterative technique offers monotone sequences which converge to the solution of the considered problem. This technique is precisely developed to ordinary differential equations by G. S. Ladde, V. Lakshmikantham and A. Vatsala (see [Ladde et al. 1985]).

In Sections 6.4 and 6.5 the method of quasilinearization is applied to first order scalar nonlinear differential equations with "maxima." The origin of this method lies in the theory of dynamic programming and was initially applied by R. Belman and R. Kalaba (see [Bellman and Kalaba 1965]). A systematic development of this method to ordinary differential equations is done by V. Lakshmikantham and A. S. Vatsala (see [Lakshmikantham and Vatsala 1998]). The main advantage of this method is the ability to find easily the successive approximations of the unknown solution as well as the quadratic convergence. In Section 6.4 an initial value problem for differential equations with "maxima" is considered. An algorithm for an appropriate construction of the initial conditions and the linear differential equations with "maxima," whose unique solutions are successive approximations, is given. The quadratic convergence of the successive approximations is proven. An example illustrates the application of the suggested algorithm. In Section 6.5 a boundary value problem is studied and a procedure for obtaining two monotone quadratically convergent sequences is suggested. Each term of both sequences is equal to the unique solution of an appropriately constructed boundary value problem for a linear differential equation with "maxima."

In Chapter 7, the averaging method is generalized to differential equations with "maxima." The study of qualitative and approximate properties of the solutions of differential equations with "maxima" is subject to specific difficulties. At the same time, the preliminary applications of the methods for theoretical approximation of the solutions may considerably simplify the problem. One very powerful method for theoretical approximation is the averaging method. In this chapter several different schemes for averaging are suggested and applied to various types of problems for differential equations with "maxima." The main characteristic of the studied differential equations is the presence of the maximum of the unknown functions as well as the maximum of its derivative in the right side of the equations. At the same time the averaged equations do not contain the maximum of the unknown function.

Sections 7.1 is devoted to the justification of the averaging method for an initial value problem for a nonlinear system of differential equations with "maxima." Section 7.2 deals with the averaging method of a multipoint boundary value problem associated to differential equation with "maxima." In both suggested schemes the averaged equations are ordinary differential equations, whose solutions could be obtained comparatively easier than the given equations with "maxima."

In Sections 7.3 and 7.4 several schemes for various types of partial averaging are suggested. Regular partial averaging, partially additive averaging, and partially multiplicative averaging method are applied to differential equations with "maxima."

Throughout the book we use the notation

$$\mathbb{R} = (-\infty, \infty), \quad \mathbb{R}_+ = [0, \infty), \quad \mathbb{R}_- = (-\infty, 0],$$
$$\mathbb{R}^n_+ = \mathbb{R}_+ \times \cdots \times \mathbb{R}_+,$$

$Dom(f)$ is the domain of the function f,

$\overline{1, n}$ is the set of all naturals from 1 to n inclusive.

Acknowledgments. The authors are immensely thankful to Prof. Dr. A. Dishliev, Prof. Dr. D. Kolev, Prof. Dr. M. Konstantinov and Prof. Dr. S. Nenov for proofreading of the manuscript and their valuable suggestions.

Chapter 2

Integral Inequalities with Maxima

Integral inequalities which provide explicit bounds for unknown functions play a fundamental role in the development of the theory of differential and integral equations. In the past few years, a number of integral inequalities have been established by many scholars who are motivated by certain applications such as existence, uniqueness, continuous dependence, comparison, perturbation, boundedness, and stability of solutions of differential and integral equations. Among these integral inequalities, we cite the famous Gronwall inequality and its different generalizations (see for example [Bainov and Simeonov 1989] and the references cited therein).

In connection with the development of the theory of differential equations with "maxima," a new type of integral inequalities is required. The main purpose of this chapter is to establish some new integral inequalities in the case when a maximum of the unknown scalar function is involved into the integral.

In this chapter we will assume that $t_0 \geq 0$ and $T \geq t_0$ are fixed points. Note that T could be equal to ∞.

2.1 Linear Integral Inequalities with Maxima for Scalar Functions of One Variable

We will solve several linear integral inequalities of Gronwall's type whose main characteristic is the presence of the maximum of the unknown function in the integral.

Theorem 2.1.1. *Let the following conditions be fulfilled:*

1. *The function $\alpha \in C^1([t_0, T), \mathbb{R}_+)$ is nondecreasing and $\alpha(t) \leq t$.*

2. *The functions $p, q \in C([t_0, T), \mathbb{R}_+)$ and $a, b \in C([\alpha(t_0), T), \mathbb{R}_+)$.*

3. *The function $\phi \in C([\alpha(t_0) - h, t_0], \mathbb{R}_+)$.*

4. *The function $u \in C([\alpha(t_0) - h, T), \mathbb{R}_+)$ satisfies the inequalities*

$$u(t) \leq k + \int_{t_0}^t [p(s)u(s) + q(s) \max_{\xi \in [s-h,s]} u(\xi)]ds$$

$$+ \int_{\alpha(t_0)}^{\alpha(t)} [a(s)u(s) + b(s) \max_{\xi \in [s-h,s]} u(\xi)]ds \quad \text{for } t \in [t_0, T),$$

$$(2.1)$$

$$u(t) \leq \phi(t) \quad \text{for } t \in [\alpha(t_0) - h, t_0], \qquad\qquad (2.2)$$

where $h = const \geq 0$, $k = const \geq 0$.

Then for $t \in [t_0, T)$ the inequality

$$u(t) \leq M \exp\left(\int_{t_0}^t \Big[p(s) + q(s)\Big] ds + \int_{\alpha(t_0)}^{\alpha(t)} \Big[a(s) + b(s)\Big] ds \right) \quad (2.3)$$

holds, where $M = \max\left(k, \max_{s \in [\alpha(t_0)-h, t_0]} \phi(s)\right)$.

Proof. Let us define a function $v : [\alpha(t_0) - h, T) \to \mathbb{R}_+$ by

$$v(t) = \begin{cases} M + \int_{t_0}^t [p(s)u(s) + q(s) \max_{\xi \in [s-h,s]} u(\xi)]ds \\ + \int_{\alpha(t_0)}^{\alpha(t)} [a(s)u(s) + b(s) \max_{\xi \in [s-h,s]} u(\xi)]ds, & t \in [t_0, T), \\ M, & t \in [\alpha(t_0) - h, t_0]. \end{cases}$$

Note the function $v(t)$ is nondecreasing, $u(t) \leq v(t)$ for $t \in [\alpha(t_0) - h, T)$ and $\max_{s \in [t-h,t]} v(s) = v(t)$ for $t \in [\alpha(t_0), T)$. Then from inequality (2.1) we get for $t \in [t_0, T)$

$$
v(t) \leq M + \int_{t_0}^{t} [p(s)v(s) + q(s) \max_{\xi \in [s-h,s]} v(\xi)]ds
$$

$$
+ \int_{\alpha(t_0)}^{\alpha(t)} [a(s)v(s) + b(s) \max_{\xi \in [s-h,s]} v(\xi)]ds \qquad (2.4)
$$

$$
= M + \int_{t_0}^{t} [p(s) + q(s)]v(s)ds + \int_{\alpha(t_0)}^{\alpha(t)} [a(s) + b(s)]v(s)ds.
$$

Change the variable $s = \alpha(\eta)$ in the second integral of inequality (2.4), use condition 1 of Theorem 2.1.1 for the function $\alpha(t)$ and we obtain

$$
v(t) \leq M + \int_{t_0}^{t} \left[p(s) + q(s) \right] v(s)ds
$$

$$
+ \int_{t_0}^{t} \left[a(\alpha(\eta)) + b(\alpha(\eta)) \right] v(\alpha(\eta))\alpha'(\eta)d\eta \qquad (2.5)
$$

$$
\leq M + \int_{t_0}^{t} \left[p(s) + q(s) + a(\alpha(s))\alpha'(s) + b(\alpha(s))\alpha'(s) \right] v(s)ds.
$$

We apply Gronwall inequality to (2.5) and obtain

$$
v(t) \leq M \exp \left(\int_{t_0}^{t} \left[p(s) + q(s) + a(\alpha(s))\alpha'(s) + b(\alpha(s))\alpha'(s) \right] ds \right)
$$

$$
(2.6)
$$

$$
= M \exp \left(\int_{t_0}^{t} \left[p(s) + q(s) \right] ds + \int_{\alpha(t_0)}^{\alpha(t)} \left[a(s) + b(s) \right] ds \right).
$$

Inequality (2.6) with $u(t) \leq v(t)$ imply the validity of (2.3).

□

As a partial case, we obtain the following result for integral inequality with "maxima:"

Corollary 2.1.1. *Let the following conditions be fulfilled:*

1. *The condition 1 of Theorem 2.1.1 is satisfied.*

2. *The functions $p, q \in C([t_0, T), \mathbb{R}_+)$.*

3. *The function $\phi \in C([\alpha(t_0) - h, t_0], \mathbb{R}_+)$.*

4. *The function $u \in C([\alpha(t_0) - h, T), \mathbb{R}_+)$ satisfies the inequalities*

$$u(t) \le k + \int_{t_0}^t [p(s)u(s) + q(s) \max_{\xi \in [s-h,s]} u(\xi)]ds \quad for \ t \in [t_0, T),$$

$$u(t) \le \phi(t) \quad for \ t \in [\alpha(t_0) - h, t_0],$$

where $h = const \ge 0$, $k = const \ge 0$ such that $k \ge \max_{s \in [\alpha(t_0)-h,t_0]} \phi(s)$.

Then for $t \in [t_0, T)$ the inequality

$$u(t) \le M \exp\left(\int_{t_0}^t \left[p(s) + q(s)\right] ds\right)$$

holds, where $M = \max\left(k, \max_{s \in [\alpha(t_0)-h,t_0]} \phi(s)\right)$.

Remark 2.1.1. *As a partial case of Theorem 2.1.1 we obtain a result for integral inequalities without maximum (see [Pachpatte 2002], Theorem 1).*

In the case when an increasing function is involved in the right-hand side of inequality (2.1) instead of a constant, we obtain the following upper bound for $u(t)$:

Theorem 2.1.2. *Let the following conditions be fulfilled:*

1. *The function $\alpha \in C^1([t_0, T), \mathbb{R}_+)$ is nondecreasing and $\alpha(t) \le t$.*

2. *The functions $p, q \in C([t_0, T), \mathbb{R}_+)$ and $a, b \in C([\alpha(t_0), T), \mathbb{R}_+)$.*

3. *The function $\phi \in C([\alpha(t_0) - h, t_0], \mathbb{R}_+)$.*

4. *The function $k \in C([t_0, T), (0, \infty))$ is nondecreasing.*

5. *The function $u \in C([\alpha(t_0) - h, T), \mathbb{R}_+)$ satisfies the inequalities*

$$u(t) \le k(t) + \int_{t_0}^t [p(s)u(s) + q(s) \max_{\xi \in [s-h,s]} u(\xi)]ds$$

$$+ \int_{\alpha(t_0)}^{\alpha(t)} [a(s)u(s) + b(s) \max_{\xi \in [s-h,s]} u(\xi)]ds \quad for \ t \in [t_0, T),$$

$$(2.7)$$

$$u(t) \le \phi(t) \quad for \ t \in [\alpha(t_0) - h, t_0], \quad (2.8)$$

where $h = const \ge 0$.

Then for $t \in [t_0, T)$ the inequality

$$u(t) \leq Mk(t) \exp\left(\int_{t_0}^{t}\left[p(s) + q(s)\right]ds + \int_{\alpha(t_0)}^{\alpha(t)}\left[a(s) + b(s)\right]ds\right)$$

$$(2.9)$$

holds, where $M = \max\left(1, \frac{\max_{\xi \in [\alpha(t_0)-h, t_0]} \phi(\xi)}{k(t_0)}\right)$.

Proof. From inequality (2.7) we obtain for $t \in [t_0, T)$ the inequality

$$\frac{u(t)}{k(t)} \leq 1 + \int_{t_0}^{t}\left[p(s)\frac{u(s)}{k(t)} + q(s)\frac{\max_{\xi \in [s-h,s]} u(\xi)}{k(t)}\right]ds$$

$$+ \int_{\alpha(t_0)}^{\alpha(t)}\left[a(s)\frac{u(s)}{k(t)} + b(s)\frac{\max_{\xi \in [s-h,s]} u(\xi)}{k(t)}\right]ds. \qquad (2.10)$$

Let us define functions $w : [\alpha(t_0) - h, T) \to \mathbb{R}_+$ and $\tilde{k} : [\alpha(t_0) - h, T) \to \mathbb{R}_+$ by

$$\tilde{k}(t) = \begin{cases} k(t), & \text{for } t \in [t_0, T), \\ k(t_0), & \text{for } t \in [\alpha(t_0) - h, t_0], \end{cases}$$

$$w(t) = \frac{u(t)}{\tilde{k}(t)} \qquad \text{for } t \in [\alpha(t_0) - h, T).$$

The function $\tilde{k}$ is continuous nondecreasing on $[\alpha(t_0) - h, T)$.

From monotonicity of $k(t)$ we obtain for $t \in [t_0, T)$ and $s \in [\alpha(t_0), t]$ the inequality

$$\frac{\max_{\xi \in [s-h,s]} u(\xi)}{k(t)} \leq \frac{\max_{\xi \in [s-h,s]} u(\xi)}{\tilde{k}(s)} = \max_{\xi \in [s-h,s]}\frac{u(\xi)}{\tilde{k}(s)} \leq \max_{\xi \in [s-h,s]}\frac{u(\xi)}{\tilde{k}(\xi)}.$$

$$(2.11)$$

From inequalities (2.10) and (2.11) and the definition of the function $w(t)$ it follows that

$$w(t) \leq 1 + \int_{t_0}^{t}\left[p(s)\varphi(s) + q(s)\max_{\xi \in [s-h,s]} w(\xi)\right]ds$$

$$+ \int_{\alpha(t_0)}^{\alpha(t)}\left[a(s)\varphi(s) + b(s)w(\xi)\right]ds, \quad \text{for } t \in [t_0, T), \qquad (2.12)$$

$$w(t) = \frac{u(t)}{k(t_0)} \leq \frac{\phi(t)}{k(t_0)}, \quad \text{for } t \in [\alpha(t_0) - h, t_0]. \qquad (2.13)$$

From the definition of the function $\tilde{k}(t)$, and inequalities (2.12) and (2.13) according to Theorem 2.1.1 it follows that for $t \in [t_0, T)$ the inequality

$$w(t) \leq M \exp \left(\int_{t_0}^t \left[p(s) + q(s) \right] ds + \int_{\alpha(t_0)}^{\alpha(t)} \left[a(s) + b(s) \right] ds \right) \quad (2.14)$$

holds.

From inequality (2.14), the definition of functions $\tilde{k}(t)$ and $w(t)$ we obtain inequality (2.9).

$\square$

Corollary 2.1.2. *Let the conditions of Theorem 2.1.2 be satisfied, and equality $k(t_0) = max_{s \in [\alpha(t_0)-h, t_0]} \phi(s)$ holds.*
Then for $t \in [t_0, T)$ the inequality

$$u(t) \leq k(t) \exp \left(\int_{t_0}^t \left[p(s) + q(s) \right] ds + \int_{\alpha(t_0)}^{\alpha(t)} \left[a(s) + b(s) \right] ds \right) \quad (2.15)$$

holds.

In the case when the function involved in the right-hand side of inequality (2.1) is not monotonic, we obtain the following result:

Theorem 2.1.3. *Let the following conditions be fulfilled:*

1. *The function $\alpha \in C^1([t_0, T), \mathbb{R}_+)$ is nondecreasing and $\alpha(t) \leq t$.*

2. *The functions $p, q \in C([t_0, T), \mathbb{R}_+)$ and $a, b \in C([\alpha(t_0), T), \mathbb{R}_+)$.*

3. *The function $\phi \in C([\alpha(t_0)-h, T), \mathbb{R}_+)$, $max_{s \in [\alpha(t_0)-h, t_0]} \phi(s) > 0$.*

4. *The function $u \in C([\alpha(t_0) - h, T), \mathbb{R}_+)$ satisfies the inequalities*

$$u(t) \leq \phi(t) + \int_{t_0}^t [p(s)u(s) + q(s) \max_{\xi \in [s-h,s]} u(\xi)] ds$$

$$+ \int_{\alpha(t_0)}^{\alpha(t)} [a(s)u(s) + b(s) \max_{\xi \in [s-h,s]} u(\xi)] ds \ \text{for} \ t \in [t_0, T),$$

$$(2.16)$$

$$u(t) \leq \phi(t) \quad \text{for} \ t \in [\alpha(t_0) - h, t_0], \quad (2.17)$$

where $h = const \geq 0$.

Then for $t \in [t_0, T)$ the inequality

$$u(t) \leq \phi(t)$$

$$+ e(t) \exp \left(\int_{t_0}^{t} \left[p(s) + q(s) \right] ds + \int_{\alpha(t_0)}^{\alpha(t)} \left[a(s) + b(s) \right] ds \right)$$

$$(2.18)$$

holds, where $e : [t_0, T) \to \mathbb{R}_+$ *is defined by*

$$e(t) = \max_{s \in [\alpha(t_0) - h, t_0]} \phi(s) + \int_{t_0}^{t} [p(s)\phi(s) + q(s) \max_{\xi \in [s-h,s]} \phi(\xi)] ds$$

$$+ \int_{t_0}^{\max(\alpha(t), t_0)} \left(a(s)\phi(s) + b(s) \max_{\xi \in [s-h,s]} \phi(\xi) \right) ds. \qquad (2.19)$$

Proof. Let us define a function $z : [\alpha(t_0) - h, T) \to \mathbb{R}_+$ by

$$z(t) = \begin{cases} \int_{t_0}^{t} \left[p(s)u(s) + q(s) \max_{\xi \in [s-h,s]} u(\xi) \right] ds \\ + \int_{\alpha(t_0)}^{\alpha(t)} \left[a(s)u(s) + b(s) \max_{\xi \in [s-h,s]} u(\xi) \right] ds, & t \in [t_0, T) \\ 0, & t \in [\alpha(t_0) - h, t_0). \end{cases} \qquad (2.20)$$

From inequality (2.16) and the definition of function $z(t)$ we have for $t \in [\alpha(t_0) - h, T)$

$$u(t) \leq \phi(t) + z(t). \qquad (2.21)$$

Let $t \in [t_0, T)$ be such that $\alpha(t) \geq t_0$. Then from inequality (2.21) it follows the validity of the inequality

$$\int_{\alpha(t_0)}^{\alpha(t)} \left[a(s)u(s) + b(s) \max_{\xi \in [s-h,s]} u(\xi) \right] ds$$

$$\leq \int_{\alpha(t_0)}^{t_0} \left[a(s)z(s) + b(s) \max_{\xi \in [s-h,s]} z(\xi) \right] ds$$

$$+ \int_{t_0}^{\alpha(t)} \left[a(s) \left(\phi(s) + z(s) \right) + b(s) \left(\max_{\xi \in [s-h,s]} \phi(\xi) + \max_{\xi \in [s-h,s]} z(\xi) \right) \right] ds$$

$$= \int_{t_0}^{\max(\alpha(t), t_0)} \left(a(s)\phi(s) + b(s) \max_{\xi \in [s-h,s]} \phi(\xi) \right) ds$$

$$+ \int_{\alpha(t_0)}^{\alpha(t)} \left(a(s)z(s) + b(s) \max_{\xi \in [s-h,s]} z(\xi) \right) ds. \qquad (2.22)$$

Let $t \in [t_0, T)$ be such that $\alpha(t) < t_0$. Then from the definition of function $z(t)$ we get

$$\int_{\alpha(t_0)}^{\alpha(t)} \left[a(s)u(s) + b(s) \max_{\xi \in [s-h,s]} u(\xi) \right] ds$$

$$= \int_{\alpha(t_0)}^{\alpha(t)} \left(a(s)z(s) + b(s) \max_{\xi \in [s-h,s]} z(\xi) \right) ds$$

$$\leq \int_{t_0}^{\max(\alpha(t),t_0)} \left(a(s)\phi(s) + b(s) \max_{\xi \in [s-h,s]} \phi(\xi) \right) ds$$

$$+ \int_{\alpha(t_0)}^{\alpha(t)} \left(a(s)z(s) + b(s) \max_{\xi \in [s-h,s]} z(\xi) \right) ds. \qquad (2.23)$$

From the definition of the function $z(t)$ and inequalities (2.22) and (2.23), it follows that

$$z(t) \leq e(t) + \int_{t_0}^{t} \left[p(s)z(s) + q(s) \max_{\xi \in [s-h,s]} z(\xi) \right] ds$$

$$+ \int_{\alpha(t_0)}^{\alpha(t)} \left[a(s)z(s) + b(s) \max_{\xi \in [s-h,s]} z(\xi) \right] ds, \quad t \in [t_0, T), \quad (2.24)$$

$$z(t) \leq \phi(t), \quad t \in [\alpha(t_0) - h, t_0], \qquad (2.25)$$

where function $e(t)$ is defined by equality (2.19). Note that function $e(t) : [t_0, T) \to (0, \infty)$ is nondecreasing for $t \in [t_0, T)$ and $e(t_0) = \max_{s \in [t_0-h,t_0]} \phi(s)$.

From inequalities (2.24) and (2.25) according to Theorem 2.1.2 we get

$$z(t) \leq e(t) \exp \left(\int_{t_0}^{t} \left[p(s) + q(s) \right] ds + \int_{\alpha(t_0)}^{\alpha(t)} \left[a(s) + b(s) \right] ds \right).$$

$$(2.26)$$

From inequalities (2.21) and (2.26) we obtain inequality (2.18). □

Remark 2.1.2. *Note that in the case when the function $\phi(t)$ is nondecreasing the inequality (2.15) gives us better than (2.18) upper bound for function $u(t)$, since inequality $e^x \leq 1 + xe^x$, $x \geq 0$ holds.*

Now we will consider an inequality in which the unknown function in the left-hand side is in a power.

Theorem 2.1.4. *Let the following conditions be fulfilled:*

1. *The conditions 1, 2, 3 of Theorem 2.1.2 are satisfied.*

2. *The function $k \in C([t_0, T), (0, \infty))$ is nondecreasing and the inequality $M = \max_{s \in [\alpha(t_0)-h, t_0]} \phi(s) \leq \sqrt[n]{k(t_0)}$ holds.*

3. *The function $u \in C([\alpha(t_0) - h, T), \mathbb{R}_+)$ satisfies the inequalities*

$$u^n(t) \leq k(t) + \int_{t_0}^{t} [p(s)u(s) + q(s) \max_{\xi \in [s-h,s]} u(\xi)]ds$$
$$+ \int_{\alpha(t_0)}^{\alpha(t)} [a(s)u(s) + b(s) \max_{\xi \in [s-h,s]} u(\xi)]ds \text{ for } t \in [t_0, T),$$

(2.27)

$$u(t) \leq \phi(t) \quad \text{for } t \in [\alpha(t_0) - h, t_0], \tag{2.28}$$

where $h \geq 0$, $n > 1$.

Then for $t \in [t_0, T)$ the inequality

$$u(t) \leq \sqrt[n]{k(t)} + \left(M + \frac{e(t)}{n\left(k(t_0)\right)^{1-\frac{1}{n}}} \right) e^{A(t)+B(t)} \tag{2.29}$$

holds, where

$$e(t) = \int_{t_0}^{t} \left(p(s)\omega(s) + q(s) \max_{\xi \in [s-h,s]} \omega(\xi) \right) ds$$
$$+ \int_{t_0}^{\max(\alpha(t),t_0)} \left(a(s)\omega(s) + b(s) \max_{\xi \in [s-h,s]} \omega(\xi) \right) ds, \tag{2.30}$$

$$A(t) = \frac{1}{n} \int_{t_0}^{t} \left(k(s) \right)^{\frac{1-n}{n}} \left[p(s) + q(s) \right] ds, \tag{2.31}$$

$$B(t) = \frac{1}{n} \int_{\alpha(t_0)}^{\alpha(t)} \left(K(s) \right)^{\frac{1-n}{n}} \left[a(s) + b(s) \right] ds, \tag{2.32}$$

$$K(t) = \begin{cases} k(t), & t \in [t_0, T), \\ k(t_0), & t \in [\alpha(t_0), t_0), \end{cases}$$

$$\omega(t) = \begin{cases} \sqrt[n]{k(t)}, & t \in (t_0, T), \\ M, & t \in [t_0 - h, t_0]. \end{cases}$$

Proof. Let us define a function $z : [\alpha(t_0) - h, T) \to \mathbb{R}_+$ by

$$
z(t) = \begin{cases}
\dfrac{\sqrt[n]{k(t)}}{nk(t)}\left(\displaystyle\int_{t_0}^{t}\left(p(s)u(s) + q(s)\max_{\xi\in[s-h,s]}u(\xi)\right)ds\right. \\
\left. \quad + \displaystyle\int_{\alpha(t_0)}^{\alpha(t)}\left(a(s)u(s) + b(s)\max_{\xi\in[s-h,s]}u(\xi)\right)ds\right), \\
\hspace{6cm} t \in [t_0, T) \\
0, \hspace{4cm} t \in [\alpha(t_0) - h, t_0).
\end{cases} \tag{2.33}
$$

From inequality (2.27) we have for $t \in [t_0, T)$

$$
u^n(t) \le k(t)\left(1 + n\frac{z(t)}{\sqrt[n]{k(t)}}\right),
$$

or

$$
u(t) \le \sqrt[n]{k(t)}\left(1 + n\frac{z(t)}{\sqrt[n]{k(t)}}\right)^{\frac{1}{n}}.
$$

Apply Bernoulli's inequality $(1 + x)^a \le 1 + ax$, where $0 < a < 1$ and $-1 < x$, and observe that

$$
u(t) \le \sqrt[n]{k(t)}\left(1 + \frac{z(t)}{\sqrt[n]{k(t)}}\right) = \sqrt[n]{k(t)} + z(t) = \omega(t) + z(t), \quad t \in [t_0, T),
$$
$$\tag{2.34}$$

and

$$
u(t) \le \phi(t) \le \phi(t) + z(t) \le \omega(t) + z(t), \quad t \in [\alpha(t_0) - h, t_0], \tag{2.35}
$$

where

$$
\omega(t) = \begin{cases}
\sqrt[n]{k(t)}, & t \in [t_0, T) \\
\sqrt[n]{k(t_0)}, & t \in [\alpha(t_0) - h, t_0).
\end{cases}
$$

Therefore

$$
\max_{\xi\in[s-h,s]}u(\xi) \le \max_{\xi\in[s-h,s]}\omega(\xi) + \max_{\xi\in[s-h,s]}z(\xi), \quad s \in [t_0, T). \tag{2.36}
$$

Let $t \in [t_0, T)$ be such that $\alpha(t) \ge t_0$. Then from inequalities (2.34) and (2.35) we get

$$
\int_{\alpha(t_0)}^{\alpha(t)} \left[a(s)u(s) + b(s) \max_{\xi \in [s-h,s]} u(\xi) \right] ds
$$
$$
\leq \int_{\alpha(t_0)}^{t_0} \left[a(s)z(s) + b(s) \max_{\xi \in [s-h,s]} z(\xi) \right] ds
$$
$$
+ \int_{t_0}^{\alpha(t)} \left[a(s)\Big(w(s) + z(s)\Big) + b(s)\Big(\max_{\xi \in [s-h,s]} w(\xi) + \max_{\xi \in [s-h,s]} z(\xi) \Big) \right] ds
$$
$$
= \int_{t_0}^{\max(\alpha(t),t_0)} \left(a(s)w(s) + b(s) \max_{\xi \in [s-h,s]} w(\xi) \right) ds
$$
$$
+ \int_{\alpha(t_0)}^{\alpha(t)} \left(a(s)z(s) + b(s) \max_{\xi \in [s-h,s]} z(\xi) \right) ds. \tag{2.37}
$$

Let $t \in [t_0, T)$ be such that $\alpha(t) < t_0$. Then from the definition of function $z(t)$ and inequalities (2.34) and (2.35) we get

$$
\int_{\alpha(t_0)}^{\alpha(t)} \left[a(s)u(s) + b(s) \max_{\xi \in [s-h,s]} u(\xi) \right] ds
$$
$$
= \int_{\alpha(t_0)}^{\alpha(t)} \left(a(s)z(s) + b(s) \max_{\xi \in [s-h,s]} z(\xi) \right) ds
$$
$$
\leq \int_{t_0}^{\max(\alpha(t),t_0)} \left(a(s)w(s) + b(s) \max_{\xi \in [s-h,s]} w(\xi) \right) ds
$$
$$
+ \int_{\alpha(t_0)}^{\alpha(t)} \left(a(s)z(s) + b(s) \max_{\xi \in [s-h,s]} z(\xi) \right) ds. \tag{2.38}
$$

Let $C = Mn \, k(t_0)^{1-\frac{1}{n}} > 0$. Note the function $v : [t_0, T) \rightarrow (0, \infty)$, $v(t) = \frac{1}{n \, k(t_0)^{1-\frac{1}{n}}} \Big(C + e(t) \Big)$ is nondecreasing and the equality $v(t_0) = \frac{1}{n \, k(t_0)^{1-\frac{1}{n}}} \Big(C + e(t_0) \Big) = M$ holds, where the function $e(t)$ is defined by (2.30). From the definition of the function $z(t)$ and inequalities (2.37) and (2.38) it follows that

$$
z(t) \leq \frac{1}{n \, k(t)^{1-\frac{1}{n}}} \left(e(t) + \int_{t_0}^{t} \left[p(s)z(s) + q(s) \max_{\xi \in [s-h,s]} z(\xi) \right] ds \right.
$$

$$
\left. + \int_{\alpha(t_0)}^{\alpha(t)} \left[a(s)z(s) + b(s) \max_{\xi \in [s-h,s]} z(\xi) \right] ds \right)
$$

$$
\leq \frac{1}{n \, k(t_0)^{1-\frac{1}{n}}} \left(C + e(t) \right)
$$

$$
+ \frac{1}{n \, k(t)^{1-\frac{1}{n}}} \int_{t_0}^{t} \left[p(s)z(s) + q(s) \max_{\xi \in [s-h,s]} z(\xi) \right] ds
$$

$$
+ \frac{1}{n \, k(t)^{1-\frac{1}{n}}} \int_{\alpha(t_0)}^{\alpha(t)} \left[a(s)z(s) + b(s) \max_{\xi \in [s-h,s]} z(\xi) \right] ds
$$

$$
\leq v(t) + \int_{t_0}^{t} \frac{1}{n} \left[\frac{p(s)}{\left(k(s)\right)^{1-\frac{1}{n}}} z(s) + \frac{q(s)}{\left(k(s)\right)^{1-\frac{1}{n}}} \max_{\xi \in [s-h,s]} z(\xi) \right] ds
$$

$$
\tag{2.39}
$$

$$
+ \int_{\alpha(t_0)}^{\alpha(t)} \frac{1}{n} \left[\frac{a(s)}{\left(K(s)\right)^{1-\frac{1}{n}}} z(s) + \frac{b(s)}{\left(K(s)\right)^{1-\frac{1}{n}}} \max_{\xi \in [s-h,s]} z(\xi) \right] ds,
$$

$$
t \in [t_0, T),
$$

$$
z(t) = 0, \quad t \in [\alpha(t_0) - h, t_0]. \tag{2.40}
$$

From inequalities (2.39) and (2.40) according to Theorem 2.1.2 we get

$$
z(t) \leq v(t) e^{A(t)+B(t)} = \left(M + \frac{e(t)}{n \, (k(t_0))^{1-\frac{1}{n}}} \right) e^{A(t)+B(t)}, \tag{2.41}
$$

where $A(t)$ and $B(t)$ are defined by (2.31) and (2.32), respectively.

Substitute bound (2.41) for $z(t)$ in the right-hand side of (2.34) and obtain inequality (2.29).

$\square$

Remark 2.1.3. *As partial cases of Theorem 2.1.3 and Theorem 2.1.4 we obtain results for integral inequality without maximum (see [Kim 2005], Theorem 2.1 and Theorem 2.2).*

2.2 Nonlinear Integral Inequalities with Maxima for Scalar Functions of One Variable

In this section some new nonlinear integral inequalities of Bihari type will be solved. The main characteristic of the considered integral inequalities is the presence of the maximum of the unknown scalar function in the integral.

Theorem 2.2.1. *Let the following conditions be fulfilled:*

1. *The function $\alpha \in C^1([t_0, T), \mathbb{R}_+)$ is nondecreasing and $\alpha(t) \leq t$.*

2. *The functions p, $q \in C([t_0, T), \mathbb{R}_+)$ and a, $b \in C([\alpha(t_0), T), \mathbb{R}_+)$.*

3. *The function $\phi \in C([\alpha(t_0) - h, t_0], \mathbb{R}_+)$.*

4. *The function $g \in C(\mathbb{R}_+, (0, \infty))$ is increasing, $\int^\infty g(s)ds = \infty$.*

5. *The function $u \in C([\alpha(t_0) - h, T), \mathbb{R}_+)$ and satisfies the inequalities*

$$u(t) \leq k + \int_{t_0}^{t} \left[p(s)g\Big(u(s)\Big) + q(s)g\Big(\max_{\xi \in [s-h,s]} u(\xi) \Big) \right] ds \quad (2.42)$$

$$+ \int_{\alpha(t_0)}^{\alpha(t)} \left[a(s)g\Big(u(s)\Big) + b(s)g\Big(\max_{\xi \in [s-h,s]} u(\xi) \Big) \right] ds$$

$$for \ t \in [t_0, T),$$

$$u(t) \leq \phi(t) \quad for \ t \in [\alpha(t_0) - h, t_0], \quad (2.43)$$

where $h, k = const \geq 0$.

Then, for $t \in [t_0, t_1)$ the inequality

$$u(t) \leq G^{-1}\left(G(M) + \int_{t_0}^{t} \Big[p(s) + q(s) \Big] ds + \int_{\alpha(t_0)}^{\alpha(t)} \Big[a(s) + b(s) \Big] ds \right),$$

$$(2.44)$$

holds, where $M = \max\left(k, \max_{t\in[\alpha(t_0)-h,t_0]} \phi(t)\right)$, *the function* G *is defined by*

$$G(r) = \int_{r_0}^r \frac{ds}{g(s)}, \quad r_0 \geq 0, \tag{2.45}$$

$$t_1 = \sup\left\{\tau \in (t_0, T) : G(M) + \int_{t_0}^t \left[p(s) + q(s)\right]ds \right.$$
$$\left. + \int_{\alpha(t_0)}^{\alpha(t)} \left[a(s) + b(s)\right]ds \in Dom\left(G^{-1}\right) \text{ for } t \in [t_0, \tau]\right\},$$

and the function G^{-1} *is the inverse function of* G.

Proof. Let us define a function $z : [\alpha(t_0) - h, T) \to \mathbb{R}_+$ by

$$z(t) = \begin{cases} M + \int_{t_0}^t \left[p(s)g\Big(u(s)\Big) + q(s)g\Big(\max_{\xi\in[s-h,s]} u(\xi)\Big)\right]ds \\ \quad + \int_{\alpha(t_0)}^{\alpha(t)} \left[a(s)g\Big(u(s)\Big) + b(s)g\Big(\max_{\xi\in[s-h,s]} u(\xi)\Big)\right]ds, \quad (2.46) \\ \qquad\qquad\qquad\qquad\qquad\qquad\qquad\qquad t \in [t_0, T) \\ M, \qquad\qquad\qquad\qquad\qquad\qquad\quad t \in [\alpha(t_0) - h, t_0). \end{cases}$$

The function $z(t)$ is nondecreasing and the inequality $u(t) \leq z(t)$ holds for $t \in [\alpha(t_0) - h, T)$. Note $\max_{s\in[t-h,t]} z(s) = z(t)$ for $t \in [\alpha(t_0), T)$. Then from inequality (2.42) we get for $t \in [t_0, T)$

$$z(t) \leq M + \int_{t_0}^t \left[p(s)g\Big(z(s)\Big) + q(s)g\Big(\max_{\xi\in[s-h,s]} z(\xi)\Big)\right]ds$$
$$+ \int_{\alpha(t_0)}^{\alpha(t)} \left[a(s)g\Big(z(s)\Big) + b(s)g\Big(\max_{\xi\in[s-h,s]} z(\xi)\Big)\right]ds$$
$$\leq M + \int_{t_0}^t \left[p(s) + q(s)\right]g\Big(z(s)\Big)ds$$
$$+ \int_{\alpha(t_0)}^{\alpha(t)} \left[a(s) + b(s)\right]g\Big(z(s)\Big)ds. \tag{2.47}$$

Define a function $w : [\alpha(t_0) - h, T) \to [M, \infty)$ by

$$w(t) = M + \int_{t_0}^t \left[p(s) + q(s)\right]g\Big(z(s)\Big)ds + \int_{\alpha(t_0)}^{\alpha(t)} \left[a(s) + b(s)\right]g\Big(z(s)\Big)ds.$$

Differentiate the function $w(t)$, use the monotonicity of $w(t)$ and obtain

$$\left(w(t)\right)' = \left[p(t) + q(t)\right]g\left(z(t)\right)$$
$$+ \left[a\left(\alpha(t)\right) + b\left(\alpha(t)\right)\right]g\left(z\left(\alpha(t)\right)\right)\left(\alpha(t)\right)'$$
$$\leq g\left(w(t)\right)\left[p(t) + q(t) + a\left(\alpha(t)\right)\left(\alpha(t)\right)' + b\left(\alpha(t)\right)\left(\alpha(t)\right)'\right].$$
$$\text{(2.48)}$$

From definition (2.45) and inequality (2.48) it follows that

$$\frac{d}{dt}G\left(w(t)\right) = \frac{\left(w(t)\right)'}{g\left(w(t)\right)}$$
$$\leq p(t) + q(t) + a\left(\alpha(t)\right)\left(\alpha(t)\right)' + b\left(\alpha(t)\right)\left(\alpha(t)\right)'. \quad \text{(2.49)}$$

We integrate inequality (2.49) from t_0 to t for $t \in [t_0, T)$, change the variable $\eta = \alpha(s)$ and we obtain

$$G\left(w(t)\right) \leq G(M) + \int_{t_0}^{t}\left[p(\eta) + q(\eta)\right]d\eta + \int_{\alpha(t_0)}^{\alpha(t)}\left[a(\eta) + b(\eta)\right]d\eta. \quad \text{(2.50)}$$

Since G^{-1} is an increasing function, we obtain from inequality (2.50) and $u(t) \leq z(t)] \leq w(t)$ the required inequality (2.44) for $t \in [t_0, t_1)$. $\qquad \square$

Remark 2.2.1. *In the case when $h = 0$ the result of the Theorem 2.2.1 reduces to the classical Bihari inequality.*

In connection with the nonlinearity of the considered integral inequality, we introduce the following class of functions.

Definition 2.2.1. *We will say that a function $g \in C(\mathbb{R}_+, \mathbb{R}_+)$ is from the class Ω if the following conditions are satisfied:*

(i) g is a nondecreasing function;

(ii) $g(x) > 0$ for $x > 0$;

(iii) $g(tx) \geq tg(x)$ for $0 \leq t \leq 1$, $x \geq 0$;

(iv) $g(x) + g(y) \geq g(x + y)$ for $x, y \geq 0$;

(v) $\displaystyle\int_1^\infty \frac{dx}{g(x)} = \infty.$

Remark 2.2.2. *Note that the function* $g(x) = \sqrt[p]{x} \in \Omega$, *where* p *is a natural number.*

Theorem 2.2.2. *Let the following conditions be fulfilled:*

1. *The function* $\alpha \in C^1([t_0, T), \mathbb{R}_+)$ *is nondecreasing and* $\alpha(t) \leq t$.

2. *The functions* $p,\ q \in C([t_0, T), \mathbb{R}_+)$ *and* $a,\ b \in C([\alpha(t_0), T), \mathbb{R}_+)$.

3. *The function* $k \in C([\alpha(t_0) - h, T), \mathbb{R}_+)$.

4. *The function* $g \in C(\mathbb{R}_+, \mathbb{R}_+)$ *and* $g \in \Omega$.

5. *The function* $u \in C([\alpha(t_0) - h, T), \mathbb{R}_+)$ *and satisfies the inequalities*

$$u(t) \leq k(t) + \int_{t_0}^t \left[p(s)g\big(u(s)\big) + q(s)g\Big(\max_{\xi \in [s-h,s]} u(\xi) \Big) \right] ds$$

(2.51)

$$+ \int_{\alpha(t_0)}^{\alpha(t)} \left[a(s)g\big(u(s)\big) + b(s)g\Big(\max_{\xi \in [s-h,s]} u(\xi) \Big) \right] ds$$

for $t \in [t_0, T)$,

$$u(t) \leq k(t) \quad \text{for} \ \ t \in [\alpha(t_0) - h, t_0]$$

(2.52)

where $h = const \geq 0$.

Then for $t \in [t_0, t_2)$ *the inequality*

$$u(t) \leq k(t) + e(t)G^{-1}\Big(G(1)$$

$$+ \int_{t_0}^t \big[p(s) + q(s) \big] ds + \int_{\alpha(t_0)}^{\alpha(t)} \big[a(s) + b(s) \big] ds \Big)$$

(2.53)

holds, where $e(t) : [t_0, T) \to \mathbb{R}_+$ *is defined by*

$$e(t) = 1 + \int_{t_0}^t \left[p(s)g\big(k(s)\big) + q(s)g\Big(\max_{\xi \in [s-h,s]} k(\xi) \Big) \right] ds$$

$$+ \int_{t_0}^{\max\,(\alpha(t),t_0)} \left[a(s)g\big(k(s)\big) + b(s)g\Big(\max_{\xi \in [s-h,s]} k(\xi) \Big) \right] ds,$$

(2.54)

the function G is defined by equality (2.45), the function G^{-1} is the inverse of G,

$$
t_2 = \sup \left\{ \tau \in (t_0, T) \ : \ G(1) + \int_{t_0}^t \Big[p(s) + q(s) \Big] ds \right.
$$
$$
\left. + \int_{\alpha(t_0)}^{\alpha(t)} \Big[a(s) + b(s) \Big] ds \in Dom(G^{-1}) \quad for \ \ t \in [t_0, \tau] \right\}.
$$

Proof. Let us define a function $z : [\alpha(t_0) - h, T) \to \mathbb{R}_+$ by

$$
z(t) = \begin{cases} \int_{t_0}^t \left[p(s)g\Big(u(s)\Big) + q(s)g\Big(\max_{\xi \in [s-h,s]} u(\xi) \Big) \right] ds \\ \quad + \int_{\alpha(t_0)}^{\alpha(t)} \left[a(s)g\Big(u(s)\Big) + b(s)g\Big(\max_{\xi \in [s-h,s]} u(\xi) \Big) \right] ds, \\ \qquad\qquad\qquad\qquad\qquad\qquad\qquad t \in (t_0, T), \\ 0, \qquad\qquad\qquad\qquad\qquad\qquad t \in [\alpha(t_0) - h, t_0]. \end{cases}
$$

From inequality (2.51) and the definition of $z(t)$ we have for $t \in [\alpha(t_0) - h, T)$

$$
u(t) \le k(t) + z(t). \tag{2.55}
$$

Let $t \in [t_0, T)$ such that $\alpha(t) \ge t_0$. Then from inequality (2.55), the definition of the function $z(t)$, and condition 4 of Theorem 2.2.2 we have the inequality

$$
\int_{\alpha(t_0)}^{\alpha(t)} \left[a(s)g\Big(u(s)\Big) + b(s)g\Big(\max_{\xi \in [s-h,s]} u(\xi) \Big) \right] ds
$$
$$
\le \int_{\alpha(t_0)}^{t_0} \left[a(s)g\Big(z(s)\Big) + b(s)g\Big(\max_{\xi \in [s-h,s]} z(\xi) \Big) \right] ds
$$
$$
+ \int_{t_0}^{\alpha(t)} \left[a(s)g\Big(k(s) + z(s)\Big) \right.
$$
$$
\left. + b(s)g\Big(\max_{\xi \in [s-h,s]} k(\xi) + \max_{\xi \in [s-h,s]} z(\xi) \Big) \right] ds
$$
$$
\le \int_{t_0}^{\max(\alpha(t), t_0)} \left[a(s)g\Big(k(s)\Big) + b(s)g\Big(\max_{\xi \in [s-h,s]} k(\xi) \Big) \right] ds
$$
$$
+ \int_{\alpha(t_0)}^{\alpha(t)} \left[a(s)g\Big(z(s)\Big) + b(s)g\Big(\max_{\xi \in [s-h,s]} z(\xi) \Big) \right] ds. \tag{2.56}
$$

Let $t \in [t_0, T)$ such that $\alpha(t) < t_0$. Then from the definition of function $z(t)$, we get

$$\int_{\alpha(t_0)}^{\alpha(t)} \left[a(s)g\Big(u(s)\Big) + b(s)g\Big(\max_{\xi \in [s-h,s]} u(\xi) \Big) \right] ds$$

$$= \int_{\alpha(t_0)}^{\alpha(t)} \left[a(s)g\Big(z(s)\Big) + b(s)g\Big(\max_{\xi \in [s-h,s]} z(\xi) \Big) \right] ds$$

$$\leq \int_{t_0}^{\max(\alpha(t),t_0)} \left[a(s)g\Big(k(s)\Big) + b(s)g\Big(\max_{\xi \in [s-h,s]} k(\xi) \Big) \right] ds$$

$$+ \int_{\alpha(t_0)}^{\alpha(t)} \left[a(s)g\Big(z(s)\Big) + b(s)g\Big(\max_{\xi \in [s-h,s]} z(\xi) \Big) \right] ds. \qquad (2.57)$$

From the definition of function $z(t)$ and inequalities (2.52), (2.56), and (2.57) it follows that

$$z(t) \leq e(t) + \int_{t_0}^{t} \left[p(s)g\Big(z(s)\Big) + q(s)g\Big(\max_{\xi \in [s-h,s]} z(\xi) \Big) \right] ds \qquad (2.58)$$

$$+ \int_{\alpha(t_0)}^{\alpha(t)} \left[a(s)g\Big(z(s)\Big) + b(s)g\Big(\max_{\xi \in [s-h,s]} z(\xi) \Big) \right] ds \quad \text{for } t \in [t_0, T),$$

$$z(t) \leq k(t) \quad \text{for } t \in [\alpha(t_0) - h, t_0), \qquad (2.59)$$

where the function $e(t)$ is defined by (2.54). Note the function $e(t)$ is nondecreasing for $t \in [t_0, T)$ and $e(t_0) = 1$.

From inequalities (2.58) and (2.59), condition 4 of Theorem 2.2.2 and $\frac{1}{e(t)} \leq 1$ we obtain for $t \in [t_0, T)$ the inequality

$$\frac{z(t)}{e(t)} \leq 1 + \int_{t_0}^{t} \left[p(s)g\Big(\frac{z(s)}{e(s)} \Big) + q(s)g\Big(\frac{\max_{\xi \in [s-h,s]} z(\xi)}{e(t)} \Big) \right] ds$$

$$+ \int_{\alpha(t_0)}^{\alpha(t)} \left[a(s)g\Big(\frac{z(s)}{e(s)} \Big) + b(s)g\Big(\frac{\max_{\xi \in [s-h,s]} z(\xi)}{e(t)} \Big) \right] ds. \quad (2.60)$$

From monotonicity of $e(t)$ we obtain for $t \in [t_0, T)$ and $s \in [\alpha(t_0), t]$ the inequality

$$\frac{\max_{\xi \in [s-h,s]} z(\xi)}{e(t)} \leq \frac{\max_{\xi \in [s-h,s]} z(\xi)}{\hat{e}(s)} = \max_{\xi \in [s-h,s]} \frac{z(\xi)}{\hat{e}(s)} \leq \max_{\xi \in [s-h,s]} \frac{z(\xi)}{\hat{e}(\xi)},$$
$$(2.61)$$

where the continuous nondecreasing function $\hat{e} : [\alpha(t_0) - h, T) \to \mathbb{R}_+$ is defined by

$$\hat{e}(t) = \begin{cases} e(t), & \text{for } t \in [t_0, T), \\ e(t_0), & \text{for } t \in [\alpha(t_0) - h, t_0]. \end{cases}$$

From (2.60) and (2.61) follows that for $t \in [t_0, T)$ the inequality

$$\frac{z(t)}{\hat{e}(t)} \leq 1 + \int_{t_0}^{t} \left[p(s)g\left(\frac{z(s)}{\hat{e}(s)}\right) + q(s)g\left(\max_{\xi \in [s-h,s]} \frac{z(\xi)}{\hat{e}(\xi)}\right) \right] ds$$
$$+ \int_{\alpha(t_0)}^{\alpha(t)} \left[a(s)g\left(\frac{z(s)}{\hat{e}(s)}\right) + b(s)g\left(\max_{\xi \in [s-h,s]} \frac{z(\xi)}{\hat{e}(\xi)}\right) \right] ds \quad (2.62)$$

holds.

Let us define a function $u_1 : [\alpha(t_0) - h, T) \to \mathbb{R}_+$ by $u_1(t) = \frac{z(t)}{\hat{e}(t)}$. Set the right-hand side of inequality (2.62) by function $z_1 : [t_0, T) \to \mathbb{R}_+$. Note that function $z_1(t)$ is increasing, $z_1(t_0) = 1$ and for $t \in [t_0, T)$ the inequality $u_1(t) \leq z_1(t)$ holds.

Therefore

$$z_1(t) \leq 1 + \int_{t_0}^{t} \left[p(s) + q(s) \right] g(z_1(s)) ds + \int_{\alpha(t_0)}^{\alpha(t)} \left[a(s) + b(s) \right] g(z_1(s)) ds.$$
$$(2.63)$$

Define a function $w : [\alpha(t_0) - h, T) \to [1, \infty)$ by

$$w(t) = 1 + \int_{t_0}^{t} \left[p(s) + q(s) \right] g(z_1(s)) ds + \int_{\alpha(t_0)}^{\alpha(t)} \left[a(s) + b(s) \right] g(z_1(s)) ds.$$

Differentiate the function $w(t)$ and we obtain

$$\left(w(t)\right)' \leq \left[p(t) + q(t) \right] g\left(z_1(t)\right)$$
$$+ \left[a(\alpha(t)) + b(\alpha(t)) \right] g\left(z_1(\alpha(t))\right)\left(\alpha(t)\right)'. \quad (2.64)$$

From inequality (2.64) and condition 1 of Theorem 2.2.2 we get

$$\left(w(t)\right)' \leq g\left(w(t)\right)\left[p(t) + q(t) + a(\alpha(t))\left(\alpha(t)\right)' \right.$$
$$\left. + b(\alpha(t))\left(\alpha(t)\right)' \right]. \quad (2.65)$$

From the definition of G and inequality (2.65) we obtain

$$\frac{d}{dt}G\Big(w(t)\Big) = \frac{\big(w(t)\big)'}{g\big(w(t)\big)} \le p(t) + q(t) + a(\alpha(t))\big(\alpha(t)\big)'$$

$$+ b(\alpha(t))\big(\alpha(t)\big)'. \tag{2.66}$$

Integrating inequality (2.66) from t_0 to t we get for $t \in [t_0, T)$ the following inequality

$$G\Big(w(t)\Big) \le G(1) + \int_{t_0}^t \Big[p(s) + q(s)\Big]ds + \int_{t_0}^t \Big[a(\alpha(s)) + b(\alpha(s))\Big]\big(\alpha(s)\big)'ds$$

$$\le G(1) + \int_{t_0}^t \Big[p(\eta) + q(\eta)\Big]d\eta + \int_{\alpha(t_0)}^{\alpha(t)} \Big[a(\eta) + b(\eta)\Big]d\eta. \tag{2.67}$$

Since $G^{-1}(t)$ is an increasing function, from inequalities (2.67) and $u_1(t) \le z_1(t) \le w(t)$ it follows that for $t \in [t_0, T)$ the inequality

$$\frac{z(t)}{\hat{e}(t)} \le G^{-1}\left(G(1) + \int_{t_0}^t \Big[p(s)+q(s)\Big]ds + \int_{\alpha(t_0)}^{\alpha(t)} \Big[a(s)+b(s)\Big]ds\right) \tag{2.68}$$

holds.

Inequalities (2.55) and (2.68) and the definition of $\hat{e}(t)$ imply the validity of inequality (2.53).

$\square$

In the nonlinear case when the unknown function is in a power, the following result is valid:

Theorem 2.2.3. *Let the following conditions be fulfilled:*

1. *The function $\alpha \in C^1([t_0, T), \mathbb{R}_+)$ is nondecreasing and $\alpha(t) \le t$.*

2. *The functions $p, q \in C([t_0, T), \mathbb{R}_+)$ and $a, b \in C([\alpha(t_0), T), \mathbb{R}_+)$.*

3. *The function $\phi \in C([\alpha(t_0) - h, t_0), \mathbb{R}_+)$.*

4. *The function $k \in C([t_0, T), (0, \infty))$ is nondecreasing and the inequality $M = \max_{s \in [\alpha(t_0)-h, t_0]} \phi(s) \ge \sqrt[n]{k(t_0)}$ holds.*

5. *The function $g \in \Omega$.*

6. *The function $u \in C([\alpha(t_0) - h, T), \mathbb{R}_+)$ and satisfies the inequalities*

$$\left(u(t) \right)^n \leq k(t)$$

$$+ \int_{t_0}^t \left[p(s)g\left(u(s) \right) + q(s)g\left(\max_{\xi \in [s-h,s]} u(\xi) \right) \right] ds$$

$$\text{for } t \in [t_0, T), \quad (2.69)$$

$$u(t) \leq \phi(t) \quad \text{for } t \in [\alpha(t_0) - h, t_0] \quad (2.70)$$

where $h = const \geq 0$, $n = const > 1$.

Then for $t \in [t_0, t_3)$ the inequality

$$u(t) \leq \sqrt[n]{k(t)} + e_1(t)\left\{ \frac{1}{n}\left(k(t) \right)^{\frac{1-n}{n}} + G^{-1}\left(G(1) + A_1(t) + B_1(t) \right) \right\}$$

$$(2.71)$$

holds, where

$$e_1(t) = 1 + \int_{t_0}^t \left[p(s)g\left(\psi(s) \right) + q(s)g\left(\max_{\xi \in [s-h,s]} \psi(\xi) \right) \right] ds$$

$$+ \int_{t_0}^{\max(\alpha(t),t_0)} \left[a(s)g\left(\psi(s) \right) ds + b(s)g\left(\max_{\xi \in [s-h,s]} \psi(s) \right) \right] ds,$$

$$(2.72)$$

$$A_1(t) = \frac{1}{n} \int_{t_0}^t \left[p(s)\left(k(s) \right)^{\frac{1-n}{n}} + q(s) \max_{\xi \in [s-h,s]} \left(k(\xi) \right)^{\frac{1-n}{n}} \right] ds, \quad (2.73)$$

$$B_1(t) = \frac{1}{n} \int_{\alpha(t_0)}^{\alpha(t)} \left[a(s)\left(K(s) \right)^{\frac{1-n}{n}} + b(s) \max_{\xi \in [s-h,s]} \left(K(\xi) \right)^{\frac{1-n}{n}} \right] ds,$$

$$(2.74)$$

$$K(t) = \begin{cases} k(t), & t \in [t_0, T) \\ k(t_0), & t \in [\alpha(t_0), t_0), \end{cases}$$

$$\psi(t) = \begin{cases} \sqrt[n]{k(t)}, & t \in (t_0, T) \\ M, & t \in [t_0 - h, t_0]. \end{cases}$$

the function G is defined by (2.45), the function G^{-1} is the inverse of G,

$$t_3 = \sup\left\{ \tau \in [t_0, T) : G(1) + A_1(t) + B_1(t) \in Dom\left(G^{-1} \right) \right.$$

$$\left. \text{for } t \in [t_0, \tau] \right\}.$$

Proof. Let us define a function $z(t) : [\alpha(t_0) - h, T) \to \mathbb{R}_+$ by

$$
z(t) = \begin{cases}
\dfrac{\sqrt[n]{k(t)}}{n \, k(t)} \left(\displaystyle\int_{t_0}^{t} \left[p(s)g\Big(u(s)\Big) + q(s)g\Big(\max_{\xi \in [s-h,s]} u(\xi) \Big) \right] ds \right. \\
\left. + \displaystyle\int_{\alpha(t_0)}^{\alpha(t)} \left[a(s)g\Big(u(s)\Big) + b(s)g\Big(\max_{\xi \in [s-h,s]} u(\xi) \Big) \right] ds \right), \\
\hspace{8cm} t \in [t_0, T), \\
0, \hspace{3cm} t \in [\alpha(t_0) - h, t_0).
\end{cases}
$$

By inequality (2.69) and the definition of $z(t)$ we have for $t \in [t_0, T)$

$$
\Big(u(t) \Big)^n \leq k(t) \left(1 + n \dfrac{z(t)}{\sqrt[n]{k(t)}} \right)
$$

or

$$
u(t) \leq \sqrt[n]{k(t)} \left(1 + n \dfrac{z(t)}{\sqrt[n]{k(t)}} \right)^{\frac{1}{n}}.
$$

Apply Bernoulli's inequality $(1+x)^a \leq 1 + ax$ where $0 < a < 1$ and $-1 < x$, and observe that

$$
u(t) \leq \sqrt[n]{k(t)} \left(1 + \dfrac{z(t)}{\sqrt[n]{k(t)}} \right) = \sqrt[n]{k(t)} + z(t)
$$

$$
= \psi(t) + z(t), \quad t \in [t_0, T), \tag{2.75}
$$

$$
u(t) \leq \phi(t) \leq \phi(t) + z(t) \leq \psi(t) + z(t), \quad t \in [\alpha(t_0) - h, t_0], \tag{2.76}
$$

where

$$
\psi(t) = \begin{cases}
\sqrt[n]{k(t)}, & t \in [t_0, T) \\
\sqrt[n]{k(t_0)}, & t \in [\alpha(t_0) - h, t_0).
\end{cases}
$$

Therefore,

$$
\max_{\xi \in [s-h,s]} u(\xi) \leq \max_{\xi \in [s-h,s]} \psi(\xi) + \max_{\xi \in [s-h,s]} z(\xi), \quad s \in [t_0, T). \tag{2.77}
$$

Let $t \in [t_0, T)$ such that $\alpha(t) \geq t_0$. Then from inequalities (2.75) and (2.76) we get

$$\int_{\alpha(t_0)}^{\alpha(t)} \left[a(s)g\Big(u(s)\Big) + b(s)g\Big(\max_{\xi \in [s-h,s]} u(\xi)\Big) \right] ds$$

$$\leq \int_{\alpha(t_0)}^{t_0} \left[a(s)g\Big(z(s)\Big) + b(s)g\Big(\max_{\xi \in [s-h,s]} z(\xi)\Big) \right] ds$$

$$+ \int_{t_0}^{\alpha(t)} \left[a(s)g\Big(\psi(s) + z(s)\Big) \right.$$

$$\left. + b(s)g\Big(\max_{\xi \in [s-h,s]} \psi(\xi) + \max_{\xi \in [s-h,s]} z(\xi)\Big) \right] ds$$

$$= \int_{t_0}^{\max(\alpha(t),t_0)} \left[a(s)g\Big(\psi(s)\Big) + b(s)g\Big(\max_{\xi \in [s-h,s]} \psi(\xi)\Big) \right] ds$$

$$+ \int_{\alpha(t_0)}^{\alpha(t)} \left[a(s)g\Big(z(s)\Big) + b(s)g\Big(\max_{\xi \in [s-h,s]} z(\xi)\Big) \right] ds. \qquad (2.78)$$

Now, let $t \in [t_0, T)$ such that $\alpha(t) < t_0$. This time, from the definition of function $z(t)$ and inequalities (2.75) and (2.76), we get

$$\int_{\alpha(t_0)}^{\alpha(t)} \left[a(s)g\Big(u(s)\Big) + b(s)g\Big(\max_{\xi \in [s-h,s]} u(\xi)\Big) \right] ds$$

$$= \int_{\alpha(t_0)}^{\alpha(t)} \left[a(s)g\Big(z(s)\Big) + b(s)g\Big(\max_{\xi \in [s-h,s]} z(\xi)\Big) \right] ds$$

$$\leq \int_{t_0}^{\max(\alpha(t),t_0)} \left[a(s)g\Big(\sqrt[n]{\psi(s)} \Big) + b(s)g\Big(\max_{\xi \in [s-h,s]} \sqrt[n]{\psi(\xi)} \Big) \right] ds$$

$$+ \int_{\alpha(t_0)}^{\alpha(t)} \left[a(s)g\Big(z(s)\Big) + b(s)g\Big(\max_{\xi \in [s-h,s]} z(\xi)\Big) \right] ds. \qquad (2.79)$$

It follows from the definition of the function $z(t)$ and inequalities

(2.78) and (2.79) that

$$z(t) \leq \frac{1}{n\left(k(t)\right)^{\frac{n-1}{n}}} \left(e_1(t) + \int_{t_0}^{t} \left[p(s)g\Big(z(s)\Big) + q(s)g\Big(\max_{\xi \in [s-h,s]} z(\xi)\Big) \right] ds \right.$$

$$\left. + \int_{\alpha(t_0)}^{\alpha(t)} \left[a(s)g\Big(z(s)\Big) + b(s)g\Big(\max_{\xi \in [s-h,s]} z(\xi)\Big) \right] ds \right)$$

$$\leq \frac{\sqrt[n]{k(t)}}{n\,k(t)} e_1(t) + \frac{1}{n} \int_{t_0}^{t} \left[p(s)\Big(k(s)\Big)^{\frac{1-n}{n}} g\Big(z(s)\Big) \right.$$

$$+ q(s)\Big(k(s)\Big)^{\frac{1-n}{n}} g\Big(\max_{\xi \in [s-h,s]} z(\xi)\Big) \bigg] ds$$

$$+ \frac{1}{n} \int_{\alpha(t_0)}^{\alpha(t)} \left[a(s)\Big(K(s)\Big)^{\frac{1-n}{n}} g\Big(z(s)\Big) \right.$$

$$+ b(s)\Big(K(s)\Big)^{\frac{1-n}{n}} g\Big(\max_{\xi \in [s-h,s]} z(\xi)\Big) \bigg] ds, \quad t \in [t_0, T), \tag{2.80}$$

$$z(t) \leq \phi(t), \quad t \in [t_0 - h, t_0]. \tag{2.81}$$

According to Theorem 2.2.2 from inequalities (2.80) and (2.81) we get

$$z(t) \leq e_1(t) \left\{ \frac{\sqrt[n]{k(t)}}{n\,k(t)} + G^{-1}\Big(G(1) + A_1(t) + B_1(t)\Big) \right\}, \tag{2.82}$$

where A_1 and B_1 are defined by equalities (2.73) and (2.74), respectively.

Substitute bound (2.82) for $z(t)$ into the right-hand side of (2.75) and obtain required inequality (2.71). $\qquad\square$

2.3 Integral Inequalities with Maxima for Scalar Functions of Two Variables

In the present section we solve some new Gronwall's type integral inequalities for continuous scalar functions of two variables. The main characteristic of the studied inequalities is the presence of the maximum of the unknown function in the integral. The obtained inequalities are used to investigate qualitative properties of the solutions of partial differential equations with "maxima."

Let points $a, b, x_0, y_0 \in \mathbb{R}_+$ be fixed such that $a > x_0$, $b > y_0$.

Let function $\alpha : \mathbb{R}_+ \to \mathbb{R}_+$ be such that $\alpha(x) \leq x$.

Consider following sets

$$
\begin{aligned}
G(x_0, y_0) &= \{(x, y) \in \mathbb{R}^2 : x \geq x_0, \ y \geq y_0\}, \\
\mathcal{G} &= \{(x, y) \in \mathbb{R}^2 : x \in [x_0, a], \ y \in [y_0, b]\}, \\
\Omega &= \{(x, y) \in \mathbb{R}^2 : x \in [\alpha(x_0) - h, x_0], \ y \geq y_0\},
\end{aligned}
\tag{2.83}
$$

where $h \geq 0$ is a constant.

We will study inequalities of the type

$$
u(x, y) \leq \varphi(x, y) + \int_{x_0}^{x} \int_{y_0}^{y} f(s, t) u^p(s, t) dt ds
$$

$$
+ \int_{\alpha(x_0)}^{\alpha(x)} \int_{y_0}^{y} g(s, t) \max_{\xi \in [s-h, s]} u^p(\xi, t) dt ds \quad \text{for } (x, y) \in G(x_0, y_0),
$$

$$
\tag{2.84}
$$

$$
u(x, y) \leq \psi(x, y) \quad \text{for } (x, y) \in \Omega,
\tag{2.85}
$$

where $u : G(\alpha(x_0) - h, y_0) \to \mathbb{R}_+$, functions $\varphi, f, g : G(x_0, y_0) \to \mathbb{R}_+$, and $\psi : \Omega \to \mathbb{R}_+$, $p \in (0, 1]$ is a real number.

We will prove some linear integral inequalities of Gronwall-Belman type for scalar continuous functions of two variables in the case when maxima of the unknown function is involved into the integral.

Initially we will study inequalities (2.84) and (2.85) in the case when $p = 1$.

Theorem 2.3.1. *Let the following conditions be fulfilled:*

1. *The functions $f, g \in C(G(x_0, y_0), \mathbb{R}_+)$.*

2. *The function $\alpha \in C([x_0, \infty), \mathbb{R}_+)$ is nondecreasing and $\alpha(x) \leq x$ for $x \geq x_0$.*

3. *The function $u \in C(G(\alpha(x_0)-h, y_0), \mathbb{R}_+)$ satisfies the inequalities*

$$u(x, y) \leq c + \int_{x_0}^{x} \int_{y_0}^{y} f(s, t) u(s, t) dt ds$$

$$+ \int_{\alpha(x_0)}^{\alpha(x)} \int_{y_0}^{y} g(s, t) \max_{\xi \in [s-h, s]} u(\xi, t) dt ds$$

$$\text{for } (x, y) \in G(x_0, y_0), \qquad (2.86)$$

$$u(x, y) \leq c \quad \text{for } (x, y) \in \Omega, \qquad (2.87)$$

where $h = const \geq 0, c = const > 0$.

Then for $(x, y) \in G(x_0, y_0)$ the inequality

$$u(x, y) \leq c e^{\int_{x_0}^{x} \int_{y_0}^{y} \{f(s,t) + g(\alpha(s), t)\alpha'(s)\} dt ds} \qquad (2.88)$$

holds.

Proof. Let us define a function $v: G(\alpha(x_0) - h, y_0) \rightarrow (0, \infty)$ by

$$v(x, y) = \begin{cases} c + \int_{x_0}^{x} \int_{y_0}^{y} f(s, t) u(s, t) dt ds \\ + \int_{\alpha(x_0)}^{\alpha(x)} \int_{y_0}^{y} g(s, t) \max_{\xi \in [s-h, s]} u(\xi, t) dt ds, & \text{for } (x, y) \in G(x_0, y_0), \\ c, & \text{for } (x, y) \in \Omega. \end{cases}$$

The function $v \in C^{2,2}[G(x_0 - h, y_0), (0, \infty)]$ is nondecreasing in both its arguments and

$$v(x_0, y) = c \text{ for } y \in [y_0, b], \qquad v(x, y_0) = c \text{ for } x \in [x_0, a], \quad (2.89)$$

$$v_x(x, y_0) = 0 \text{ for } x \in [x_0, a], \qquad v_y(x_0, y) = 0 \text{ for } y \in [y_0, b], \quad (2.90)$$

$$v_x(x, y) \geq 0, \qquad v_y(x, y) \geq 0 \text{ on } G(x_0, y_0). \qquad (2.91)$$

From the definition of the function v it follows that for $(x, y) \in G(x_0, y_0)$ the inequalities $u(x, y) \leq v(x, y)$ and

$$\max_{\xi \in [x-h, x]} u(\xi, y) \leq \max_{\xi \in [x-h, x]} v(\xi, y) = v(x, y)$$

hold.

Then from (2.86), (2.91) and the definition of function $v(x, y)$ we obtain for $(x, y) \in G(x_0, y_0)$

$$v_{xy}(x, y) \leq \left(f(x, y) + g(\alpha(x), y)\alpha'(x) \right) v(x, y)$$

$$\leq \left(f(x, y) + g(\alpha(x), y)\alpha'(x) \right) v(x, y) + \frac{v_x(x, y)v_y(x, y)}{v(x, y)}$$

$$\tag{2.92}$$

or

$$\frac{\partial}{\partial y} \left(\frac{v_x(x, y)}{v(x, y)} \right) \leq f(x, y) + g(\alpha(x), y)\alpha'(x). \tag{2.93}$$

Integrate inequality (2.93) from y_0 to y, use (2.90) and obtain

$$\frac{v_x(x, y)}{v(x, y)} \leq \int_{y_0}^{y} \left(f(x, t) + g(\alpha(s), t)\alpha'(s) \right) dt. \tag{2.94}$$

Integrate inequality (2.94) from x_0 to x, use (2.89) and obtain

$$\ln v(x, y) - \ln v(x_0, y) \leq \int_{x_0}^{x} \int_{y_0}^{y} \left(f(s, t) + g(\alpha(s), t)\alpha'(s) \right) dt ds. \tag{2.95}$$

From inequality (2.95) it follows that

$$v(x, y) \leq c e^{\int_{x_0}^{x} \int_{y_0}^{y} \{f(s,t) + g(\alpha(s),t)\alpha'(s)\} dt ds}. \tag{2.96}$$

Inequality (2.96) proves the validity of (2.88).

$$\square$$

In the case when the constant c in the inequalities (2.86) and (2.87) is zero, the following result is true:

Corollary 2.3.1. *Let the following conditions be fulfilled:*

1. *The conditions 1, 2 of Theorem 2.3.1 are satisfied.*

2. *The function $u \in C[G(\alpha(x_0) - h, y_0), \mathbb{R}_+]$ satisfies*

$$u(x, y) \leq \int_{x_0}^{x} \int_{y_0}^{y} f(s, t)u(s, t)dt ds$$

$$+ \int_{\alpha(x_0)}^{\alpha(x)} \int_{y_0}^{y} g(s, t) \max_{\xi \in [s-h, s]} u(\xi, t)dt ds \quad for \ (x, y) \in G(x_0, y_0),$$

$$\tag{2.97}$$

$$u(x, y) = 0 \quad for \ (x, y) \in \Omega, \tag{2.98}$$

where $h = const \geq 0$.

Then

$$u(x,y) \equiv 0 \quad for \ (x,y) \in G(x_0,y_0). \tag{2.99}$$

Proof. Let $\epsilon > 0$ be an arbitrary number. Then function $u(x,y)$ satisfies inequalities (2.86) and (2.87) for the constant $c = \epsilon$.

According to Theorem 2.3.1, the inequality (2.88) is true. Since $\epsilon > 0$ is an arbitrary number, we could take $\epsilon \to 0$ that proves the validity of (2.99).

$\square$

Corollary 2.3.2. *Let the following conditions be fulfilled:*

1. *The conditions 1, 2 of Theorem 2.3.1 are satisfied.*

2. *The function $\varphi \in C(G(x_0,y_0),(0,\infty))$ is nondecreasing in both its arguments.*

3. *The function $\psi \in C(\Omega,\mathbb{R}_+)$ and $\psi(x,y) \leq \psi(x_0,y) \leq \varphi(x_0,y)$ for $(x,y) \in \Omega$.*

4. *The function $u \in C(G(\alpha(x_0)-h,y_0),\mathbb{R}_+)$ satisfies the inequalities*

$$u(x,y) \leq \varphi(x,y) + \int_{x_0}^{x}\int_{y_0}^{y} f(s,t)u(s,t)dtds$$

$$+ \int_{\alpha(x_0)}^{\alpha(x)}\int_{y_0}^{y} g(s,t) \max_{\xi \in [s-h,s]} u(\xi,t)dtds \quad for \ (x,y) \in G(x_0,y_0),$$
$$\tag{2.100}$$

$$u(x,y) \leq \psi(x,y), \quad for \ (x,y) \in \Omega, \tag{2.101}$$

where $h = const \geq 0$.

Then for $(x,y) \in G(x_0,y_0)$ the inequality

$$u(x,y) \leq \varphi(x,y)e^{\int_{x_0}^{x}\int_{y_0}^{y}\{f(s,t)+g(\alpha(s),t)\alpha'(s)\}dtds} \tag{2.102}$$

holds.

Proof. We define a function $m : G(\alpha(x_0)-h,y_0) \to \mathbb{R}_+$ by

$$m(x,y) = \begin{cases} \dfrac{u(x,y)}{\varphi(x,y)} & for \ (x,y) \in G(x_0,y_0), \\ \dfrac{u(x,y)}{\varphi(x_0,y)} & for \ (x,y) \in \Omega. \end{cases}$$

Let $(x, y) \in G(x_0, y_0)$ and $s \in [x_0, x]$, $t \in [y_0, y]$.

If $s > x_0 + h$ then from the monotonicity of the function $\varphi(x, y)$ we obtain

$$\frac{\max_{\xi \in [s-h,s]} u(\xi, t)}{\varphi(x, y)} = \max_{\xi \in [s-h,s]} \frac{u(\xi, t)}{\varphi(x, y)} \leq \max_{\xi \in [s-h,s]} m(\xi, t). \qquad (2.103)$$

If $s \in [x_0, x_0 + h]$ then we obtain

$$\frac{\max_{\xi \in [s-h,s]} u(\xi, t)}{\varphi(x, y)} = \max \left(\frac{\max_{\xi \in [s-h,x_0]} u(\xi, t)}{\varphi(x, y)}, \frac{\max_{\xi \in [x_0,s]} u(\xi, t)}{\varphi(x, y)} \right)$$

$$\leq \max \left(\max_{\xi \in [s-h,x_0]} \frac{u(\xi, t)}{\varphi(x, y)}, \max_{\xi \in [x_0,s]} \frac{u(\xi, t)}{\varphi(x, y)} \right)$$

$$\leq \max_{\xi \in [s-h,s]} m(\xi, t). \qquad (2.104)$$

From inequalities (2.100), (2.103) and (2.104) we obtain for $(x, y) \in G(x_0, y_0)$ the validity of the inequality

$$m(x, y) \leq 1 + \int_{x_0}^{x} \int_{y_0}^{y} f(s, t) m(s, t) dt ds$$

$$+ \int_{\alpha(x_0)}^{\alpha(x)} \int_{y_0}^{y} g(s, t) \max_{\xi \in [s-h,s]} m(\xi, t) dt ds. \qquad (2.105)$$

Let $(x, y) \in \Omega$. Then from condition 3 and inequality (2.101) follows

$$m(x, y) = \frac{u(x, y)}{\varphi(x_0, y)} \leq \frac{\psi(x, y)}{\varphi(x_0, y)} \leq 1. \qquad (2.106)$$

From Theorem 2.3.1 and inequalities (2.105) and (2.106) we obtain the validity of inequality (2.102).

$\square$

Now we will solve the nonlinear integral inequalities (2.84) and (2.85) in the case when $0 < p < 1$, i.e., in the case when the unknown function and its maximum are involved nonlinearly into the integrals.

Theorem 2.3.2. *Let the following conditions be fulfilled:*

1. *The conditions 1, 2 of Theorem 2.3.1 are satisfied.*

2. *The function $u \in C(G(\alpha(x_0)-h, y_0), \mathbb{R}_+)$ satisfies the inequalities*

$$u(x,y) \le c + \int_{x_0}^{x} \int_{y_0}^{y} f(s,t)u^p(s,t)dsdt$$

$$+ \int_{\alpha(x_0)}^{\alpha(x)} \int_{y_0}^{y} g(s,t) \max_{\xi \in [s-h,s]} u^p(\xi,t)dsdt \quad for \quad (x,y) \in G(x_0, y_0),$$

$$(2.107)$$

$$u(s,y) \le c \quad for \quad (s,y) \in \Omega, \tag{2.108}$$

where $0 < p < 1$, $h = const \le 0$, $c = const > 0$.

Then for $(x,y) \in G(x_0, y_0)$ the inequality

$$u(x,y) \le \left(c^{1-p} + (1-p)\int_{x_0}^{x}\int_{y_0}^{y}\{f(s,t) + g(\alpha(s),t)\alpha'(s)\}dtds\right)^{\frac{1}{1-p}}$$

$$(2.109)$$

holds.

Proof. We define a function $v : G(\alpha(x_0) - h, y_0) \to (0, \infty)$ by

$$v(x,y) =$$

$$\begin{cases} c + \int_{x_0}^{x}\int_{y_0}^{y} f(s,t)u^p(s,t)dsdt + \int_{\alpha(x_0)}^{\alpha(x)}\int_{y_0}^{y} g(s,t)\max_{\xi\in[s-h,s]} u^p(\xi,t)dsdt \\ \qquad\qquad\qquad\qquad\qquad\qquad\qquad\qquad for \quad (x,y) \in G(x_0, y_0), \\ c \quad for \quad (s,y) \in \Omega. \end{cases}$$

The function $v \in C^{2,2}(G(x_0, y_0), (0, \infty))$ is nondecreasing in both its arguments and

$$v(x_0, y) = c \ for \ y \in [y_0, b], \qquad v(x, y_0) = c \ for \ x \in [x_0, a], \quad (2.110)$$

$$v_x(x, y_0) = 0 \ for \ x \in [x_0, a], \qquad v_y(x_0, y) = 0 \ for \ y \in [y_0, b], \quad (2.111)$$

$$v_x(x, y) \ge 0, \quad v_y(x, y) \ge 0 \ on \ G(x_0, y_0). \tag{2.112}$$

From the definition of the function v it follows that the function $v^p(x,y)$ is nondecreasing and for $(x,y) \in G(x_0, y_0)$ the inequalities $u(x,y) \le v(x,y)$, $u^p(x,y) \le v^p(x,y)$ and $\max_{\xi\in[x-h,x]} u^p(\xi,y) \le \max_{\xi\in[x-h,x]} v^p(\xi,y) = v^p(x,y)$ hold.

Then from the inequality (2.107) we obtain

$$v_{xy}(x,y) \le \left(f(x,y) + g(\alpha(x),y)\alpha'(x)\right)v^p(x,y), \quad (x,y) \in G(x_0, y_0). \tag{2.113}$$

From inequalities (2.112) and (2.113) it follows that

$$\frac{v_{xy}(x,y)}{v^p(x,y)} - \frac{pv_y(x,y)v_x(x,y)}{v^{p+1}(x,y)} \le f(x,y) + g(\alpha(x),y)\alpha'(x)$$

or

$$\frac{\partial}{\partial y}\left(\frac{v_x(x,y)}{v^p(x,y)}\right) \le f(x,y) + g(\alpha(x),y)\alpha'(x), \qquad (x,y) \in G(x_0,y_0).$$
(2.114)

Integrate inequality (2.114) from y_0 to y, use (2.111) and obtain

$$\frac{v_x(x,y)}{v^p(x,y)} - \frac{v_x(x_0,y)}{v^p(x_0,y)} \le \int_{y_0}^{y}\left(f(x,t) + g(\alpha(x),t)\alpha'(x)\right)dt$$

or

$$\frac{v_x(x,y)}{v^p(x,y)} \le \int_{y_0}^{y}\left(f(x,t) + g(\alpha(x),t)\alpha'(x)\right)dt. \tag{2.115}$$

Integrate inequality (2.115) from x_0 to x, use (2.110) and obtain

$$\frac{1}{1-p}\left(v^{1-p}(x,y) - v^{1-p}(x_0,y)\right) \le \int_{x_0}^{x}\int_{y_0}^{y}\left(f(s,t) + g(\alpha(s),t)\alpha'(s)\right)dtds,$$

or

$$\frac{1}{q}\left(v^q(x,y) - c^q\right) \le \int_{x_0}^{x}\int_{y_0}^{y}\left(f(s,t) + g(\alpha(s),t)\alpha'(s)\right)dtds, \tag{2.116}$$

where $p + q = 1$

From inequality (2.116) follows that

$$v(x,y) \le \left(c^q + (1-p)\int_{x_0}^{x}\int_{y_0}^{y}\{f(s,t) + g(\alpha(s),t)\alpha'(s)\}dtds\right)^{\frac{1}{1-p}}.$$
(2.117)

Inequality (2.117) proves the validity of (2.109).

$\square$

In the case when constant c in inequalities (2.107) and (2.108) is zero, we obtain the following result:

Corollary 2.3.3. *Let the following conditions be fulfilled:*

1. *The conditions 1,2 of Theorem 2.3.1 are satisfied.*

2. *The function $u \in C(G(\alpha(x_0)-h, y_0), \mathbb{R}_+)$ satisfies the inequalities*

$$u(x,y) \leq \int_{x_0}^{x} \int_{y_0}^{y} f(s,t)u^p(s,t)dtds$$

$$+ \int_{\alpha(x_0)}^{\alpha(x)} \int_{y_0}^{y} g(s,t) \max_{\xi \in [s-h,s]} u^p(\xi,t)dtds$$

$$\text{for} \quad (x,y) \in G(x_0, y_0), \tag{2.118}$$

$$u(s,y) \equiv 0 \quad \text{for} \quad (s,y) \in \Omega, \tag{2.119}$$

where $h = const \geq 0$, $p \in (0,1)$.

Then for $(x,y) \in G(x_0, y_0)$ the inequality

$$u(x,y) \leq \left\{ (1-p) \int_{x_0}^{x} \int_{y_0}^{y} \{f(s,t) + g(\alpha(s),t)\alpha'(s)\}dtds \right\}^{\frac{1}{1-p}} \tag{2.120}$$

holds.

Proof. Choose an arbitrary number $\epsilon > 0$. Then the function $u(x,y)$ satisfies inequalities (2.107) and (2.108) for $c = \epsilon$.

According to Theorem 2.3.2 we obtain the validity of the inequality

$$v(x,y) \leq \left(\epsilon^{1-p} + (1-p) \int_{x_0}^{x} \int_{y_0}^{y} \{f(s,t) + g(\alpha(s),t)\alpha'(s)\}dtds \right)^{\frac{1}{1-p}}.$$
$$\tag{2.121}$$

Since ϵ is an arbitrary number, it follows from (2.121) the validity of inequality (2.120) for $(x,y) \in G(x_0, y_0)$.

□

Corollary 2.3.4. *Let the following conditions be fulfilled:*

1. *The conditions 1, 2, 3 of Corollary 2.3.2 are satisfied.*

2. *The function $u \in C(G(\alpha(x_0)-h, y_0), \mathbb{R}_+)$ satisfies the inequalities*

$$u(x,y) \leq \varphi(x,y) + \int_{x_0}^{x} \int_{y_0}^{y} f(s,t)u^p(s,t)dtds$$

$$+ \int_{\alpha(x_0)}^{\alpha(x)} \int_{y_0}^{y} g(s,t) \max_{\xi \in [s-h,s]} u^p(\xi,t)dtds$$

$$\text{for} \quad (x,y) \in G(x_0, y_0), \tag{2.122}$$

$$u(s,y) \leq \psi(s,y) \quad \text{for} \quad (s,y) \in \Omega, \tag{2.123}$$

where $h = const \geq 0$, $p \in (0,1)$.

Then for $(x, y) \in G(x_0, y_0)$ *the inequality*

$$u(x, y) \leq \varphi(x, y)$$

$$\times \left\{ 1 + (1 - p) \int_{x_0}^{x} \int_{y_0}^{y} \left(\frac{f(s, t)}{\varphi^{1-p}(s, t)} + \frac{g(\alpha(s), t)}{\varphi^{1-p}(\alpha(s), t)} \alpha'(s) \right) dt ds \right\}^{\frac{1}{1-p}}$$

$$(2.124)$$

holds.

Proof. We define functions $m(x, y) : G(\alpha(x_0) - h, y_0) \to \mathbb{R}_+$ and $F, P : G(x_0, y_0) \to \mathbb{R}_+$ by

$$F(x, y) = \frac{f(x, y)}{\varphi^{1-p}(x, y)}, \qquad P(x, y) = \frac{g(x, y)}{\varphi^{1-p}(x, y)},$$

and

$$m(x, y) = \begin{cases} \dfrac{u(x, y)}{\varphi(x, y)} & \text{for} \quad (x, y) \in G(x_0, y_0), \\ \dfrac{u(x, y)}{\varphi(x_0, y)} & \text{for} \quad (x, y) \in \Omega. \end{cases}$$

From inequalities (2.122) and (2.123), using the monotonicity of the function $\varphi(x, y)$ we obtain the inequalities

$$m(x, y) \leq 1 + \int_{x_0}^{x} \int_{y_0}^{y} F(s, t) m^p(s, t) dt ds$$

$$+ \int_{\alpha(x_0)}^{\alpha(x)} \int_{y_0}^{y} P(s, t) \max_{\xi \in [s-h, s]} m^p(\xi, t) dt ds \quad \text{for} \quad (x, y) \in G(x_0, y_0),$$

$$(2.125)$$

$$m(s, y) \leq 1 \quad \text{for} \quad (s, y) \in \Omega. \qquad (2.126)$$

Applying Theorem 2.3.2 to inequalities (2.125) and (2.126), we obtain the inequality

$$m(x, y) \leq \left(1 + (1 - p) \int_{x_0}^{x} \int_{y_0}^{y} \{ F(s, t) + P(\alpha(s), t) \alpha'(s) \} dt ds \right)^{\frac{1}{1-p}}$$

$$(2.127)$$

Inequality (2.127) proves the validity of (2.124).

□

2.4 Applications of the Integral Inequalities with Maxima

We will apply some of the proved above inequalities to study various properties of solutions of differential equations with "maxima."

Consider the following system of differential equations with "maxima"

$$x' = f\left(t, x(t), \max_{s \in [\beta(t), \alpha(t)]} x(s)\right) \quad \text{for} \quad t \geq t_0 \tag{2.128}$$

with initial condition

$$x(t) = \phi(t) \quad \text{for} \quad t \in [\alpha(t_0) - h, t_0], \tag{2.129}$$

where $x \in \mathbb{R}^n$, $\phi : [\alpha(t_0) - h, t_0] \to \mathbb{R}^n$, $f : [0, \infty) \times \mathbb{R}^n \times \mathbb{R}^n \to \mathbb{R}^n$, $h > 0$ is a constant, $t_0 \geq 0$.

Theorem 2.4.1. *[Uniqueness] Let the following conditions be fulfilled:*

1. *The functions $\alpha, \beta \in C([t_0, \infty), \mathbb{R}_+)$, $\alpha(t)$ is an increasing function, $\beta(t) \leq \alpha(t) \leq t$, and $\alpha(t) - \beta(t) \leq h$ for $t \geq t_0$.*

2. *The function $f \in C([t_0, \infty) \times \mathbb{R}^n \times \mathbb{R}^n, \mathbb{R}^n)$ and satisfies for $t \geq t_0$ and $x_i, y_i \in \mathbb{R}^n$, $i = 1, 2$ the condition*

$$\|f(t, x_1, y_1) - f(t, x_2, y_2)\| \leq g(t) \|x_1 - x_2\| + r(t) \|y_1 - y_2\|,$$

where $g(t), r(t) \in C([t_0, \infty), \mathbb{R}_+)$.

3. *For any function $\phi \in C([\alpha(t_0) - h, t_0], \mathbb{R}^n)$ the initial value problem (2.128), (2.129) has at least one solution $x(t; t_0, \phi)$ defined for $t \geq \alpha(t_0) - h$.*

Then the initial problem (2.128), (2.129) has an unique solution.

Proof. Let $\phi \in C([\alpha(t_0) - h, t_0], \mathbb{R}^n)$ be a fixed initial function. Assume there exist two different solutions $u(t) = u(t; t_0, \phi)$ and $v(t) = v(t; t_0, \phi)$ of the initial value problems (2.128) and (2.129), which are defined for $t \geq \alpha(t_0) - h$. Both functions $u(t)$ and $v(t)$ satisfy the integral equations

$$u(t) = \phi(t_0) + \int_{t_0}^t f(s, u(s), \max_{\xi \in [\beta(s), \alpha(s)]} u(\xi)) ds \quad \text{for} \quad t \geq t_0,$$

$$v(t) = \phi(t_0) + \int_{t_0}^t f(s, v(s), \max_{\xi \in [\beta(s), \alpha(s)]} v(\xi)) ds \quad \text{for} \quad t \geq t_0$$

and $u(t) = v(t) = \phi(t)$ for $t \in [\alpha(t_0) - h, t_0]$.

Then the norm of the difference of both solutions $u(t)$ and $v(t)$ satisfies the inequalities

$$\|u(t) - v(t)\|$$

$$\leq \int_{t_0}^{t} \|f(s, u(s), \max_{\xi \in [\beta(s), \alpha(s)]} u(\xi)) - f(s, v(s), \max_{\xi \in [\beta(s), \alpha(s)]} v(\xi))\| ds$$

$$\leq \int_{t_0}^{t} \left(g(s)\|u(s) - v(s)\| ds + r(s)\| \max_{\xi \in [\beta(s), \alpha(s)]} u(\xi) - \max_{\xi \in [\beta(s), \alpha(s)]} v(xi)\| \right) ds$$

$$\leq \int_{t_0}^{t} g(s) \|u(s) - v(s)\| \, ds + \int_{t_0}^{t} r(s) \max_{\xi \in [\beta(s), \alpha(s)]} \|u(\xi) - v(\xi)\| ds,$$

$$t \geq t_0, \qquad (2.130)$$

$$\|u(t) - v(t)\| = 0, \quad t \in [\alpha(t_0) - h, t_0].$$

Set $p(t) = \|u(t) - v(t)\|$ for $t \in [t_0 - h, \infty)$, change the variable $\eta = \alpha(s)$ in the second integral of (2.130), use the inequality

$$\max_{\xi \in [\beta(t), \alpha(t))]} p(\xi) \leq \max_{\xi \in [\alpha(t) - h, \alpha(t))]} p(\xi)$$

that follows from condition 1 of Theorem 2.4.1 and obtain the inequality

$$p(t) \leq \int_{t_0}^{t} g(\eta)p(\eta)d\eta + \int_{\alpha(t_0)}^{\alpha(t)} r(\alpha^{-1}(\eta))(\alpha^{-1}(\eta))' \max_{\xi \in [\eta - h, \eta]} p(\xi)d\eta,$$

$$t \geq t_0. \quad (2.131)$$

According to Theorem 2.1.1 from inequality (2.131) and $p(t) \equiv 0$, $t \in [\alpha(t_0) - h, t_0]$ we obtain $p(t) \leq 0$ for $t \geq t_0$ that proves the validity of equality $\|u(t) - v(t)\| = 0$ for $t \geq t_0$ or $u(t) \equiv v(t)$. $\qquad \square$

Now we will obtain bounds for the solution of the initial problem (2.128), (2.129) in the case when the right-hand side is nonlinear.

Theorem 2.4.2. *[Bounds] Let the following conditions be fulfilled:*

1. *The functions $\alpha, \beta \in C^1([t_0, \infty), \mathbb{R}_+)$, $\alpha(t)$ is a nondecreasing function, $\beta(t) \leq \alpha(t) \leq t$ and $\alpha(t) - \beta(t) \leq h$ for $t \geq t_0$.*

2. *The function $f \in C([t_0, \infty) \times \mathbb{R}^n \times \mathbb{R}^n, \mathbb{R}^n)$ and satisfies for $t \geq t_0$ and $x, y \in \mathbb{R}^n$ the condition*

$$\left\| f(t, x, y) \right\| \leq P(t)\sqrt{\|x\|} + B(t)\sqrt{\|y\|}$$

where $P(t)$, $B(t) \in C([t_0, \infty), \mathbb{R}_+)$.

3. The function $\phi \in C([\alpha(t_0) - h, t_0], \mathbb{R}_+)$.

4. The function $u(t; t_0, \phi)$ is the solution of initial value problem (2.128), (2.129) defined for $t \geq \alpha(t_0) - h$.

Then the solution of initial value problem (2.128), (2.129) satisfies for $t \geq t_0$ the inequality

$$||u(t)|| \leq \frac{1}{4}\left(2\sqrt{||\phi(t_0)||} + \int_{t_0}^{t}\Big[P(s) + B(s)\Big]ds\right)^2. \qquad (2.132)$$

Proof. The solution $u(t) = u(t; t_0, \phi)$ of initial value problem (2.128), (2.129) satisfies the integral equations

$$u(t) = \phi(t_0) + \int_{t_0}^{t} f\left(s, \; u(s), \; \max_{\xi \in [\beta(s), \alpha(s)]} u(\xi)\right)ds \quad \text{for} \quad t \geq t_0,$$

and

$$u(t) = \phi(t) \quad \text{for} \quad t \in [\alpha(t_0) - h, t_0].$$

Then for the norm of the solution $u(t)$ we obtain

$$||u(t)|| \leq ||\phi(t_0)|| + \int_{t_0}^{t}\left\|f\left(s, \; u(s), \; \max_{\xi \in [\beta(s), \alpha(s)]} u(\xi)\right)\right\|ds$$

$$\leq ||\phi(t_0)|| + \int_{t_0}^{t}\left(P(s)\sqrt{||u(s)||} + B(s)\sqrt{\left\|\max_{\xi \in [\beta(s), \alpha(s)]} u(\xi)\right\|}\right)ds$$

$$\leq ||\phi(t_0)|| + \int_{t_0}^{t} P(s)\sqrt{||u(s)||}ds$$

$$+ \int_{t_0}^{t} B(s)\sqrt{\max_{\xi \in [\beta(s), \alpha(s)]} ||u(\xi)||}ds \quad \text{for } t \geq t_0, \qquad (2.133)$$

and

$$||u(t)|| = ||\phi(t)|| \quad \text{for} \quad t \in [\alpha(t_0) - h, t_0]. \qquad (2.134)$$

Set $\psi(t) = ||u(t)||$ for $t \in [\alpha(t_0) - h, \infty)$. Then we get for $t \geq t_0$.

$$\psi(t) \leq ||\phi(t_0)|| + \int_{t_0}^{t} P(s)\sqrt{\psi(s)}ds + \int_{t_0}^{t} B(s)\sqrt{\max_{\xi \in [\beta(s), \alpha(s)]} \psi(\xi)}ds.$$

$$(2.135)$$

Change the variable $\eta = \alpha(s)$ into the second integral of (2.135), use the inequality $\max_{\xi \in [\beta(t), \alpha(t))]} \psi(\xi) \leq \max_{\xi \in [\alpha(t)-h, \alpha(t)]} \psi(\xi)$ that follows from condition 1 of Theorem 2.4.2 and obtain

$$\psi(t) \leq ||\phi(t_0)|| + \int_{t_0}^{t} P(\eta)\sqrt{\psi(\eta)}d\eta$$
$$+ \int_{\alpha(t_0)}^{\alpha(t)} B(\alpha^{-1}(\eta))\left(\alpha^{-1}(\eta)\right)' \sqrt{\max_{\xi \in [\eta-h,\eta]} \psi(\xi)}d\eta. \quad (2.136)$$

Note conditions of Theorem 2.2.1 are satisfied for $k = ||\varphi(t_0)||$ and $p(t) \equiv P(t)$, $q(t) \equiv 0$ for $t \in [t_0, \infty)$, $a(t) \equiv 0$, $b(s) \equiv B(\alpha^{-1}(s))(\alpha^{-1}(s))'$ for $t \in [\alpha(t_0), \infty)$, $g(u) = \sqrt{u}$, $G(u) = 2\sqrt{u}$, $G^{-1}(u) = \frac{1}{4}u^2$, $Dom(G^{-1}) = \mathbb{R}_+$ and $t_1 = \infty$.

According to Theorem 2.2.1 from inequality (2.136) we obtain for $t \geq t_0$

$$\psi(t) \leq \frac{1}{4}\left(2\sqrt{||\phi(t_0)||} + \int_{t_0}^{t}\Big[P(s) + B(s)\Big]ds\right)^2. \quad (2.137)$$

Substitute the bound (2.137) for $\psi(t)$ into the right-hand side of equality $||u(t)|| = \psi(t)$ and obtain the required inequality (2.132). $\qquad\square$

Consider the following scalar differential equations with "maximum"

$$x\,x' = f\Big(t, x(t), \max_{s \in [\beta(t), \alpha(t)]} x(s)\Big) \quad \text{for} \quad t \geq t_0 \quad (2.138)$$

with initial condition

$$x(t) = \varphi(t) \quad \text{for} \quad t \in [\alpha(t_0) - h, t_0], \quad (2.139)$$

where $x \in \mathbb{R}$, $\varphi : [t_0 - h, t_0] \to \mathbb{R}$, $f : [0, \infty) \times \mathbb{R} \times \mathbb{R} \to \mathbb{R}$, $h > 0$ is a constant.

Theorem 2.4.3. *[Bounds] Let the following conditions be fulfilled:*

1. *The functions $\alpha, \beta \in C([t_0, \infty), \mathbb{R}_+)$, $\alpha(t)$ is an increasing function, $\beta(t) \leq \alpha(t) \leq t$ and $\alpha(t) - \beta(t) \leq h$ for $t \geq t_0$.*

2. *The function* $f \in C([t_0, \infty) \times \mathbb{R} \times \mathbb{R}, \mathbb{R})$, $f(t, 0, 0) \equiv 0$ *and satisfies for* $t \geq t_0$, $x_i, y_i \in \mathbb{R}$, $(i = 1, 2)$ *the condition*

$$|f(t, x_1, y_1) - f(t, x_2, y_2)| \leq g(t)|x_1 - x_2| + r(t)|y_1 - y_2|,$$

where $g(t), r(t) \in C([t_0, \infty), \mathbb{R}_+)$.

3. *The function* $\varphi \in C([\alpha(t_0) - h, t_0], \mathbb{R})$, $|\varphi(t_0)| > 0$.

4. *The solution* $x(t; t_0, \varphi)$ *of initial value problem* (2.138), (2.139) *is defined for* $t \geq \alpha(t_0) - h$.

Then for $t \geq t_0$ *the inequality*

$$|x(t; t_0, \varphi)| \leq |\varphi(t_0)|$$
$$+ M\left(1 + \frac{1}{|\varphi(t_0)|} \int_{t_0}^t (g(s) + r(s))ds\right) e^{\frac{1}{|\varphi(t_0)|} \int_{t_0}^t \left(g(s) + r(s)\right)ds} \quad (2.140)$$

holds, where $M = \max_{s \in [\alpha(t_0) - h, t_0]} |\varphi(s)|$.

Proof. The solution $x(t) = x(t; t_0, \varphi)$ satisfies the following integral equation

$$\left(x(t)\right)^2 = \left(\varphi(t_0)\right)^2 + \int_{t_0}^t 2f(s, x(s), \max_{\xi \in [\beta(s), \alpha(s)]} x(\xi))ds \quad \text{for} \quad t \geq t_0,$$
$$x(t) = \varphi(t), \quad t \in [\alpha(t_0) - h, t_0].$$

According to condition 2 of Theorem 2.4.3 we get

$$|x(t)|^2$$
$$\leq |\varphi(t_0)|^2 + 2\int_{t_0}^t |f(s, x(s), \max_{\xi \in [\beta(s), \alpha(s)]} x(\xi))|ds$$
$$\leq |\varphi(t_0)|^2 + 2\int_{t_0}^t g(s)|x(s)|ds + 2\int_{t_0}^t r(s) \max_{\xi \in [\beta(s), \alpha(s)]} |x(\xi)|ds,$$
$$t \geq t_0. \quad (2.141)$$

Set $u(t) = |x(t)|$ for $t \in [t_0 - h, \infty)$, change the variable $s = \alpha^{-1}(\eta)$ in the second integral of (2.141), use the inequality

$$\max_{\xi \in [\beta(t), \alpha(t))]} u(\xi) \leq \max_{\xi \in [\alpha(t) - h, \alpha(t)]} u(\xi)$$

that follows from condition 1 of Theorem 2.4.3 and obtain for $t \geq t_0$ the following inequality

$$\left(u(t)\right)^2 \leq |\varphi(t_0)|^2 + \int_{t_0}^{t} 2g(s)u(s)ds$$

$$+ \int_{\alpha(t_0)}^{\alpha(t)} 2r(\alpha^{-1}(\eta))(\alpha^{-1}(\eta))' \max_{\xi \in [\eta-h,\eta]} u(\xi)d\eta. \quad (2.142)$$

From inequality (2.142) according to Theorem 2.1.4 for $n = 2$, $\phi(t) = |\varphi(t)|$ on $[t_0 - h, t_0]$, $k(t) = |\varphi(t_0)|^2 > 0$, $p(t) = 2g(t)$, $q(t) \equiv 0$ on $[t_0, \infty)$, and $a(t) \equiv 0$, $b(t) = 2r(\alpha^{-1}(t))(\alpha^{-1}(t))'$ on $[\alpha(t_0), \infty)$ we obtain the inequality

$$u(t) \leq |\phi(t_0)| + \left(M + \frac{e(t)}{2|\phi(t_0)|}\right)e^{\frac{1}{|\phi(t_0)|}\int_{t_0}^{t}\left(g(s)+r(s)\right)ds}, \quad (2.143)$$

hold, where

$$e(t) \leq 2M \int_{t_0}^{t}\left(g(s)+r(s)\right)ds.$$

Inequality (2.143) proves the validity of inequality (2.140).

$\square$

We apply some of the solved integral inequalities in Section 2.3 to investigate qualitative properties of partial differential equations with "maxima".

Consider the nonlinear partial differential equation with "maxima"

$$u_{xy}(x,y) = F(x, y, u(x,y), \max_{\xi \in [\alpha(x),\beta(x)]} u(\xi, y)) \quad \text{for} \quad (x,y) \in \mathcal{G}$$

$$(2.144)$$

with initial conditions

$$u(x_0, y) = \varphi_1(y), \quad u_x(x_0, y) = \varphi_2(y) \quad \text{for} \quad y \in [y_0, b],$$

$$u(x, y) = \varphi_3(x, y) \quad \text{for} \quad x \in [\beta(x_0) - h, x_0], \quad y \in [y_0, b], \quad (2.145)$$

and boundary conditions

$$u(x, y_0) = \psi_1(x), \quad u(x, b) = \psi_2(x) \quad \text{for } x \in [x_0, a], \quad (2.146)$$

where $u : \mathbb{R}^2 \to \mathbb{R}$, $h = sup\{\beta(s) - \alpha(s) : s \in [x_0, a]\}$, and the sets $G(x_0, y_0), \mathcal{G}, \Omega$ are defined by (2.83).

We will assume that the solution of (2.144)–(2.146) exists on the rectangular $\mathcal{G}$.

We will say that conditions (**H2.4**) are satisfied if

H2.4.1. Functions $\alpha, \beta \in C([x_0, a], \mathbb{R})$ are such that $\alpha(x) \le \beta(x) \le x$ for $x \in [x_0, a]$, $\beta(x)$ is a nondecreasing function;

H2.4.2. Functions $\varphi_1, \varphi_2 \in C^1([y_0, b], \mathbb{R})$;

H2.4.3. Functions $\psi_1, \psi_2 \in C^1([x_0, a], \mathbb{R})$ and $\varphi_1(y_0) = \psi_1(x_0)$;

H2.4.4. Function $\varphi_3(x, y) \in C([x_0 - h, x_0] \times [y_0, b], \mathbb{R})$ and the equality $\varphi_3(x_0, y) = \varphi_1(y)$ holds for $y \in [y_0, b]$.

H2.4.5. The mixed problem (2.144)-(2.146) has a solution, defined in $\mathcal{G}$.

If conditions (H2.4) are satisfied, then the solution $u(x, y)$ of (2.144)-(2.146) satisfies for $(x, y) \in \mathcal{G}$ the integral equation

$$u(x, y) = \varphi_1(y) + \psi_1(x) - \psi_1(x_0)$$
$$+ \int_{x_0}^{x} \int_{y_0}^{y} F(s, t, u(s, t), \max_{\xi \in [\alpha(s), \beta(s)]} u(\xi, t)) dt ds \quad (2.147)$$

$$u(x, y) = \varphi_3(x, y) \text{ for } x \in [\beta(x_0) - h, x_0]. \quad (2.148)$$

We will apply directly the above-solved integral inequalities involving a maximum of the unknown function to obtain some qualitative properties of the solutions of the partial differential equations (2.144)-(2.146).

Initially we will prove the uniqueness of the solution of (2.144)-(2.146).

Theorem 2.4.4. *[Uniqueness] Let the following conditions be fulfilled:*

1. *The function $F(x, y, u, v) \in C(\mathcal{G} \times \mathbb{R}^2, \mathbb{R})$ and there exists $f, g \in C(\mathcal{G}, \mathbb{R}_+)$ such that $|F(x, y, u_1, v_1) - F(x, y, u_2, v_2)| \le f(x, y)|u_1 - u_2| + g(x, y)|v_1 - v_2|$ for $(x, y) \in \mathcal{G}$ and $u_1, u_2, v_1, v_2 \in \mathbb{R}$.*

2. *The conditions H2.4 are satisfied.*

Then the mixed problem for the partial differential equations with "maxima" (2.144)-(2.146) has an unique solution.

Proof. Assume that there are two solutions $u(x,y), v(x,y)$ of (2.144)-(2.146) defined in $\mathcal{G}$. Both functions satisfy the integral equations (2.147) and (2.148). Then we obtain the inequalities

$$|u(x,y) - v(x,y)|$$

$$\leq \int_{x_0}^x \int_{y_0}^y |F(s,t,u(s,t), \max_{\xi \in [\alpha(s),\beta(s)]} u(\xi,t))$$

$$- F(s,t,v(s,t), \max_{\xi \in [\alpha(s),\beta(s)]} v(\xi,t))|dtds$$

$$\leq \int_{x_0}^x \int_{y_0}^y \Big(f(x,y)|u(s,t) - v(s,t)|$$

$$+ g(x,y)| \max_{\xi \in [\alpha(s),\beta(s)]} u(\xi,t) - \max_{\xi \in [\alpha(s),\beta(s)]} v(\xi,t)| \Big) dtds.$$

From the continuity of the function $u(x,y)$ it follows that for any fixed points $s \in [x_0, a]$ and $t \in [y_0, y]$ there exists a point $\eta \in [\alpha(s), \beta(s)]$ such that the inequality $\max_{\xi \in [\alpha(s),\beta(s)]} u(\xi,t) = u(\eta,t)$ holds and therefore

$$| \max_{\xi \in [\alpha(s),\beta(s)]} u(\xi,t) - \max_{\xi \in [\alpha(s),\beta(s)]} v(\xi,t)| = |u(\eta,t) - \max_{\xi \in [\alpha(s),\beta(s)]} v(\xi,t)|$$

$$\leq |u(\eta,t) - v(\eta,t)| \leq \max_{\xi \in [\alpha(s),\beta(s)]} |u(\xi,t) - v(\xi,t)|. \quad (2.149)$$

Then we obtain

$$|u(x,y) - v(x,y)| \leq \int_{x_0}^x \int_{y_0}^y f(x,y)|u(s,t) - v(s,t)|$$

$$+ \int_{x_0}^x \int_{y_0}^y g(x,y) \max_{\xi \in [\alpha(s),\beta(s)]} |u(\xi,t) - v(\xi,t)|dtds \quad (2.150)$$

We change the variable $\eta = \beta(s)$ in the second integral of (2.150), use the inequality

$$\max_{\xi \in [\alpha(s),\beta(s)]} |u(\xi,t) - v(\xi,t)| \leq \max_{\xi \in [\eta-h,\eta]} |u(\xi,t) - v(\xi,t)|,$$

and obtain

$$|u(x,y) - v(x,y)| \leq \int_{x_0}^x \int_{y_0}^y f(x,y)|u(s,t) - v(s,t)|$$

$$+ \int_{\beta(x_0)}^{\beta(x)} \int_{y_0}^y g(\beta^{-1}(\eta), y)(\beta^{-1}(\eta))' \max_{\xi \in [\eta-h,\eta]} |u(\xi,t) - v(\xi,t)|dtd\eta.$$

$$(2.151)$$

From inequality (2.151) and Corollary 2.3.1 we obtain $|u(x,y) - v(x,y)| \equiv 0$.

$\square$

We will obtain some bounds of the solution $u(x,y)$ of (2.144)-(2.146).

Theorem 2.4.5. *Let the following conditions be fulfilled:*

1. *The function $F(x,y,u,v) \in C(\mathcal{G} \times \mathbb{R}^2, \mathbb{R})$ and there exists $f,g \in C(\mathcal{G}, \mathbb{R}_+)$ such that*

$$|F(x,y,u,v)| \leq f(x,y)|u| + g(x,y)|v| \quad for \quad (x,y) \in \mathcal{G}, \quad u,v \in \mathbb{R}.$$

2. *The conditions H2.4 are satisfied.*

Then for $(x,y) \in \mathcal{G}$ the solution $u(x,y)$ of the partial differential equation with "maxima" (2.144)-(2.146) satisfies the inequality

$$|u(x,y)| \leq |\varphi_1(y) + \psi_1(x) - \psi_1(x_0)| e^{\int_{x_0}^{x} \int_{y_0}^{y} \{f(s,t) + g(s,t)\gamma'(\beta(s))\beta'(s)\} dt ds},$$
(2.152)

where $\gamma(s)$ is the inverse of $\beta(s)$.

Proof. The solution $u(x,y)$ of (2.144)-(2.146) satisfies on $\mathcal{G}$ the inequality

$$|u(x,y)| \leq |\varphi_1(y) + \psi_1(x) - \psi_1(x_0)| + \int_{x_0}^{x} \int_{y_0}^{y} f(s,t)|u(s,t)| dt ds$$

$$+ \int_{x_0}^{x} \int_{y_0}^{y} g(s,t)| \max_{\xi \in [\alpha(s),\beta(s)]} |u(\xi,t)| dt ds. \quad (2.153)$$

From the continuity of the function $u(x,y)$ follows that for any fixed points $s \in [x_0, a]$ and $t \in [y_0, y]$ there exists a point $\eta \in [\alpha(s), \beta(s)]$ such that

$$| \max_{\xi \in [\alpha(s),\beta(s)]} u(\xi,t)| = |u(\eta,t)| \leq \max_{\xi \in [\alpha(s),\beta(s)]} |u(\xi,t)|$$

$$\leq \max_{\xi \in [\beta(s)-h,\beta(s)]} |u(\xi,t)|. \quad (2.154)$$

From inequalities (2.153) and (2.154) we obtain

$$|u(x,y)| \leq |\varphi_1(y) + \psi_1(x) - \psi_1(x_0)| + \int_{x_0}^{x} \int_{y_0}^{y} f(s,t)|u(s,t)| dt ds$$

$$+ \int_{\beta(x_0)}^{\beta(x)} \int_{y_0}^{y} g(\gamma(s),t)\gamma'(s)| \max_{\xi \in [s-h,s]} |u(\xi,t)| dt ds. \quad (2.155)$$

From inequality (2.155) and Corollary 2.3.4 follows the validity of (2.152).

$\qquad\qquad\qquad\qquad\qquad\qquad\qquad\qquad\qquad\qquad\qquad\qquad$ $\square$

Theorem 2.4.6. *Let the following conditions be fulfilled:*

1. *The function $F(x, y, u, v) \in C(\mathcal{G} \times \mathbb{R}^2, \mathbb{R})$ and there exists $f, g \in C(\mathcal{G}, \mathbb{R}_+)$ such that*

$$|F(x, y, u, v)| \le f(x, y)|u|^p + g(x, y)|v|^p \quad for \quad (x, y) \in \mathcal{G}, \; u, v \in \mathbb{R},$$

where $0 < p < 1$.

2. *The conditions H2.4 are satisfied.*

Then for $(x, y) \in \mathcal{G}$ the solution $u(x, y)$ of the partial differential equation with "maxima" (2.144)-(2.146) satisfies the inequality

$$|u(x, y)| \le q(x, y)\Big\{1$$

$$+ (1-p) \int_{x_0}^x \int_{y_0}^y \frac{f(s, t)}{q(s, t)^{1-p}} + \frac{g(s, t)}{q(\beta(s), t)^{1-p}} \gamma'(\beta(s))\beta'(s) dt ds \Big\}^{\frac{1}{1-p}},$$

$$\tag{2.156}$$

where $q(x, y) = |\varphi_1(y) + \psi_1(x) - \psi_1(x_0)|$, $\gamma(s)$ is the inverse of $\beta(s)$.

The proof of inequality (2.156) follows from the Corollary 2.3.4 and equalities (2.147) and (2.148).

Chapter 3

General Theory

Some fundamental concepts and theorems in the qualitative analysis of various types of differential equations with "maxima" will be introduced in this chapter.

3.1 Existence Theory for Initial Value Problems

We will prove some existence results for differential equations with "maxima." The main characteristic of the considered equations is the presence of maximum of the derivative of the unknown function in the right side of the equation. Both scalar and multidimensional cases are studied.

Consider the scalar differential equation with "maxima"

$$y'(t) = F\left(t, \max_{s \in [p(t),\, q(t)]} y(s), \max_{s \in [\beta(t),\, \alpha(t)]} y'(s)\right) \quad \text{for} \quad t \geq 0 \qquad (3.1)$$

with initial condition

$$y(t) = \psi(t), \quad y'(t) = \psi'(t) \quad \text{for} \quad t \leq 0, \qquad (3.2)$$

where $y \in \mathbb{R}$, $\beta(t) \leq \alpha(t) \leq t$ and $p(t) \leq q(t) \leq t$ for $t \geq 0$.

Let $\mathcal{B}_i$ be Banach spaces with norms $\|.\|_i$, $i = 1, 2$.

In our further investigations we will use the following existence results:

Proposition 3.1.1. *(see [Angelov and Bainov 1981]) Let the following conditions be fulfilled:*

1. *On $\mathcal{B}_i$ the nonlinear continuous operators $N_i : \mathcal{B}_i \to \mathcal{B}_i$ satisfy*

$$\left\| \Big((1 + \lambda\gamma - \lambda\alpha)I + \lambda N_i \Big)x - \Big((1 + \lambda\gamma - \lambda\alpha)I + \lambda N_i \Big)y \right\|_i$$
$$\geq \left\| x - y \right\|_i$$

    for some $\gamma, \alpha > 0$ and $x, y \in \mathcal{B}_i$.

2. *The linear mapping $j : \mathcal{B}_1 \to \mathcal{B}_2$ has the property $j(N_1 x) = N_2(j)$, $x \in \mathcal{B}_1$.*

Then for every $x \in \mathcal{B}_1$ and $y \in \mathcal{B}_2$ for which $jx = (N_2 + \gamma I)y$, there exists a unique element $z \in \mathcal{B}_1$ satisfying the inequalities $(N_1 + \gamma I)z = x$, $jz = y$.

Proposition 3.1.2. *(see [Angelov and Bainov 1981]) Under the assumptions of Proposition 3.1.1, if*

$$jx = (N_2 + \gamma I)y, \quad j\bar{x} = (N_2 + \gamma I)\bar{y}$$

and

$$(N_1 + \gamma I)z = x, \quad jz = y, \quad (N_1 + \gamma I)\bar{z} = \bar{x}, \quad j\bar{z} = \bar{y},$$

then

$$\left\| z - \bar{z} \right\|_1 \leq \frac{1}{\alpha} \left\| x - \bar{x} \right\|_1.$$

After the change of variables $x(t) = y'(t)$ for $t > 0$ and $\varphi(t) = \psi'(t)$ for $t \leq 0$ in (3.1), (3.2), we obtain the following initial value problem

$$x(t) = F\left(t, \max_{s \in [p(t), q(t)]} \int_0^s x(\theta)d\theta, \max_{s \in [u\beta(t), \alpha(t)]} x(s) \right), \quad t > 0, \quad (3.3)$$
$$x(t) = \varphi(t), \quad t \leq 0. \tag{3.4}$$

Remark 3.1.1. *Note the function $M(t) = \max_{s \in [p(t), q(t)]} f(s)$ is continuous if $f, p, q \in C(\mathbb{R}, \mathbb{R})$. In order to emphasize this fact we shall introduce the following metric in the set of all intervals*

$$J = \{[p, q] : p, q \in \mathbb{R}; p \leq q\},$$
$$\rho([p, q], [\bar{p}, \bar{q}]) = \max \{|p - \bar{p}|, |q - \bar{q}|\}.$$

It is now obvious that the map $P : \mathbb{R}_+ \to J$ such that $P(t) = [p(t), q(t)]$ is continuous when $p(t)$ and $q(t)$ are continuous. In an analogous way we can establish the continuity of the map $Q : J \to \mathbb{R}$ such that $Q([p, q]) = \max_{s \in [p, q]} f(s)$. Then $M(t)$ is the superposition of P and Q, i.e., $M(t) = Q(P(t))$.

Theorem 3.1.1. *Let the following conditions be fulfilled:*

1. *The functions $p, q, \alpha, \beta : \mathbb{R}_+ \to \mathbb{R}$ are continuous, $p(t) \leq q(t) \leq t$ and $\beta(t) \leq \alpha(t) \leq t$.*

2. *The function $F(t, x, y) : \mathbb{R}_+ \times \mathbb{R}^2 \to \mathbb{R}$ is continuous and there exist constants $\gamma > m > 0$ such that*

$$\left|F(t, x, y)\right| \leq \frac{1}{\gamma}\, \mu(t, |x|, |y|),$$

$$\left|F(t, x, y) - F(t, \bar{x}, \bar{y})\right| \leq \frac{m}{\gamma}\lambda(t, |x - \bar{x}|, |y - \bar{y}|),$$

   where $\mu(t, u, v)$, $\lambda(t, u, v) : \mathbb{R}_+^3 \to \mathbb{R}_+$ are nondecreasing in u, v, the function $\bar{\mu}(t) = \mu(t, q(t)u, u)$ is bounded in t and $\lambda(t, q(t)u, u) \leq u$ for every $u \in \mathbb{R}_+$.

3. *The initial function $\varphi(t) : \mathbb{R}_- \to \mathbb{R}$ is bounded, continuous and satisfies the condition*

$$\varphi(0) = F\left(0, \max_{s \in [p(0), q(0)]} \int_0^s \varphi(\theta)d\theta, \max_{s \in [\beta(0), \alpha(0)]} \varphi(s)\right).$$

Then there exists a unique continuous and bounded solution of the initial value problem (3.3), (3.4).

Proof. Let $\mathcal{B}_1$ be Banach space of all bounded and continuous functions $f(t) : \mathbb{R} \to \mathbb{R}$ with supremum norm and $\mathcal{B}_2$ be Banach space of all bounded and continuous functions $g(t) : \mathbb{R}_- \to \mathbb{R}$ with supremum norm.

Define the operators $N_i : \mathcal{B}_i \to \mathcal{B}_i$, $(i = 1, 2)$, by equalities

$$(N_1 f)(t) = \begin{cases} -\gamma F\left(t, \displaystyle\max_{s \in [p(t), q(t)]} \int_0^s f(\tau)d\tau, \max_{s \in [\beta(t), \alpha(t)]} f(s)\right), & t > 0 \\[2mm] -\gamma F\left(0, \displaystyle\max_{s \in [p(0), q(0)]} \int_0^s f(\tau)d\tau, \max_{s \in [\beta(0), \alpha(0)]} f(s)\right), & t \leq 0, \end{cases}$$

for $f \in \mathcal{B}_i$ and

$$(N_2 g)(t) = -\gamma F\left(0, \max_{s \in [p(0), q(0)]} \int_0^s g(\tau)d\tau, \max_{s \in [\beta(0), \alpha(0)]} g(s)\right), \quad t \leq 0.$$

The linear map $j : \mathcal{B}_1 \to \mathcal{B}_2$ is defined as a restriction of the function $f \in \mathcal{B}_1$ on the semi-axis $\mathbb{R}_-$.

Then the operators $N_i : \mathcal{B}_i \to \mathcal{B}_i$. Indeed, if follows from the estimates

$$\left|(N_1 f)(t)\right| \leq \mu\left(t,\ q(t)||f||_1,\ ||f||_1\right), \quad t > 0$$
$$\left|(N_1 f)(t)\right| \leq \mu\left(0,\ q(0)||f||_1,\ ||f||_1\right), \quad t \leq 0$$

that $(N_1 f)(t)$ is bounded. The continuity of the functions

$$M_f(t) = \max_{s \in [p(t),\ q(t)]} \int_0^s f(\tau)d\tau,$$
$$P_f(t) = \max_{s \in [\beta(t), \alpha(t)]} f(s)$$

implies the continuity of the function $(N_1 f)(t)$ that is $(N_1 f)(t) \in \mathcal{B}_1$.

Let $f,\ g \in \mathcal{B}_1$. Then for $t > 0$ we obtain

$$\left|(N_1 f)(t) - (N_1 g)(t)\right|$$
$$\leq m\lambda\left(t,\ \max_{s \in [p(t),q(t)]} \int_0^s |f(\tau) - g(\tau)|d\tau,\ \max_{s \in [\beta(t),\alpha(t)]} |f(s) - g(s)|\right)$$
$$\leq m\lambda\left(t,\ q(t)||f - g||_1,\ ||f - g||_1\right) \leq m||f - g||_1$$

and for $t \leq 0$

$$\left|(N_1 f)(t) - (N_1 g)(t)\right| \leq m\lambda\left(0,\ q(0)||f - g||_1,\ ||f - g||_1\right)$$
$$\leq m||f - g||_1$$

that is

$$\left\|N_1 f - N_1 g\right\|_1 \leq m||f - g||_1.$$

It is easy to verify that the operator $N_i + (\gamma - \alpha)I$ is an accretive one where $\alpha = \gamma - m$.

Finally, define the function

$$h(t) = \begin{cases} 0, & t > 0, \\ (N_2 + \gamma I)\varphi(t), & t \leq 0. \end{cases}$$

Then Proposition 3.1.1 implies an existence of a unique function $x(t) \in \mathcal{B}_1$ such that

$$(N_1 + \gamma I)x(t) = h(t)$$

and $jx(t) = \varphi(t)$, i.e., the initial value problem (3.3), (3.4) has a unique solution.

Thus Theorem 3.1.1 is proved.

$\square$

As a consequence of Proposition 3.1.2, we obtain:

Theorem 3.1.2. *Let $x(t; \varphi_1)$ and $x(t; \varphi_2)$ be two solutions of the initial value problem (3.1),(3.2).*

Then

$$\left| x(t; \varphi_1) - x(t; \varphi_2) \right| \le \frac{\gamma}{\gamma - m} \left[\left| \varphi_1(0) - \varphi_2(0) \right| + \sup_{t \in \mathbb{R}_-} \left| \varphi_1(t) - \varphi_2(t) \right| \right].$$

Now we will consider a scalar differential equation with "maxima" which is more general than (3.1) and (3.2):

$$y'(t) = F\Big(t, \max_{s \in [p_1(t), q_1(t)]} y(s), \ldots, \max_{s \in [p_m(t), q_m(t)]} y(s), $$

$$\max_{s \in [\alpha_1(t), \beta_1(t)]} y'(s), \ldots, \max_{s \in [\alpha_n(t), \beta_n(t)]} y'(s) \Big), \quad t > 0, \tag{3.5}$$

$$y(t) = \psi(t), \quad y'(t) = \psi'(t), \quad t \le 0, \tag{3.6}$$

where $y \in \mathbb{R}$, p_k, q_k, α_j, $\beta_j : \mathbb{R}_+ \to \mathbb{R}$, $(k = 1, 2, \ldots, m; \ j = 1, 2, \ldots, n)$, $p_k(t) \le q_k(t) \le t$, $\alpha_j(t) \le \beta_j(t) \le t$.

Changing the variables $x(t) = y'(t)$ and $\varphi(t) = \psi'(t)$, we obtain the following initial value problem

$$x(t) = F\Big(t, \max_{s \in [p_1(t), \ q_1(t)]} \int_0^s x(\tau) d\tau, \ldots, \max_{s \in [p_m(t), \ q_m(t)]} \int_0^s x(\tau) d\tau, $$

$$\max_{s \in [\beta_1(t), \alpha_1(t)]} x(s), \ldots, \max_{s \in [\beta_n(t), \alpha_n(t)]} x(s) \Big), \quad t > 0, \tag{3.7}$$

$$x(t) = \varphi(t), \quad t \le 0. \tag{3.8}$$

The analogous result is valid.

Theorem 3.1.3. *Let the following conditions be fulfilled:*

1. *The functions $p_i(t)$, $q_i(t)$, $\alpha_i(t)$, $\beta_i(t) : \mathbb{R}_+ \to \mathbb{R}$ are continuous, $p_i(t) \le q_i(t) \le t$, $\beta_i(t) \le \alpha_i(t) \le t$.*

2. *The functions* $F\big(t,\, x_1,\ldots,\, x_m,\, y_1,\ldots,\, y_n\big) : \mathbb{R}_+ \times \mathbb{R}^{m+n} \to \mathbb{R}$
are continuous and

$$\Big|F\big(t,\, x_1,\ldots,\, x_m,\, y_1,\ldots,\, y_n\big)\Big|$$
$$\leq \frac{1}{\gamma}\mu\Big(t,\, |x_1|,\ldots,\, |x_m|,\, |y_1|,\ldots,\, |y_n|\Big),$$

$$\Big|F\big(t,\, x_1,\ldots,\, x_m,\, y_1,\ldots,\, y_n\big) - F\big(t,\, \bar{x}_1,\ldots,\, \bar{x}_m,\, \bar{y}_1,\ldots,\, \bar{y}_n\big)\Big|$$
$$\leq \frac{m}{\gamma}\lambda\Big(t,\, |x_1-\bar{x}_1|,\ldots,\, |x_m-\bar{x}_m|,\, |y_1-\bar{y}_1|,\ldots,\, |y_n-\bar{y}_n|\Big)$$

for some $\gamma > m > 0$ *where the function* $\bar{\mu}(t) = \mu\big(t,\, q_1(t)u,\ldots,\, q_m(t)u,\, u,\ldots,\, u\big)$ *is bounded in* t *and* $\lambda\big(t,\, q_1(t)u,\ldots,\, q_m(t)u,\, u,\ldots,\, u\big) \leq u$ *for every* $u \in \mathbb{R}_+$.

3. *The initial function* $\varphi(t) : \mathbb{R}_- \to \mathbb{R}$ *satisfies condition 3 of Theorem 3.1.1 (with obvious modification on the conformity condition).*

Then there exists a unique continuous and bounded solution of the initial value problem (3.7), (3.8).

The proof is analogous to the one of Theorem 3.1.1.

Theorem 3.1.4. *Let* $x(t;\varphi_1)$ *and* $x(t;\varphi_2)$ *be two solutions of the initial value problem (3.7), (3.8).*
Then

$$|x(t;\varphi_1) - x(t;\varphi_2)| \leq \frac{\gamma}{\gamma-m}\Big[|\varphi_1(0)-\varphi_2(0)| + \sup_{t\in\mathbb{R}_-}|\varphi_1(t)-\varphi_2(t)|\Big].$$

Now we will consider the following system of differential equations with "maxima:"

$$y_k'(t) = F_k\Big(t,\, \max_{s\in[p_1(t),q_1(t)]}y_1(s),\ldots,\, \max_{s\in[p_n(t),q_n(t)]}y_n(s),$$
$$\max_{s\in[\beta_1(t),\alpha_1(t)]}y_1'(s),\ldots,\, \max_{s\in[\beta_n(t),\alpha_n(t)]}y_n'(s)\Big), \quad t>0, \quad (3.9)$$
$$y_k(t) = \psi_k(t), \quad y_k'(t) = \psi_k'(t), \quad t\leq 0, \qquad (3.10)$$

where $y_k \in \mathbb{R}$, p_i, q_i, α_j, $\beta_j : \mathbb{R}_+ \to \mathbb{R}$, $(k, i, j = 1, 2, \ldots, n)$, $p_k(t) \leq q_k(t) \leq t$, $\beta_j(t) \leq \alpha_j(t) \leq t$.

Change the variables in the initial value problem (3.9), (3.10) and obtain the system

$$x_k(t) = F_k \left(t, \max_{s \in [p_1(t), q_1(t)]} \int_0^s x_1(\tau) d\tau, \ldots, \max_{s \in [p_n(t), q_n(t)]} \int_0^s x_n(\tau) d\tau, \right.$$

$$\left. \max_{s \in [\beta_1(t), \alpha_1(t)]} x_1(s), \ldots, \max_{s \in [\beta_n(t), \alpha_n(t)]} x_n(s) \right), \quad t > 0, \quad (3.11)$$

with initial condition

$$x_k(t) = \varphi_k(t), \quad t \leq 0. \tag{3.12}$$

Theorem 3.1.5. *Let the following conditions be fulfilled:*

1. *The functions $p_k(t)$, $q_k(t)$, $\alpha_k(t)$, $\beta_k(t) : \mathbb{R}_+ \to \mathbb{R}$ are continuous, and $p_k(t) \leq q_k(t) \leq t$, $\beta_k(t) \leq \alpha_k(t) \leq t$, $(k = 1, 2, \ldots, n)$.*

2. *The functions $F_k(t, x_1, \ldots, x_n, y_1, \ldots, y_n) : \mathbb{R}_+ \times \mathbb{R}^{2n} \to \mathbb{R}$ are continuous and*

$$\left| F_k(t, x_1, \ldots, x_n, y_1, \ldots, y_n) \right| \leq \frac{1}{\gamma} \mu_k \left(t, |x_1|, \ldots, |x_n|, |y_1|, \ldots, |y_n| \right),$$

$$\left| F_k(t, x_1, \ldots, x_n, y_1, \ldots, y_n) - F_k(t, \bar{x}_1, \ldots, \bar{x}_n, \bar{y}_1, \ldots, \bar{y}_n) \right|$$

$$\leq \frac{m}{\gamma} \lambda_k \left(t, |x_1 - \bar{x}_1|, \ldots, |x_n - \bar{x}_n|, |y_1 - \bar{y}_1|, \ldots, |y_n - \bar{y}_n| \right)$$

for some $\gamma > m > 0$;

$$\bar{\mu}_k(t) = \mu_k \left(t, q_1(t)y, \ldots, q_n(t)y, y, \ldots, y \right)$$

is bounded in t and

$$\sum_{k=1}^{n} \lambda_k \left(t, q_1(t)y_k, \ldots, q_n(t)y_k, y_k, \ldots, y_k \right) \leq \sum_{k=1}^{n} y_k.$$

3. *The initial functions $\varphi_k(t) : \mathbb{R}_- \to \mathbb{R}$ are continuous, bounded and satisfy the condition*

$$\varphi_k(0) = F_k \left(0, \max_{s \in [p_1(0), q_1(0)]} \int_0^s \varphi_1(\tau) d\tau, \ldots, \max_{s \in [\beta_n(0), \alpha_n(0)]} \varphi_n(s) \right).$$

Then there exists a unique continuous and bounded solution of the initial value problem (3.11), (3.12).

Proof. Let $\mathcal{B}_1$ be Banach space $CB(\mathbb{R}) \times \cdots \times CB(\mathbb{R})$ with norm

$$\|\{f_1,\ldots,f_n\}\|_1 = \|f_1\|_{CB} + \cdots + \|f_n\|_{CB}$$

where $CB(\mathbb{R})$ is Banach space of all continuous and bounded functions $f(t): \mathbb{R} \to \mathbb{R}$ with supremum norm $\|\cdot\|_{CB}$. Let $\mathcal{B}_2$ be Banach space $CB(\mathbb{R}) \times CB(\mathbb{R}) \times \ldots \times CB(\mathbb{R}_-)$ with the corresponding norm. Define the operators $N_i : \mathcal{B}_i \to \mathcal{B}_i$ $(i = 1,\ 2)$ by the formulas

$$N_1\{f_1,\ldots,\ f_n\} = \{h_1,\ldots,\ h_n\}, \quad \{f_1,\ldots,\ f_n\} \in \mathcal{B}_1,$$

$$h_k(t) = \begin{cases} -\gamma F_k\Big(t, \max\limits_{s\in[p_1(t),q_1(t)]}\int_0^s f_1(\tau)d\tau, \ldots, \max\limits_{s\in[p_n(t),q_n(t)]}\int_0^s f_n(\tau)d\tau, \\ \max\limits_{s\in[\beta_1(t),\alpha_1(t)]} f_1(s), \ldots, \max\limits_{s\in[\beta_n(t),\alpha_n(t)]} f_n(s)\Big), \quad t > 0, \\ -\gamma F_k\Big(0, \max\limits_{s\in[p_1(0),q_1(0)]}\int_0^s f_1(\tau)d\tau, \ldots, \max\limits_{s\in[p_n(0),q_n(0)]}\int_0^s f_n(\tau)d\tau, \\ \max\limits_{s\in[\beta_1(0),\alpha_1(0)]} f_1(s), \ldots, \max\limits_{s\in[\beta_n(0),\alpha_n(0)]} f_n(s)\Big), \quad t \leq 0 \end{cases}$$

$$(k = 1,\ 2,\ldots,\ n);$$

$$N_2\{g_1,\ldots,\ g_n\} = \{\bar{h}_1,\ldots,\ \bar{h}_n\},$$

where

$$\bar{h}_k(t) = -\gamma F_k\Big(t, \max\limits_{s\in[p_1(0),\ q_1(0)]}\int_0^s g_1(\tau)d\tau, \ldots, \max\limits_{s\in[p_n(0),\ q_n(0)]}\int_0^s g_n(\tau)d\tau,$$

$$\max\limits_{s\in[\beta_1(0),\alpha_1(0)]} g_1(s), \ldots, \max\limits_{s\in[\beta_n(0),\alpha_n(0)]} g_n(s)\Big), \quad t \leq 0,$$

$\{g_1,\ldots,\ g_n\} \in \mathcal{B}_2$.

The linear map $j : \mathcal{B}_1 \to \mathcal{B}_2$ is defined as a restriction of the function $f \in \mathcal{B}_1$ on the semi-axis $\mathbb{R}_-$.

As in the proof of Theorem 3.1.1, it can be established that $N_i : \mathcal{B}_i \to \mathcal{B}_i$. We shall show only the Lipschitz continuity of N_1.

Indeed, for $t > 0$ and $\{f_1, \dots, f_n\}, \{\bar{f}_1, \dots, \bar{f}_n\} \in \mathcal{B}_1$ we have

$$\left| N_1 \{f_1, \dots, f_n\} - N_1 \{\bar{f}_1, \dots, \bar{f}_n\} \right|$$

$$\leq \sum_{k=1}^{n} \left| h_k(t) - \bar{h}_k(t) \right|$$

$$\leq m \sum_{k=1}^{n} \lambda_k \Big(t, \ q_1(t) \big| \big| f_k - \bar{f}_k \big| \big|, \dots, q_n(t) \big| \big| f_k - \bar{f}_k \big| \big|,$$

$$\big| \big| f_k - \bar{f}_k \big| \big|, \dots, \big| \big| f_k - \bar{f}_k \big| \big| \Big)$$

$$\leq m \sum_{k=1}^{n} \big| \big| f_k - \bar{f}_k \big| \big|$$

$$= m \big| \big| \{f_1, \dots, f_n\} - \{\bar{f}_1, \dots, \bar{f}_n\} \big| \big|_1.$$

An analogous estimate for $t \leq 0$ together with the last inequalities implies

$$\left| \left| N_1 \{f_1, \dots, f_n\} - N_1 \{\bar{f}_1, \dots, \bar{f}_n\} \right| \right|_1 \leq$$
$$m \big| \big| \{f_1, \dots, f_n\} - \{\bar{f}_1, \dots, \bar{f}_n\} \big| \big|_1$$

which completes the proof of Theorem 3.1.5.

$\square$

As a consequence of Proposition 3.1.2 we obtain:

Theorem 3.1.6. *Let the conditions of Theorem 3.1.5 be satisfied and the functions $\{x_1(\varphi), \dots, x_n(\varphi)\}$ and $\{x_1(\bar{\varphi}), \dots, x_n(\bar{\varphi})\}$ be two solutions of the initial value problem (3.11), (3.12).*
Then we have

$$\sum_{k=1}^{n} \left| x_k(\varphi)(t) - x_k(\bar{\varphi})(t) \right|$$

$$\leq \frac{\gamma}{\gamma - m} \left[\sum_{k=1}^{n} \left| \varphi_k(0) - \bar{\varphi}_k(0) \right| + \sum_{k=1}^{n} \sup_{t \in \mathbb{R}_-} \left| \varphi_k(t) - \bar{\varphi}_k(t) \right| \right],$$

where $\varphi = \{\varphi_1, \dots, \varphi_n\}$, $\bar{\varphi} = \{\bar{\varphi}_1, \dots, \bar{\varphi}_n\}$.

3.2   Existence Theory for Boundary Value Problems

In this section the method of a priori estimates based on the Leray-Schauder topological degree theory is developed to establish the existence of solutions of general boundary value problems for differential equations with "maxima."

Consider the following differential equation with "maxima:"

$$x' = f\big(t,\ x(t),\ \max_{\tau \in S(t)} x(\tau)\big), \qquad t \in [a, b],$$

$$x(t) = 0, \qquad t \notin [a, b], \tag{3.13}$$

$$\varphi(x) = 0,$$

where $x :\in \mathbb{R}^n$, $S(t) \subset (-\infty, t]$, $f : [a, b] \times \mathbb{R}^n \times \mathbb{R}^n \to \mathbb{R}^n$, φ is a vector functional over a space of vector functions on $[a, b]$ to be specified in the sequel, which represents the boundary conditions.

We emphasize that the additional assumption $x(t) = 0$ for $t \notin [a, b]$ is unnecessary if $S(t) \subset [a, b]$ for all $t \in [a, b]$. In the opposite case, however, one can expect different types of such additional assumption depending on the concrete applications. For example, instead of $x = 0$ it can be required that $x(t) = \varphi(t)$ outside the interval $[a, b]$, where $\varphi(.) \in \mathbb{R}^n$ is some given vector function, or $x(t) = x(a)$ for $t \leq a$ and $x(t) = x(b)$ for $t \geq b$. Furthermore, the right side of the differential equation in (3.13) can also be more complex, e.g., there can occur the dependence on maximum values of different components of the state vector x on different time intervals.

Initially we will develop the Leray-Schauder topological degree theory, which will be the mathematical tool in the existence results for different types of boundary value problems for differential equations with "maxima."

Let $L^p(a, b)$ stand for the standard Lebesgue space of functions integrable on the interval (a, b) with exponent $1 \leq p < \infty$ (or essentially bounded when $p = \infty$). In the sequel we make an extensive use of the space $L^p\big((a, b);\ \mathbb{R}^n\big)$ (denoted by L^p_n for brevity) of functions with values in $\mathbb{R}^n$ with components in $L^p(a, b)$, equipped with its usual norm

$$\|x\|_p := \begin{cases} \left(\displaystyle\int_a^b |x(\tau)|^p d\tau \right)^{1/p}, & 1 \leq p < \infty, \\[2mm] \operatorname*{ess\,sup}_{(a,b)} |x(t)|, & p = \infty, \end{cases}$$

where $|\cdot|$ stands for the norm in $\mathbb{R}^n$. We denote by AC_n^p the space of vector functions with absolutely continuous components and with derivatives in L_n^p, equipped with the norm

$$||x||_{AC_n^p} := |x(a)| + ||x'||_p.$$

Also we denote by $C_n[a, b]$ (C_n, for short) the space of all $\mathbb{R}^n$-valued functions with continuous components, equipped with the usual maximum norm.

Suppose we have to solve a nonlinear boundary value problem

$$\begin{cases} Lx = F(x), \\ \phi(x) = 0, \end{cases} \tag{3.14}$$

where $x \in X$ is an unknown, $L : X \to Y$ is some linear operator, $F : X \to Y$ is a nonlinear operator, $\phi : X \to \mathbb{R}^n$ is a vector functional, and X and Y are given Banach spaces.

Let X be isomorphic to the direct product of some Banach space E and the finite-dimensional space $\mathbb{R}^n$ by an isomorphism $J : E \times \mathbb{R}^n \to X$, by $J(u, \lambda) = \Lambda u + D\lambda$, where $\Lambda : E \to X$ and $D : \mathbb{R}^n \to X$. Assuming that $Q = L\Lambda : E \to E$ is invertible, we can then reduce the problem (3.14) to the form

$$\begin{cases} u = \mathcal{F}(u, \lambda), \\ \mathbf{D}(u, \lambda) = 0, \end{cases} \tag{3.15}$$

where $u \in E$ and $\lambda \in \mathbb{R}^n$ are unknowns, $\mathcal{F} : E \times \mathbb{R}^n \to E$ is a nonlinear operator and $\mathbf{D} : E \times \mathbb{R}^n \to \mathbb{R}^n$ a nonlinear vector functional. To study the solvability of the latter we apply the topological degree theory.

For the purpose of the investigations in this section, the Leray-Schauder topological degree theory will be applied to mappings of the form

$$\Psi(u, \lambda) = \begin{Bmatrix} u - \mathcal{F}(u, \lambda) \\ \mathbf{D}(u, \lambda) \end{Bmatrix}. \tag{3.16}$$

Introduce regions (open bounded subsets) $\Omega_1 \subset E$ and $\Omega_2 \subset \mathbb{R}^n$ and denote their boundaries by $\partial\Omega_1$ and $\partial\Omega_2$, respectively. Also, let $\Omega = \Omega_1 \times \Omega_2$.

To develop the Leray-Schauder topological degree theory, one needs the compactness of Ψ, which is provided by the following assumption:

H3.2.1. $\mathcal{F}$ is a compact and continuous operator, and $\mathbf{D}$ is a continuous vector functional which maps bounded sets into bounded sets.

In fact, the condition $(H3.2.1)$ clearly implies that the mapping Ψ is a compact perturbation of the identity in $E \times \mathbb{R}^n$. Then the additional requirement of nondegeneracy of the mapping Ψ on $\partial\Omega$ is enough to define correctly the topological degree $\deg(\Psi,\ \Omega,\ 0)$, having all the ordinary properties (see [Nirenberg 1974]). We say that the mapping Ψ (the system (3.15)) is *topologically nontrivial* on the regions $\Omega_1 \subset E$ and $\Omega_2 \subset \mathbb{R}^n$ (is $V(\Omega_1,\Omega_2)$, for short) whenever $\deg(\Psi,\ \Omega,\ 0) \neq 0$. Thus, if the system (3.15) is $V(\Omega_1,\Omega_2)$, then it admits at least one solution $(u,\lambda) \in \Omega$.

To calculate the topological degree of Ψ, we will use the following result that extends the analogous statements by S.A. Vavilov ([Vavilov 1993]).

Consider an auxiliary finite-dimensional continuous vector field $\mathbf{D}_0(\lambda) : \mathbb{R}^n \to \mathbb{R}^n$. In the sequel we will always take $\mathbf{D}_0(\lambda) := \mathbf{D}(0, \lambda)$.

Theorem 3.2.1. *Suppose $\Omega_1 \subset E$ is convex, $\mathcal{F}$ is a compact and continuous operator, $\mathbf{D}$ is a continuous vector functional which maps bounded sets into bounded sets, and:*

(i) $\mathcal{F}(\partial\Omega_1, cl\Omega_2) \subset \Omega_1$;

(ii) for all $u \in cl\Omega_1$ and for all $\lambda \in \partial\Omega_2$ the inequality $0 \leq |\mathbf{D}(u,\lambda) - \mathbf{D}_0(\lambda)| < |\mathbf{D}_0(\lambda)|$ holds.

Then $\deg(\Psi,\ \Omega,\ 0) = \deg(\mathbf{D}_0,\ \Omega_2,\ 0)$. In particular, when the latter is not zero, then the system (3.15) has at least one solution $(u',\lambda') \in \Omega$. The set of such solutions can be approximated by the Galerkin numerical scheme applied to (3.15).

Proof. Choose some $u_0 \in \Omega_1$ and consider the homotopy

$$\Psi_t(u, \lambda) = \left\{ \begin{array}{c} u - (1-t)u_0 - t\mathcal{F}(u,\lambda) \\ \mathbf{D}_0(\lambda) + t(\mathbf{D}(u,\lambda) - \mathbf{D}_0(\lambda)) \end{array} \right\},\quad t \in [0,1].$$

Since $\partial\Omega = \partial\Omega_1 \times cl\Omega_2 \cup cl\Omega_1 \times \partial\Omega_2$, we observe that $\Psi_t \neq 0$ on $\partial\Omega$ for all $t \in [0,1]$. Furthermore, clearly $\deg(\Psi_0,\ \Omega,\ 0) = \deg(\mathbf{D}_0,\ \Omega_2,\ 0)$ by the properties of the degree [Krasnoselskii and Zabreiiko 1975]. Thus, noting the compactness of the above homotopy and applying the homotopy invariance of the degree, we conclude the proof.

$\square$

Remark 3.2.1. *Suppose $0 \in \Omega_1$. The statement of Theorem 3.2.1 remains valid if (i) is replaced by*

$$\lambda \in cl\Omega_2, \quad t \in [0, 1], \quad u = t\mathcal{F}(u, \lambda) \Rightarrow u \notin \partial\Omega_1.$$

If, in particular, Ω_1 is an open ball with center zero, then the latter condition follows from the existence of a uniform a priori estimate for $u \in E$ from the first equation of (3.15).

To calculate the topological degree of Ψ we will also use the iterated mapping

$$\Psi^{(1)}(u, \lambda) = \left\{ \begin{array}{c} u - \mathcal{F}(u, \lambda) \\ \mathbf{D}\big(\mathcal{F}(u, \lambda), \lambda\big) \end{array} \right\}. \tag{3.17}$$

Theorem 3.2.2. *If both Ψ and $\Psi^{(1)}$ have the compactness property (C), and one of them is nondegenerate on $\partial\Omega$, then so is the other. Moreover, in this case*

$$\deg\big(\Psi, \ \Omega, \ 0\big) = \deg\big(\Psi^{(1)}, \ \Omega, \ 0\big).$$

Proof. The first part of the statement is obvious, for the sets of zeros of Ψ and $\Psi^{(1)}$ coincide. To prove the second part, note that the vector fields Ψ and $\Psi^{(1)}$ can have opposite directions on $\partial\Omega$ only if $u = \mathcal{F}(u, \lambda)$. But in this case their finite-dimensional components are equal and nonzero due to nondegeneracy. Therefore Ψ and $\Psi^{(1)}$ never have opposite directions, which implies the statement. $\qquad\square$

Now we will apply the obtained above theoretical result to investigate the solvability of the system (3.13) with a boundary condition

$$\phi\left(\lambda + \int_{\alpha}^{t} x(s)ds\right) = 0, \tag{3.18}$$

where the boundary condition (3.18) is represented by a nonlinear vector functional $\phi : AC_n^q \to \mathbb{R}^n$, $1 < q < \infty$. Applying the isomorphism between AC_n^q and $L_n^q \times \mathbb{R}^n$ given by the formula

$$J : L_n^q \times \mathbb{R}^n \ni (u, \lambda) \mapsto x = \int_a^t u(\tau)d\tau + \lambda \quad \in AC_n^q,$$

we reduce the original problem (3.13), (3.18) to the following system
of type (3.15):

$$u(t) = f\left(t, \ \lambda + \int_a^t u(\tau)d\tau, \ \max\left(0_S, \ \lambda + \max_{\tau \in \tilde{S}(t)} \int_a^\tau u(s)ds\right)\right),$$

$$\phi\left(\lambda + \int_a^t u(\tau)d\tau\right) = 0.$$

$$(3.19)$$

Here and in the sequel we write for brevity $\tilde{S}(t) := S(t) \cap [a, b]$ and

$$0_S(t) = \begin{cases} 0, & S(t) \neq \tilde{S}(t), \\ -\infty, & S(t) = \tilde{S}(t). \end{cases}$$

Assume, further, that the set of functions $S(.)$ takes closed values and
is measurable in the sense that for any open $V \subset \mathbb{R}$ the set

$$S^{-1}(V) := \{t \mid S(t) \cap V \neq \emptyset\}$$

is measurable. Note that there are various measurability criteria for
set functions (see, for instance, Chapter III of [Castaing and Valadier
1977]). In particular, our assumptions hold when $S(t) := [r(t), s(t)]$,
where $r(.)$ and $s(.)$ are measurable (not necessarily almost everywhere
finite) functions. Finally, define

$$\gamma := \left(\frac{\sup \bigcup_{t \in [a,b]} \tilde{S}(t) - a}{b - a}\right)^{(q-1)/q}.$$

To get a solvability result for the original problem, we can now use the
topological degree theory by applying Theorems 3.2.1 and 3.2.2.

We now pass to important particular examples of the boundary
value problem (3.13).

Consider a general nonlinear two-point boundary value problem for
a differential equation with "maxima:"

$$x' = f\left(t, \ x(t), \ \max_{\tau \in S(t)} x(\tau)\right), \qquad t \in [a, b],$$

$$x(t) = 0, \qquad t \notin [a, b],$$

$$h\left(x(a), \ x(b)\right) = 0,$$

$$(3.20)$$

where $h: \mathbb{R}^{2n} \to \mathbb{R}^n$.

After the above reduction, the system (3.20) could be written in the form

$$u(t) = f\left(t, \ \lambda + \int_a^t u(\tau)d\tau, \ \max\left(0_S, \ \lambda + \max_{\tau \in \tilde{S}(t)} \int_a^\tau u(s)ds\right)\right),$$

$$h\left(\lambda, \ \lambda + \int_a^b u(\tau)d\tau\right) = 0. \tag{3.21}$$

Theorem 3.2.3. *Let the following conditions be fulfilled:*

(i) $f(t, \ x, \ y)$ *is a Carathéodory vector function (i.e., continuous in* $(x, y) \in \mathbb{R}^{2n}$ *for a.e.* $t \in [a, b]$ *and measurable in* t *for each* (x, y)*), while for a.e.* $t \in [a, b]$ *the inequalities* $|x| \leq U, \ |y| \leq V$ *imply*

$$|f(t, \ x, \ y)| \leq \alpha(t, \ U, \ V), \quad \|\alpha(\cdot, \ U, \ V)\|_q \leq \Phi(U, V),$$

$$\left.\begin{array}{l} |x_1| \leq U, \ |x_2| \leq U \\ |y_1| \leq V, \ |y_2| \leq V \end{array}\right\}$$

$$\Rightarrow \left|\int_a^b \Big(f(t, \ x_1, \ y_1) - f(t, \ x_2, \ y_2)\Big)dt\right| \leq \Phi_1(U, V).$$

(ii) *The vector function* $h(x, y)$ *is continuous, and*

$$|x| \leq U, \ |y_1| \leq V, \ |y_2|$$
$$\leq V \Rightarrow |h(x, y_1) - h(x, y_2)| \leq \delta(U, V)|y_1 - y_2|.$$

(iii) $|\mathbf{D}_0(\lambda)| \geq \beta(\rho_2) > 0$ *if* $|\lambda| = \rho_2$*, while* $\deg\left(\mathbf{D}_0, \ |\lambda| < \rho_2, \ 0\right) \neq 0$*, where*

$$\mathbf{D}_0(\lambda) = h\left(\lambda, \ \lambda + \int_a^b f\Big(\tau, \ \lambda, \ \max\left(0_S(\tau), \ \lambda\right)\Big)d\tau\right).$$

(iv) *We have*

$$\Phi(\rho_2 + \rho_1, \ \rho_2 + \gamma\rho_1) < \rho_1/(b - a)^{(q-1)/q},$$
$$\delta(\rho_2, \ \rho_2 + \Phi(\rho_2 + \rho_1, \ \rho_2 + \gamma\rho_1))\Phi_1(\rho_2 + \rho_1, \ \rho_2 + \gamma\rho_1) < \beta(\rho_2).$$
$$\tag{3.22}$$

Then the two-point boundary value problem (3.20) has at least one solution $x \in AC_n^q$, $1 < q < \infty$, satisfying $||x'||_q < \rho_1/(b-a)^{(q-1)/q}$ and $|x(a)| < \rho_2$.

Proof. Substituting the first equation of (3.21) into the second, we obtain the system

$$u(t) = f\left(t, \ \lambda + \int_a^t u(\tau)d\tau, \ \max\left(0_S, \ \lambda + \max_{\tau \in \tilde{S}(t)} \int_a^\tau u(s)ds\right)\right),$$

$$h\left(\lambda, \ \lambda + \int_a^b f\left(t, \ \lambda + \int_a^t u(\tau)d\tau, \max\left(0_S, \lambda + \max_{\tau \in \tilde{S}(t)} \int_a^\tau u(s)ds\right)\right)dt\right) = 0.$$

$$(3.23)$$

Therefore, to any solution $(u, \lambda) \in L_n^q \times \mathbb{R}^n$ of the latter there corresponds a solution $x \in AC_n^q$ of the original problem (3.20). Consider the open balls $B_1 \subset L_n^q$ and $B_2 \subset \mathbb{R}^n$,

$$B_1 := \left\{ u \ \Big| \ ||u||_q < \rho_1/(b-a)^{(q-1)/q} \right\}, \quad B_2 := \left\{ \lambda \ \Big| \ |\lambda| < \rho_2 \right\}.$$

The statement will be proven if we show that the system (3.23) is $V(B_1, B_2)$. For this purpose we apply Theorem 3.2.1.

To verify the compactness assumption (C), note that according to (i) the Nemytskiĭ operator

$$N : L_n^\infty \times L_n^\infty \ni (v_1, v_2) \mapsto f(\cdot, \ v_1(.), \ v_2(.)) \in L_n^q$$

is continuous and maps bounded sets into bounded sets. Now define formally an operator M on the space C_n by

$$(Mv)(t) := \max_{\tau \in \tilde{S}(t)} v(t).$$

For any $v \in C_n$ clearly $(Mv)(t)$ is measurable due to the Krasnosel'skiĭ-Ladyzhenskiĭ lemma (see Theorem 6.2 of [Appell and Zabrejko 1990]) and bounded. Thus $M : C_n \to L_n^\infty$ and obviously it maps bounded sets into bounded sets. The continuity of M follows from the fact that if a sequence of continuous functions converges uniformly on $[a, b]$, then their maxima on any compact subset of $[a, b]$ converge to the maximum of the limit function on the same subset, the rate of the latter convergence being independent of the choice of the subset. Hence one easily shows that the compactness of the nonlinear operator on the right side of the first equation of (3.23) is ensured by the compactness of the imbedding $AC_n^q \subset C_n$, while the continuity of the vector functional corresponding to the second equation is provided by (ii). To conclude the proof it remains to observe that conditions (i) and (ii) of Theorem 3.2.1 are ensured by (3.22).

$\square$

Remark 3.2.2. *In* (i) *one may choose* $\Phi_1(U, V) := 2(b - a)^{(q-1)/q} \Phi(U, V)$. *However, in applications one can often get sharper estimates.*

Remark 3.2.3. *The theorem also opens the way to numerical treatment of the original value problem (3.20). Namely, the set of pairs* (u, λ) *in appropriate regions (see the proof), which corresponds to the solutions of (3.20) found in the theorem, can be approximated by the Galerkin numerical scheme applied to (3.21). The same refers to all the statements below.*

As an important particular case consider the boundary value problem

$$
\begin{aligned}
x' &= f\big(t,\ x(t),\ \max_{\tau \in S(t)} x(\tau)\big), \qquad t \in [a, b], \\
x(t) &= 0, \qquad t \notin [a, b], \\
x(a) &= x(b) + \xi h_1(x(a), x(b)),
\end{aligned}
\tag{3.24}
$$

where $h_1 : \mathbb{R}^{2n} \to \mathbb{R}^n$ and $\xi \in \mathbb{R}$ is a perturbation parameter. For $\xi = 0$ this is a periodic-type boundary value problem (however, one should not speak of periodic solutions unless $S(t) \subset (-\infty, t]$). One is therefore interested both in existence of solutions for $\xi = 0$ and in whether they do not disappear for small values of ξ.

Theorem 3.2.4. *Let the following conditions be fulfilled:*

(i) $f(t,\ x,\ y)$ *is a Carathéodory vector function, and for a.e.* $t \in [a, b]$ *the inequalities* $|x| \leq U,\ |y| \leq V$ *imply*

$$
|f(t,\ x,\ y)| \leq \alpha(t,\ U,\ V), \qquad \|\alpha(\cdot,\ U,\ V)\|_q \leq \Phi(U, V),
$$

and the inequalities

$$
\left.
\begin{aligned}
|x_1| \leq U,\ \ |x_2| \leq U \\
|y_1| \leq V,\ \ |y_2| \leq V
\end{aligned}
\right\}
$$

imply

$$\left| \int_a^b \Big(f(t, \ x_1, \ y_1) - f(t, \ x_2, \ y_2) \Big) dt \right| \le \Phi_1(U, V).$$

(ii) *The vector function* $h_1(x, y)$ *is continuous.*

(iii) $|\mathbf{D}_0(\lambda)| \ge \beta(\rho_2) > 0$ *if* $|\lambda| = \rho_2$, *while* $\deg \big(\mathbf{D}_0, \ |\lambda| < \rho_2, \ 0\big) \ne 0$, *where*

$$\mathbf{D}_0(\lambda) = \int_a^b f\Big(\tau, \ \lambda, \ \max\big(0_S(\tau), \lambda\big)\Big) d\tau.$$

(iv) *We have*

$$\begin{cases} \Phi(\rho_2 + \rho_1, \ \rho_2 + \gamma\rho_1) < \rho_1/(b-a)^{(q-1)/q}, \\[2mm] \Phi_1(\rho_2 + \rho_1, \ \rho_2 + \gamma\rho_1) < \beta(\rho_2). \end{cases} \tag{3.25}$$

Then there is $\xi^* > 0$ *such that for each* ξ *with* $|\xi| \le \xi^*$ *the boundary value problem (3.24) has at least one solution* $x_\xi \in AC_n^q$, $1 < q < \infty$, *satisfying* $\|x_\xi'\|_q < \rho_1/(b-a)^{(q-1)/q}$ *and* $|x_\xi(a)| < \rho_2$.

Proof. Write out the system of type (3.15) corresponding to the original problem:

$$u(t) = f\Big(t, \ \lambda + \int_a^t u(\tau) d\tau, \ \max\big(0_S, \ \lambda + \max_{\tau \in \tilde{S}(t)} \int_a^\tau u(s) ds\big)\Big),$$

$$\int_a^b u(\tau) d\tau + \xi h_1\Big(\lambda, \ \lambda + \int_a^b u(\tau) d\tau\Big) = 0, \tag{3.26}$$

and consider the iteration, as it was introduced by (3.17), of the system obtained by setting $\xi = 0$:

$$u(t) = f\Big(t, \ \lambda + \int_a^t u(\tau) d\tau, \ \max\big(0_S, \ \lambda + \max_{\tau \in \tilde{S}(t)} \int_a^\tau u(s) ds\big)\Big),$$

$$\int_a^b f\Big(t, \ \lambda + \int_a^t u(\tau) d\tau, \ \max\big(0_S, \ \lambda + \max_{\tau \in \tilde{S}(t)} \int_a^\tau u(s) ds\big)\Big) dt = 0. \tag{3.27}$$

Let B_1 and B_2 be the balls introduced in the proof of Theorem 3.2.3. The assumptions imply that (3.27) is $V(B_1, B_2)$ according to

Theorem 3.2.1. Applying the iteration Theorem 3.2.2, one observes that the system (3.26) for $\xi = 0$ is also $V(B_1, B_2)$. The desired conclusion follows now from the stability of topological degree with respect to small perturbations of the mapping.

$\square$

Remark 3.2.4. *As will be clear from the proof, one can in fact also assert that as $\xi \to 0$, the sets of solutions to (3.24) found in the theorem are uniformly attracted to the set of solutions to the unperturbed problem with $\xi = 0$.*

Another classical application is the Cauchy problem

$$x' = f\big(t,\ x(t),\ \max_{\tau \in S(t)} x(\tau)\big), \quad t \in [a, b],$$
$$x(t) = 0, \qquad t \notin [a, b], \tag{3.28}$$
$$x(a) = \lambda_0,$$

where $\lambda_0 \in \mathbb{R}^n$. It is worth noting that the system (3.21) in this case is reduced to a single equation:

$$u(t) = f\Bigg(t,\ \lambda_0 + \int_a^t u(\tau)d\tau,\ \max\Big(0_S,\ \lambda_0 + \max_{\tau \in \tilde{S}(t)} \int_a^\tau u(s)ds\Big)\Bigg).$$
$$\tag{3.29}$$

The following statement can be obtained either as a simple corollary of Theorem 3.2.3 or independently by applying the Schauder fixed point principle to (3.29).

Theorem 3.2.5. *Let the following conditions be fulfilled:*

(i) $f(t,\ x,\ y)$ *is a Carathéodory vector function, and for a.e. $t \in [a, b]$ the inequalities $|x| \leq U,\ |y| \leq V$ implies*

$$\big|f(t,\ x,\ y)\big| \leq \alpha(t,\ U,\ V), \quad \big\|\alpha(\cdot,\ U,\ V)\big\|_q \leq \Phi(U, V).$$

(ii) *We have*

$$\Phi\Big(|\lambda_0| + \rho_1,\ |\lambda_0| + \gamma\rho_1\Big) < \rho_1/(b - a)^{(q-1)/q}. \tag{3.30}$$

Then the Cauchy problem (3.28) admits at least one solution $x \in AC_n^q$, $1 < q < \infty$, satisfying $\|x'\|_q < \rho_1/(b - a)^{(q-1)/q}$.

Note that the operator M introduced in the proof of Theorem 3.2.3 is a Volterra operator. To show this, we consider the following simple example which rather frequently appears in applications:

$$x' = \mathcal{F}(t) + \mathcal{B}(t) \max_{\tau \in S(t)} x(\tau), \qquad t \in [a, b],$$
$$x(t) = x(a) = x(b), \qquad t \le a, \tag{3.31}$$

where $\mathcal{B}(.)$ is an $n \times n$ matrix function, and $\mathcal{F}(.)$ is a vector function. For brevity set $B := \int_a^b \mathcal{B}(t)dt$, $F := \int_a^b \mathcal{F}(t)dt$, $\beta_1 := \int_a^b |||\mathcal{B}(t)|||dt$, $f(t) := |\mathcal{F}(t)|$, $b(t) := |||\mathcal{B}(t)|||$, where $||| \cdot |||$ stands for the matrix norm compatible with the norm in $\mathbb{R}^n$. Also, let

$$\mu(t) := b(t) \exp \left(\int_a^t b(\tau)d\tau \right).$$

Theorem 3.2.6. *Let $f \in L_1^q$, $b \in L_1^q$ and $\theta := \inf_{|\lambda|=1} |B\lambda| > 0$. Then the problem (3.31) admits at least one solution $x \in AC_n^q$, $1 < q < \infty$, provided that*

$$(b-a)^{(q-1)/q}||\mu||_q \beta_1 < \theta.$$

Proof. Write out the system of equations of type (3.15) corresponding to the problem (3.31):

$$u(t) = \mathcal{F}(t) + \mathcal{B}(t)\left(\lambda + \max \left(0_S, \max_{\tau \in \tilde{S}(t)} \int_a^\tau u(s)ds \right) \right),$$
$$F + B\lambda + \int_a^b \mathcal{B}(t) \max \left(0_S, \max_{\tau \in \tilde{S}(t)} \int_a^\tau u(s)ds \right) dt = 0. \tag{3.32}$$

We will prove that for some $B_1 \subset L_n^q$ and $B_2 \subset \mathbb{R}^n$ this system is $V(B_1, B_2)$ according to Theorem 2.2.1, and thus show the statement. In fact, let B_1 be as in the proof of Theorem 2.2.3 and $B_2 \subset \mathbb{R}^n$ be the ball $|\lambda - \lambda^*| < \rho_2$, $\lambda^* := -B^{-1}F$. Consider the auxiliary vector field

$$\mathbf{D}_0(\lambda) := F + B\lambda.$$

It is clear that $\deg(\mathbf{D}_0, B_2, 0) \neq 0$ and $|\mathbf{D}_0(\lambda)| \ge \theta \rho_2$ when $|\lambda - \lambda^*| = \rho_2$. Condition (ii) of Theorem 2.2.1 holds provided that

$$\beta_1 \rho_1 < \theta \rho_2. \tag{3.33}$$

Now turn to the first equation of (3.32). To abbreviate the notation, let $y(t) := |u(t)|$ and $y_0 := |\lambda|$. Obviously this equation implies

$$y(t) \leq f(t) + b(t)\left(y_0 + \int_a^t y(\tau)d\tau\right),$$

and constructing the Cauchy majorant problem and using a Chaplygin type result on differential inequalities one easily concludes that

$$y(t) \leq f(t) + \mu(t)\left(y_0 + \int_a^t \left(f(\tau)/\mu(\tau)\right)d\tau\right).$$

Hence if $\lambda \in cl B_2$, then $\|u\|_q \leq c_1 + \|\mu\|_q (\rho_2 + c_2)$, where c_1, c_2 are some positive constants. Condition (i) of Theorem 3.2.1 will hold provided that

$$c_1 + \|\mu\|_q (\rho_2 + c_2) < \rho_1/(b-a)^{(q-1)/q}. \tag{3.34}$$

It remains to note that, under the conditions of the theorem being proved, the relations (3.33) and (3.34) are valid simultaneously for sufficiently large $\rho_2 > 0$.

$\square$

Analogous results can be easily provided for more general problems

$$x' = \mathcal{F}(t) + \mathcal{A}(t)x(t) + \mathcal{B}(t)\max_{\tau \in S(t)} x(\tau), \quad t \in [a, b],$$
$$x(t) = x(a) = x(b), \quad t \leq a, \tag{3.35}$$

where $\mathcal{A}(.)$ is an $n \times n$ matrix function. In this case seeking solutions in the form

$$x(t) = X(t)\lambda + X(t)\int_a^t X^{-1}(\tau)u(\tau)d\tau,$$

where $X(t)$ is the fundamental matrix of the system $x' - \mathcal{A}(t)x = 0$, we obtain an auxiliary system of type (3.32) to be analyzed,

$$u(t) = \mathcal{F}(t) + \mathcal{B}(t)\max\left(\lambda, X(t)\lambda + \max_{\tau \in \tilde{S}(t)} X(\tau)\int_a^\tau X^{-1}(s)u(s)ds\right),$$

$$\tilde{F} + \left(X(b) - I_n\right)\lambda + X(b)\int_a^b \mathcal{B}(t)\max\left(\lambda, \ X(t)\lambda\right.$$

$$+ \left.\max_{\tau \in \tilde{S}(t)} X(\tau)\int_a^\tau X^{-1}(s)u(s)ds\right)dt = 0,$$

where $\tilde{F} = X(b)\int_a^b X^{-1}(t)\mathcal{F}(t)dt$ and I_n is the $n \times n$ identity matrix.

3.3 Differential Equations with "Maxima" via Weakly Picard Operator Theory

Consider the following initial value problem for a scalar differential equation with "maxima"

$$x'(t) = f\left(t,\ x(t),\ \max_{\xi\in[a,t]} x(\xi)\right), \quad t \in [a,b], \tag{3.36}$$

$$x(a) = \alpha, \tag{3.37}$$

where $x \in \mathbb{R}$, $a,\ b \in \mathbb{R}$, $a < b$, $\alpha \in \mathbb{R}$.

We will say that conditions **H3.3** are satisfied if:

H3.3.1 The function $f \in C([a,b] \times \mathbb{R}^2, \mathbb{R})$;

H3.3.2 There exists $L > 0$ such that

$$\left|f(t,\ u_1,\ u_2) - f(t,\ v_1,\ v_2)\right| \le L\max\left(\left|u_1 - v_1\right|, \left|u_2 - v_2\right|\right)$$

for all $t \in [a,b]$ and $u_i,\ v_i \in \mathbb{R}$, $i = 1,\ 2$;

H3.3.3 The inequality $L(b-a) < 1$ holds.

Note the initial value problem (3.36), (3.37) is equivalent to the integral equation

$$x(t) = \alpha + \int_a^t f\left(s,\ x(s),\ \max_{\xi\in[a,s]} x(\xi)\right)ds, \quad t \in [a,b]. \tag{3.38}$$

Let us consider the operators $B_f,\ E_f : C[a,b] \to C[a,b]$ defined by

$$B_f(x(t)) = \alpha + \int_a^t f\left(s,\ x(s),\ \max_{\xi\in[a,s]} x(\xi)\right)ds$$

and

$$E_f(x(t)) = x(a) + \int_a^t f\left(s,\ x(s),\ \max_{\xi\in[a,s]} x(\xi)\right)ds.$$

For any $\alpha \in \mathbb{R}$, we consider the set $X_\alpha = \{x \in C[a,b] : x(a) = \alpha\}$. We note that $C[a,b] = \cup_{\alpha\in\mathbb{R}} X_\alpha$ is a partition of $C[a,b]$.

Lemma 3.3.1. *Let conditions H3.3 be fulfilled.*
   Then

(a) $B_f(C[a,b]) \subset X_\alpha$ and $E_f(X_\alpha) \subset X_\alpha$, $\forall \alpha \in \mathbb{R}$;

(b) $B_f|_{X_\alpha} = E_f|_{X_\alpha}$, $\forall \alpha \in \mathbb{R}$.

In this section we shall prove that if conditions H3.3.1 and H3.3.2 are satisfied and if L is small enough, then the operator E_f is weakly Picard operator in $(C[a,b], \|.\|)$ where $\|x\| := \max_{t \in [a,b]} x(t)$, and we study the equation (3.36) in the terms of the weakly Picard operator theory.

Let (X,d) be a metric space and $A : X \to X$ be an operator. We shall use the following notations:

$F_A = \{x \in X \mid A(x) = x\}$ – the fixed point set of A;

$I(A) = \{Y \subset X \mid A(Y) \subset Y,\ Y \neq \emptyset\}$ – the family of the nonempty invariant subsets of A.

Consider the Pompeiu-Housdorff functional $H : P(X) \times P(X) \to \mathbb{R}_+ \cup \{+\infty\}$ defined by

$$H(Y,Z) = \max\left\{ \sup_{y \in Y} \inf_{z \in Z} d(y,z), \sup_{z \in Z} \inf_{y \in Y} d(y,z) \right\}.$$

Definition 3.3.1. *([Rus 2001]) Let (X,d) be a metric space. An operator $A : X \to X$ is called a Picard operator (PO) if there exists $x^* \in X$ such that:*

(i) $F_A = \{x^*\}$;

(ii) the sequence $\left(A^n(x_0)\right)_{n \in \mathbb{N}}$ approaches x^* for all $x_0 \in X$.

Definition 3.3.2. *([Rus 2001]). Let (X,d) be a metric space. An operator $A : X \to X$ is a weakly Picard operator (WPO) if the sequence $\left(A^n(x)\right)_{n \in \mathbb{N}}$ converges for all $x \in X$, and its limit (which may depend on x) is a fixed point of A.*

Let A be a weakly Picard operator. Consider the operator $A^\infty : X \to X$ defined by $A^\infty(x) := \lim_{n \to \infty} A^n(x)$ ([Rus 2001]).

It is clear that $A^\infty(X) = F_A$.

Definition 3.3.3. *([Rus 2001]) Let A be a weakly Picard operator and $c > 0$. The operator A is called a c-weakly Picard operator if $d(x, A^\infty(x)) \leq cd(x, A(x))$, $\forall x \in X$.*

Now we will prove the following existence result:

Theorem 3.3.1. *Let conditions H3.3 be fulfilled.*

Then the problem (3.36), (3.37) has a unique solution in $C[a,b]$ and this solution is the uniform limit of the successive approximations.

Proof. The problem (3.36), (3.37) is equivalent to the following operator equation

$$B_f(x) = x, \quad x \in C[a,b]$$

and the solution of (3.36), (3.37) is a fixed point of the above equation.

On the other hand we have

$$\left| B_f(x)(t) - B_f(y)(t) \right|$$
$$\leq L_f \int_a^t \max\left(|x(s) - y(s)|, \left| \max_{\xi \in [a,s]} x(\xi) - \max_{\xi \in [a,s]} y(\xi) \right| \right) ds.$$

But

$$\max_{s \in [a,b]} \left| \max_{\xi \in [a,s]} x(\xi) - \max_{\xi \in [a,s]} y(\xi) \right| \leq \max_{s \in [a,b]} |x(s) - y(s)|.$$

So,

$$\left\| B_f(x) - B_f(y) \right\| \leq L_f(b-a) \|x - y\|, \quad \forall x, y \in C[a,b],$$

i.e., B_f is a contraction w.r.t. Chebyshev norm on $C[a,b]$. The proof follows from the contraction principle.

$\square$

Remark 3.3.1. *The operator B_f in Theorem 3.3.1 is PO. But*

$$B_f\big|_{X_\alpha} = E_f\big|_{X_\alpha}, \quad \forall \alpha \in \mathbb{R}.$$

Hence, the operator E_f is WPO and $F_{E_f} \cap X_\alpha = \{x_\alpha^*\}$, $\forall \alpha \in \mathbb{R}$, where x_α^* is the unique solution of the problem (3.36), (3.37).

We will prove the following comparison result:

Theorem 3.3.2. *Let the following conditions be fulfilled:*

1. *The conditions H3.3 are satisfied.*

2. *The function $f : [a,b] \times \mathbb{R}^2 \to \mathbb{R}$ is increasing in its second and third argument.*

3. *The function* $x(t)$ *is a solution of the equation (3.36) and* $y(t)$ *is a solution of the inequality*

$$y'(t) \le f\left(t, \ y(t), \ \max_{\xi \in [a,t]} y(\xi)\right), \quad t \in [a,b]. \qquad (3.39)$$

Then $y(a) \le x(a)$ *implies that* $y(t) \le x(t)$ *for* $t \in [a,b]$.

Proof. In the terms of the operator E_f, the equation (3.36) and inequality (3.39) could be written in the form

$$x = E_f(x) \quad \text{and} \quad y \le E_f(y).$$

and

$$x(a) \le y(a).$$

From the conditions H3.3 we have that the operator E_f is WPO. From the condition 2, E_f^∞ is increasing (see [Rus 2001]). If $\alpha \in \mathbb{R}$, then we denote by $\tilde{\alpha}$ the following function

$$\tilde{\alpha} : [a,b] \to \mathbb{R}, \quad \tilde{\alpha}(t) = \alpha, \quad \forall t \in [a,b].$$

We have

$$y \le E_f(y) \le \dots \le E_f^\infty(y) = E_f^\infty(\tilde{y}(a)) \le E_f^\infty(\tilde{x}(a)) = x.$$

□

In our further investigations, we need the following result:

Lemma 3.3.2. *(Comparison principle, see [Rus 2001]) Let* $(X, \ d, \ \le)$ *be an ordered metric space and* $A, \ B, \ C : X \to X$ *be such that:*

1. $A \le B \le C$;

2. *The operators* $A, \ B, \ C$ *are* $WPOs$;

3. *The operator* B *is increasing.*

Then $x \le y \le z$ *imply that* $A^\infty(x) \le B^\infty(y) \le C^\infty(z)$.

From the above result we obtain:

Theorem 3.3.3. *Let the following conditions be fulfilled:*

1. *The conditions H3.3 are satisfied for the functions* $f_i : [a,b] \times \mathbb{R}^2 \to \mathbb{R}$, $i = 1, 2, 3$ *and*

   (i) $f_1(t, x, y) \le f_2(t, x, y) \le f_3(t, x, y)$ *for any* $(t, x, y) \in [a,b] \times \mathbb{R}^2$;

   (ii) *the function* $f_2(t, x, y)$ *is increasing in its second and third arguments;*

2. *The functions* $x_i \in C^1([a,b], \mathbb{R})$, $(i = 1, 2, 3)$, *are solutions of the equations*

$$x_i'(t) = f_i\left(t, \ x_i(t), \max_{\xi \in [a,t]} x_i(\xi)\right), \quad \text{for } t \in [a,b], \quad i = 1, 2, 3.$$

Then the inequalities $x_1(a) \le x_2(a) \le x_3(a)$ *imply* $x_1(t) \le x_2(t) \le x_3(t)$ *for* $t \in [a,b]$.

Proof. From Theorem 3.3.1 we have that the operators E_{f_i}, $i = 1, 2, 3$, are $WPOs$. From the condition (ii) the operator E_{f_2} is monotone increasing. From the condition (i) it follows that $E_{f_1} \le E_{f_2} \le E_{f_3}$.

Let $\tilde{x}_i(a) \in C[a,b]$ be defined by $\tilde{x}_i(a)(t) = x_i(a)$, $\forall t \in [a,b]$. It is clear that

$$\tilde{x}_1(a)(t) \le \tilde{x}_2(a)(t) \le \tilde{x}_3(a)(t), \quad \forall t \in [a,b].$$

From Lemma 3.3.2 we obtain that

$$E_{f_1}^\infty\big(\tilde{x}_1(a)\big) \le E_{f_2}^\infty\big(\tilde{x}_2(a)\big) \le E_{f_3}^\infty\big(\tilde{x}_3(a)\big). \tag{3.40}$$

Since $x_i = E_{f_i}^\infty\big(\tilde{x}_i(a)\big)$ we obtain from the inequalities (3.40) $x_1(t) \le x_2(t) \le x_3(t)$ for $t \in [a,b]$.

$\square$

Denote by $x^*(\,\cdot\,;\, \alpha, f)$ the solution of the Cauchy problem (3.36), (3.37).

We need the following well-known result:

Theorem 3.3.4. *([Rus 2001]). Let* (X, d) *be a complete metric space and* A, $B : X \to X$ *two operators. We suppose that*

(i) *the operator* A *is an* α-contraction;

(ii) $F_B \ne \emptyset$;

(iii) there exists $\eta > 0$ such that

$$d\big(A(x), B(x)\big) \le \eta, \quad \forall x \in X.$$

Then, if $F_A = \{x_A^*\}$ and $x_B^* \in F_B$, we have

$$d\big(x_A^*, x_B^*\big) \le \frac{\eta}{1 - \alpha}.$$

We will prove the following result:

Theorem 3.3.5. *Let the following conditions be fulfilled:*

1. *Conditions H3.3 are satisfied for the functions f_i, $i = 1, 2$.*

2. *There exists $\eta > 0$ such that*

$$\big|f_1(t, u_1, u_2) - f_2(t, u_1, u_2)\big| \le \eta_2 \quad for\ t \in [a, b],\ u_i \in \mathbb{R}, i = 1, 2.$$

Then

$$\big|\big|x_1^*(t;\ \alpha_1, f_1) - x_2^*(t;\ \alpha_2, f_2)\big|\big| \le \frac{\eta_1 + (b - a)\eta_2}{1 - L_f(b - a)}$$

where $x_i^*(t;\ \alpha_i, f_i)$, $i = 1,\ 2$ are the solution of the problem (3.36), (3.37) with respect to α_i, f_i and $L = \max\big(L_1, L_2\big)$.

Proof. Consider the operators $B_{\alpha_i,\ f_i}$, $i = 1,\ 2$. According to Theorem 3.3.1 both operators are contractions.

Additionally

$$\|B_{\alpha_1,\ f_1}(x) - B_{\alpha_2,\ f_2}(x)\| \le \eta_1 + (b - a)\eta_2, \quad \forall x \in C[a, b].$$

Now the proof follows from Theorem 3.3.4, with $A = B_{\alpha_1,\ f_1}$, $B = B_{\alpha_2,\ f_2}$, $\eta = \eta_1 + (b - a)\eta_2$ and $\alpha = L(b - t_0)$, where $L = \max\big(L_1, L_2\big)$. $\square$

Theorem 3.3.6. *(see [Rus 2001]) Let (X, d) be a metric space and $A_i : X \to X$, $i = 1,\ 2$. Suppose that*

(i) the operator A_i is a c_i-weakly Picard operator, $i = 1,\ 2$;

(ii) there exists $\eta > 0$ such that

$$d\big(A_1(x), A_2(x)\big) \le \eta, \quad \forall x \in X.$$

Then $H\left(F_{A_1}, F_{A_2}\right) \leq \eta \max\left(c_1, c_2\right)$.

We obtain the following result:

Theorem 3.3.7. *Let the following conditions be fulfilled:*

1. *The functions $f_i \in C([a,b] \times \mathbb{R}^2, \mathbb{R})$, $i = 1,\ 2$, satisfy the conditions H3.3.*

2. *$S_{E_{f_1}}$, $S_{E_{f_2}}$ be the solution set of system (3.36) corresponding to f_1 and f_2.*

3. *There exists $\eta > 0$, such that*

$$\left|f_1(t,\ u_1,\ u_2) - f_2(t,\ u_1,\ u_2)\right| \leq \eta \qquad (3.41)$$

for all $t \in [a,b]$, $u_i \in \mathbb{R}$, $i = 1,\ 2$.

Then

$$H_{\|\cdot\|_C}\left(S_{E_{f_1}}, S_{E_{f_2}}\right) \leq \frac{(b-a)\eta}{1 - L_f(b-a)}$$

where $L_f = \max\left(L_{f_1}, L_{f_2}\right)$ and $H_{\|\cdot\|_C}$ denotes the Pompeiu-Housdorff functional with respect to $\|\cdot\|_C$ on $C[a,b]$.

Proof. According to condition 1 the operators E_{f_1} and E_{f_2} are c_i-weakly Picard operators, $i = 1,\ 2$.

Consider the set of functions $X_\alpha := \left\{x \in C[a,b] \ : \ x(a) = \alpha\right\}$.
It is clear that $E_{f_1}\big|_{X_\alpha} = B_{f_1}$, $E_{f_2}\big|_{X_\alpha} = B_{f_2}$. Therefore

$$\left|E_{f_1}^2(x) - E_{f_1}(x)\right| \leq L_{f_1}(b-a)\left|E_{f_1}(x) - x\right|,$$

$$\left|E_{f_2}^2(x) - E_{f_2}(x)\right| \leq L_{f_2}(b-a)\left|E_{f_2}(x) - x\right|,$$

for all $x \in C[a,b]$.

Now, choosing

$$\alpha_1 = L_{f_1}(b-a) \quad \text{and} \quad \alpha_2 = L_{f_2}(b-a),$$

we get that E_{f_1} and E_{f_2} are c_i-weakly Picard operators, $i = 1,2$ with $c_1 = (1-\alpha_1)^{-1}$ and $c_2 = (1-\alpha_2)^{-1}$. From (3.41) we obtain that

$$\|E_{f_1}(x) - E_{f_2}(x)\|_C \leq (b-a)\eta, \quad \forall x \in C[a,b].$$

Applying Theorem 3.3.6 we have that

$$H_{\|\cdot\|_C}\left(S_{E_{f_1}}, S_{E_{f_2}}\right) \leq \frac{(b-a)\eta}{1 - L_f(b-a)}$$

where $L_f = \max\left(L_{f_1}, L_{f_2}\right)$ and $H_{\|\cdot\|_C}$ is the Pompeiu-Housdorff functional with respect to $\|\cdot\|_C$ on $C[a,b]$.

$\square$

Chapter 4

Stability Theory and Lyapunov Functions

In this chapter the basic results in the stability theory for differential equations with "maxima" are presented. The basic methods of investigations are the method of Lyapunov functions and its modification of Razumikhin. Different types of stability are defined and sufficient conditions are obtained.

Some stability properties of solutions of differential equations with "maxima" are studied in [Ivanov et al. 2002], [Magomedov 1991], [Magomedov 1983b], [Magomedov 1981], [Magomedov and Nabiev 1986], [Magomedov and Ryabov 1991], [Nabiev 1985], [Nabiev 1984a], [Nabiev 1984b], [Pinto and Trofimchuk 2000], [Voulov 1995], [Voulov 1992a], [Voulov 1992b], [Voulov 1991], [Voulov and Bainov 1992], [Voulov and Bainov 1991], [Yuldashev and Kuldashev 1994], and [Zhang and Cheng 1999].

The main object of investigations in this chapter is the following nonlinear differential equations with "maxima"

$$x' = F(t, x(t), \max_{s \in [t-r,t]} x(s)) \quad \text{for} \quad t \geq t_0 \tag{4.1}$$

with initial condition

$$x(t + t_0) = \phi(t), \quad t \in [-r, 0], \tag{4.2}$$

where $x \in \mathbb{R}^n$, $F : \mathbb{R}_+ \times \mathbb{R}^n \times \mathbb{R}^n \to \mathbb{R}^n$, $F = (F_1, F_2, \ldots, F_n)$, $r > 0$ is a given fixed number, $t_0 \in \mathbb{R}_+$, and $\phi \in C([-r, 0], \mathbb{R}^n)$.

We denote by $x(t; t_0, \phi)$ the solution of the initial value problem (4.1), (4.2). In our further investigations we will assume that solution $x(t; t_0, \phi)$ is defined on $[t_0 - r, \infty)$ for any initial function $\phi \in C([-r, 0], \mathbb{R}^n)$.

Introduce the following set of functions:

$$K = \{a \in C(\mathbb{R}_+, \mathbb{R}_+) \ : \ a(s) \text{ is strictly increasing and } a(0) = 0\}.$$
$$(4.3)$$

4.1 Stability and Uniform Stability

The problems of stability of solutions of differential equations via Lyapunov functions have been successfully investigated in the past. One type of stability, very useful in real world problems, deals with two different measures. Stability in terms of two measures of differential equations has been studied by means of various types of Lyapunov functions (see [Lakshmikantham and Liu 1993] and references cited therein). In this section the stability in terms of two measures will be defined and studied for differential equations with "maxima."

We will introduce the class Λ of Lyapunov functions:

Definition 4.1.1. *We will say that the function $V(t, x) : \Omega \times \mathbb{R}^n \to \mathbb{R}_+$, $\Omega \subset \mathbb{R}_+$, belongs to class Λ if*

1. *$V(t, x)$ is a continuous function in $\Omega \times \mathbb{R}^n$;*

2. *Function $V(t, x)$ is Lipschitz with respect to its second argument.*

Let the function $V \in \Lambda$, $t \in \Omega$, and $\phi \in C([t - r, t], \mathbb{R}^n)$. We define derivative of the function $V(t, x)$ along the trajectory of solution of (4.1) as follows

$$D_{(4.1)}V(t, \phi(t)) = \limsup_{\epsilon \to 0} \frac{1}{\epsilon} \{V(t + \epsilon, \phi(t)$$
$$+ \epsilon F(t, \phi(t), \max_{s \in [-r, 0]} \phi(t + s))) - V(t, \phi(t))\}. \quad (4.4)$$

Note that $V(t, x)$ is a function, and its derivative defined by (4.4) is a functional.

We will study the stability in the regular case as well as in the case when two different measures are used for the initial conditions and for the solutions. We will obtain some sufficient conditions in both cases.

4.1.1   Stability in Terms of Two Measures

Consider the set of measures

$$\Gamma = \{h \in C([-r, \infty) \times \mathbb{R}^n, \mathbb{R}_+) \; : \; \inf_{x \in \mathbb{R}^n} h(t, x) = 0 \text{ for each } t \in [-r, \infty)\}. \tag{4.5}$$

Let $\rho >$ be a constant, $h \in \Gamma$. Define sets:

$$
\begin{aligned}
S(h, \rho) &= \{(t, x) \in \mathbb{R}_+ \times \mathbb{R}^n \; : \; h(t, x) < \rho\}; \\
S^C(h, \rho) &= \{(t, x) \in \mathbb{R}_+ \times \mathbb{R}^n \; : \; h(t, x) \geq \rho\}.
\end{aligned}
\tag{4.6}
$$

Definition 4.1.2. *Let $h, h_0 \in \Gamma$. The system of differential equations with "maxima" (4.1), (4.2) is said to be*

(S4.1.1) *equi-stable in terms of measures (h_0, h) if for every $\epsilon > 0$ and for any $t_0 \geq 0$, there exists $\delta = \delta(t_0, \epsilon) > 0$ such that for any $\phi \in C([-r, 0], \mathbb{R}^n)$ inequality $\max_{s \in [-r, 0]} h_0(t_0 + s, \phi(s)) < \delta$ implies $h(t, (x(t; t_0, \phi))) < \epsilon$ for $t \geq t_0$, where $x(t; t_0, \phi)$ is a solution of the initial value problem for differential equations with "maxima" (4.1), (4.2);*

(S4.1.2) *uniformly stable in terms of measures (h_0, h) if δ in (S4.1.1) is independent on t_0;*

In our further investigations we will use a comparison scalar ordinary differential equation

$$u' = g_1(t, u), \tag{4.7}$$

where $u \in \mathbb{R}$.

Definition 4.1.3. *([Lakshmikantham and Liu 1993]). Let $h \in \Gamma$. Function $V(t, x) \in \Lambda$ is said to be h-decrescent if there exists a constant $\delta > 0$ and a function $a \in K$ such that $h(t, x) < \delta$ implies that $V(t, x) \leq a(h(t, x))$.*

Definition 4.1.4. *([Lakshmikantham and Liu 1993]). Let $h, h_0 \in \Gamma$. Function $h_0(t, x)$ is uniformly finer than $h(t, x)$, if there exists a constant $\delta > 0$ and a function $a \in K$ such that $h_0(t, x) < \delta$ implies that $h(t, x) \leq a(h_0(t, x))$.*

In our further investigations, we need the following comparison result:

Lemma 4.1.1. *Let the following conditions be fulfilled:*

1. *The function $F \in C([t_0, T] \times \mathbb{R}^n \times \mathbb{R}^n, \mathbb{R}^n)$, where to, $T \in \mathbb{R}_+$, $t_0 < T$.*

2. *The function $V : [t_0, T] \times \mathbb{R}^n \to \mathbb{R}_+$, $V \in \Lambda$ and*

   (i) *for any number $t \in [t_0, T]$ and any function $\psi \in C([t - r, t], \mathbb{R}^n)$ such that $V(t, \psi(t)) > V(t + s, \psi(t + s))$ for $s \in [-r, 0)$ the inequality*

$$D_{(4.1)} V(t, \psi(t)) \leq g_1(t, V(t, \psi(t)))$$

   holds, where $g_1 \in C([t_0, T] \times \mathbb{R}_+, \mathbb{R}_+)$, $g_1(t, 0) \equiv 0$.

3. *The function $x(t; t_0, \phi)$ is a solution of (4.1) with initial condition (4.2), which is defined for $t \in [t_0 - r, T]$, where $\phi \in C([-r, 0], \mathbb{R}^n)$.*

4. *The function $u^*(t) = u^*(t; t_0, u_0)$ is the maximal solution of (4.7) with initial condition $u^*(t_0) = u_0$, which is defined for $t \in [t_0, T]$.*

Then the inequality $max_{s \in [-r, 0]} V(t_0 + s, \phi(s)) \leq u_0$ implies the validity of the inequality $V(t, x(t; t_0, \phi)) \leq u^*(t)$ for $t \in [t_0, T]$.

Proof. Let $u_0 \in \mathbb{R}_+$ and $\phi \in C([-r, 0], \mathbb{R}^n)$ be such that $max_{s \in [-r, 0]} V(t_0 + s, \phi(s)) \leq u_0$. Let n be a natural number and $v_n(t)$ be the maximal solution of the initial value problem

$$u' = g_1(t, u) + \frac{1}{n},$$

$$u(t_0) = u_0 + \frac{1}{n}.$$

Define a function $m(t) \in C([t_0 - r, T], \mathbb{R}_+)$: $m(t) = V(t, x(t; t_0, \phi))$.

Because of the fact that $u^*(t; t_0, u_0) = lim_{n \to \infty} v_n(t)$, it is enough to prove that for any natural number n the inequality

$$m(t) \leq v_n(t) \quad \text{for} \quad t \in [t_0, T] \tag{4.8}$$

holds.

Note that for any natural number n inequality $m(t_0) < v_n(t_0)$ holds.

Assume inequality (4.8) is not true. Let n be a natural number such that there exists a point $\eta \in (t_0, T]$: $m(\eta) > v_n(\eta)$. Let $t^* = sup\{t \in$

$[t_0, T]$: $m(s) < v_n(s)$ for $s \in [t_0, t)\}$. According to the assumption $t^* < T$.

Therefore

$$m(t^*) = v_n(t^*), \qquad m(t) < v_n(t) \qquad \text{for} \quad t \in [t_0, t^*), \qquad (4.9)$$
$$m(t) \geq v_n(t) \qquad \text{for} \quad t \in (t^*, t^* + \delta),$$

where $\delta > 0$ is a small enough number.

From inequality (4.9) it follows that

$$m'(t^*) \geq v'_n(t^*) = g_1(t^*, v_n(t^*)) + \frac{1}{n} = g_1(t^*, m(t^*)) + \frac{1}{n}. \qquad (4.10)$$

From $g_1(t, u) + \frac{1}{n} > 0$ on $[t^* - r, t^*] \cap [t_0, T]$ it follows the function $v_n(t)$ is nondecreasing on $[t^* - r, t^*] \cap [t_0, T]$.

If $t^* - r \geq t_0$ then $m(t^*) = v_n(t^*) \geq v_n(s) > m(s)$ for $s \in [t^* - r, t^*)$.

If $t^* - r < t_0$, then as above $m(t^*) > m(s)$ for $s \in [t_0, t^*)$ and $m(t^*) = v_n(t^*) \geq v_n(t_0) = u_0 + \frac{1}{n} > u_0 \geq sup_{s \in [-r,0]} V(t_0 + s, \phi(s)) \geq m(s)$ for $s \in [t^* - r, t_0)$.

Therefore $m(t^*) > m(s)$ for $s \in [t^* - r, t^*)$.

According to condition (i) of Lemma 4.1.1 and definition of function $m(t)$ we get $m'(t^*) \leq g_1(t^*, m(t^*)) < g_1(t^*, m(t^*)) + \frac{1}{n}$ that contradicts (4.10).

Therefore the inequality (4.8) holds and hence the conclusion of Lemma 4.1.1 follows.

$\square$

We will illustrate the importance of the condition $\max_{s \in [-r,0]} V(t_0 + s, \phi(s)) \leq u_0$ for the claim of Lemma 4.1.1.

Example 4.1.1. *Consider the scalar differential equation with "maxima"*

$$x'(t) = \max_{t \in [t - \frac{\pi}{2}, t]} x(s) \quad \text{for } t \geq 0, \qquad (4.11)$$

with the initial condition

$$x(s) = -a \sin(t) \quad \text{for} \quad t \in [-\frac{\pi}{2}, 0], \qquad (4.12)$$

where $a > 0$ is a constant.

The solution of the inital value problem (4.11), (4.12) is given by

$$x(t) = \begin{cases} -a\sin(t) & \text{for } t \in [-\frac{\pi}{2}, 0], \\ a\sin(t) & \text{for } t \in [0, \frac{\pi}{4}], \\ \dfrac{a\sqrt{2}}{2}\, e^{t-\frac{\pi}{4}} & \text{for } t \geq \frac{\pi}{4}. \end{cases} \qquad (4.13)$$

Let $V(t,x) = x^2$, $t \in [0, \frac{\pi}{2}]$ and the function $\psi \in C([t - \frac{\pi}{2}, t], \mathbb{R})$ be such that $(\psi(t))^2 > (\psi(t+s))^2$ for $s \in [-\frac{\pi}{2}, 0)$. Then the inequality

$$D_{(4.11)}V(t, \psi(t)) = 2\psi(t) \max_{s \in [t-\frac{\pi}{2}, t]} \psi(s) \leq 2(\psi(t))^2 = g_1(t, V(t, \psi(t)))$$

holds, where $g_1(t, u) \equiv 2u$.

Consider the scalar differential equation $u' = 2u$ with the initial condition $u(0) = u_0$, which solution is $u(t) = u_0 e^{2t}$.

It is easy to check that if $V(0, a\sin 0) = 0 < u_0$ and $a^2 > u_0$, then the inequality $(-a\sin(t))^2 \leq u_0$ does not hold for $t \in [-\frac{\pi}{2}, 0)$. At the same time the inequality $V(t, x(t)) = (a\sin t)^2 \leq u_0 e^{2t}$ does not hold for $t \in [0, \frac{\pi}{4}]$, and the conclusion ot Lemma 4.1.1 is not true, i.e., the initial inequality is necessary to be satisfied on the whole initial interval, not only at one single point.

We will give some sufficient conditions for stability of the considered differential equations with "maxima" based on the applications of Lyapunov functions and the Razumikhin method.

Theorem 4.1.1. *Let the following conditions be fulfilled:*

1. *The function $F \in C(\mathbb{R}_+ \times \mathbb{R}^n \times \mathbb{R}^n, \mathbb{R}^n)$, $F(t, 0, 0) \equiv 0$.*

2. *The function $g_1 \in C(\mathbb{R}_+ \times \mathbb{R}, \mathbb{R}_+)$, $g_1(t, 0) \equiv 0$.*

3. *The functions $h_0, h \in \Gamma$, h_0 is uniformy finer than h.*

4. *There exists a function $V(t, x) \in \Lambda$ such that:*

    (i) *for any number $t \in \mathbb{R}_+$ and any function $\psi \in C([t-r, t], \mathbb{R}^n)$ such that $(t, \psi(t)) \in S(h, \rho)$ and $V(t, \psi(t)) > V(t + s, \psi(t + s))$ for $s \in [-r, 0)$ the inequality*

    $$D_{(4.1)}V(t, \psi(t)) \leq g_1(t, V(t, \psi(t)))$$

    holds.

(ii) $b(h(t,x)) \leq V(t,x) \leq a(h_0(t,x))$ *for* $(t,x) \in S(h,\rho),$ *where functions* $a,b \in K$ *and the constant* $\rho > 0.$

5. *The zero solution of the scalar differential equation (4.7) is equi-stable.*

Then differential equations with "maxima" (4.1) is equi-stable in terms of measures $(h_0, h).$

Proof. Let $\epsilon > 0$ be a number such that $\epsilon < \rho.$

From condition 5 of Theorem 4.1.1 follows that there exists $\delta_1 = \delta_1(t_0, \epsilon) > 0$ such that $|v_0| < \delta_1$ implies

$$|v(t;t_0,v_0)| < b(\epsilon), \quad \text{for} \quad t \geq t_0, \qquad (4.14)$$

where $v(t;t_0,v_0)$ is a solution of the scalar ordinary differential equation (4.7) with initial condition $v(t_0) = v_0.$

From condition 3 of Theorem 4.1.1 follows that there exists $\delta_2 > 0$ and a function $\psi_1 \in K$ such that the inequality

$$h_0(t,x) < \delta_2 \qquad (4.15)$$

implies

$$h(t,x) \leq \psi_1(h_0(t,x)). \qquad (4.16)$$

Since $a \in K$ and $\psi_1 \in K$ we can find $\delta_3 = \delta_3(t_0,\epsilon) > 0,$ $\delta_3 < \delta_2$ such that the inequalities

$$a(\delta_3) < \delta_1, \quad \psi_1(\delta_3) < \epsilon \qquad (4.17)$$

hold.

Let $t_0 \in \mathbb{R}_+$ and $\varphi \in C([-r,0],\mathbb{R}^n)$ be such that

$$\max_{s\in[-r,0]} h_0(t_0 + s, \varphi(s)) < \delta_3. \qquad (4.18)$$

From (4.16) and (4.18) we get

$$h(t_0 + s, \varphi(s)) \leq \psi_1(h_0(t_0 + s, \varphi(s))) < \psi_1(\delta_3) < \epsilon, \quad \text{for } s \in [-r,0]. \qquad (4.19)$$

We will prove that inequality

$$h(t, x(t;t_0,\varphi)) < \epsilon \text{ for } t \geq t_0 \qquad (4.20)$$

holds, where $x(t; t_0, \varphi)$ is a solution of the initial value problem (4.1), (4.2) with the chosen-above initial function φ.

Assume that inequality (4.20) is not true. From the assumption and inequality (4.19) follows that there exists a point $t^* > t_0$ such that

$$h(t^*, x(t^*; t_0, \varphi)) = \epsilon, \quad h(t, x(t; t_0, \varphi)) < \epsilon, \quad t \in [t_0 - r, t^*). \quad (4.21)$$

Therefore $(t, x(t)) \in S(h, \epsilon)$ for $t \in [t_0, t^*)$. Since $\epsilon < \rho$ the inclusion $(t, x(t)) \in S(h, \rho)$ is valid for $t \in [t_0 - r, t^*]$.

Let $v^*(t; t_0, v_0)$ be the maximal solution of (4.7) with initial condition $v(t_0) = v_0$, where $v_0 = \max_{s \in [-r,0]} V(t_0 + s, \varphi(s))$. Therefore, from Lemma 4.1.1 we obtain

$$V(t, x(t; t_0, \varphi)) \leq v^*(t; t_0, v_0), \quad t \in [t_0, t^*]. \quad (4.22)$$

From condition (ii) of Theorem 4.1.1 and inequalities (4.17) and (4.18) follows

$$v_0 = \max_{s \in [-r,0]} V(t_0 + s, \varphi(s)) \leq a(h_0(t_0 + s, \varphi(s))) \leq a(\delta_3) < \delta_1. \quad (4.23)$$

From inequality (4.23) according to (4.14) we get

$$|v^*(t; t_0, v_0)| < b(\epsilon) \quad \text{for } t \geq t_0. \quad (4.24)$$

From inequalities (4.22) and (4.24), the choice of the point t^*, and condition (ii) of Theorem 4.1.1 follows

$$b(\epsilon) = b(h(t^*, x(t^*; t_0, \varphi))) \leq V(t^*, x(t^*; t_0, \varphi)) \leq v^*(t^*; t_0, v_0) < b(\epsilon).$$

The obtained contradiction proves the validity of the inequality (4.20), which proves equi-stability in terms of measures (h_0, h) of the considered differential equations with "maxima."

$$\square$$

Sufficient conditions for uniform stability in terms of two measures for diferential equations with "maxima" are given in the following theorem:

Theorem 4.1.2. *Let the conditions 1, 2, 3, 4 of Theorem 4.1.1 be fulfilled and the zero solution of the scalar differential equation (4.7) be uniform-stable.*

Then differential equations with "maxima" (4.1) is uniformly stable in terms of measures (h_0, h).

The proof of Theorem 4.1.2 is similar to the one for Theorem 4.1.1, where δ_1 depends only on ϵ but not on t_0.

The most difficult part in the application of the method of Lyapunov and its modification of Razimikhin is the construction of Lyapunov functions, provided the necessary conditions. In connection with this, sometimes it is more applicable to use sufficient conditions employing so-called perturbing Lyapunov function. In this case we will use the comparison scalar ordinary differential equation (4.7) combined with another comparison differential equation

$$v' = g_2(t, v), \tag{4.25}$$

where $v \in \mathbb{R}$, $g_2(t, 0) \equiv 0$.

Theorem 4.1.3. *Let the following conditions be fulfilled:*

1. *The function $F \in C(\mathbb{R}_+ \times \mathbb{R}^n \times \mathbb{R}^n, \mathbb{R}^n)$, $F(t, 0, 0) \equiv 0$.*

2. *The functions $g_1, g_2 \in C(\mathbb{R}_+ \times \mathbb{R}, \mathbb{R}_+)$, $g_1(t, 0) \equiv 0$, $g_2(t, 0) \equiv 0$.*

3. *The functions $h_0, h \in \Gamma$, h_0 is uniformy finer than h.*

4. *There exists a function $V_1 \in \Lambda$ which is h_0-decrescent and*

   (i) for $t \geq 0$ and $\psi \in C([t-r, t], \mathbb{R}^n)$ such that $V_1(t, \psi(t)) > V_1(t+ s, \psi(t + s))$ for $s \in [-r, 0)$ and $(t, \psi(t)) \in S(h, \rho)$ the inequality

   $$D_{(4.1)}V_1(t, \psi(t)) \leq g_1(t, V_1(t, \psi(t)))$$

   holds, where $\rho > 0$ is a constant.

5. *For any number $\mu > 0$ there exists a function $V_2^{(\mu)} \in \Lambda$ such that*

   (ii) $b(h(t, x)) \leq V_2^{(\mu)}(t, x) \leq a(h_0(t, x))$ for $(t, x) \in S(h, \rho) \cap S^C(h_0, \mu)$,

   where $a, b \in K$.

   (iii) for any number $t \geq 0$ and for any function $\psi \in C([t - r, t], \mathbb{R}^n)$ such that $(t, \psi(t)) \in S(h, \rho) \cap S^C(h_0, \mu)$ and

   $$V_1(t, \psi(t)) + V_2^{(\mu)}(t, \psi(t)) > V_1(t+s, \psi(t+s)) + V_2^{(\mu)}(t+s, \psi(t+s))$$

for $s \in [-r, 0)$ the inequality

$$D_{(4.1)}V_1(t, \psi(t)) + D_{(4.1)}V_2^{(\mu)}(t, \psi(t))$$
$$\leq g_2\left(t, V_1(t, \psi(t)) + V_2^{(\mu)}(t, \psi(t))\right)$$

holds;

6. *The zero solution of scalar differential equation (4.7) is equi-stable.*

7. *The zero solution of scalar differential equation (4.25) is uniformly stable.*

Then differential equations with "maxima" (4.1) is equi-stable in terms of measures (h_0, h).

Proof. Since function $V_1(t, x)$ is h_0-decrescent, there exists a constant $\rho_1 > 0$ and a function $\psi_1 \in K$ such that $h_0(t, x) < \rho_1$ implies that

$$V_1(t, x) \leq \psi_1(h_0(t, x)). \tag{4.26}$$

Without loss of generality we assume that $\rho_1 < \rho$.

Since $h_0(t, x)$ is uniformly finer than $h(t, x)$, there exists a constant $\rho_0 > 0$ and a function $\psi_2 \in K$ such that inequality $h_0(t, x) < \rho_0$ implies $h(t, x) \leq \psi_2(h_0(t, x))$. We will assume that $\rho_0 < \rho$ and $\psi_2(\rho_0) < \rho_1$.

Let $\epsilon > 0$ be a fixed number, $\epsilon < \rho$, and $t_0 \geq 0$ be a fixed point.

Since the zero solution of scalar impulsive differential equation (4.25) is uniformly stable there exists $\delta_1 = \delta_1(\epsilon) \in K$ such that the inequality $|v_0| < \delta_1$ implies

$$|v(t; t_0, v_0)| < b(\epsilon), \quad t \geq t_0, \tag{4.27}$$

where $v(t; t_0, v_0)$ is a solution of equation (4.25) with initial condition $v(t_0) = v_0$.

Since the functions $a \in K$ and $\psi_2 \in K$ we can find $\delta_2 = \delta_2(\epsilon) > 0$, $\delta_2 < \rho_0$ such that the inequalities

$$a(\delta_2) < \frac{\delta_1}{2}, \quad \psi_2(\delta_2) < \epsilon \tag{4.28}$$

hold.

Since the zero solution of scalar impulsive differential equation (4.7) is equi-stable there exists $\delta_3 = \delta_3(t_0, \epsilon) > 0$ such that inequality $|u_0| < \delta_3$ implies

$$|u(t; t_0, u_0)| < \frac{\delta_1}{2}, \quad t \geq t_0, \qquad (4.29)$$

where $u(t; t_0, u_0)$ is a solution of (4.7) with initial condition $u(t_0) = u_0$.

Since the function $\psi_1 \in K$ there exists $\delta_4 = \delta_4(t_0, \epsilon) > 0$ such that for $|u| < \delta_4$ the inequality

$$\psi_1(u) < \delta_3 \qquad (4.30)$$

holds.

Now let function $\phi \in C([-r, 0], \mathbb{R}^n)$ be such that

$$\max_{s \in [-r, 0]} h_0(t_0 + s, \phi(s)) < \delta_6, \qquad (4.31)$$

where $\delta_6 = \min\{\delta_2, \delta_4, \rho_1\}$, $\delta_6 = \delta_6(t_0, \epsilon) > 0$.

From inequality (4.31) follows that $h_0(t_0 + s, \phi(s)) < \delta_6 \leq \rho_1$ for $s \in [-r, 0]$. From inequalities (4.26) and (4.30) we obtain

$$V_1(t_0 + s, \phi(s)) \leq \psi_1(h_0(t_0 + s, \phi(s))) < \delta_3 \quad \text{for } s \in [-r, 0]. \qquad (4.32)$$

From condition 2, inequality (4.31), and the choice of δ_2 and δ_6 follows that

$$h(t_0+s, \phi(s)) < \psi_2(h_0(t_0+s, \phi(s))) < \psi_2(\delta_6) \leq \psi_2(\delta_2) < \epsilon, \quad s \in [-r, 0]. \qquad (4.33)$$

We will prove that if inequality (4.31) is satisfied, then

$$h(t, x(t; t_0, \phi)) < \epsilon, \quad t \geq t_0, \qquad (4.34)$$

where $x(t; t_0, \phi)$ is a solution of initial value problem (4.1), (4.2).

Suppose inequality (4.34) is not true. Therefore there exists a point $t^* > t_0$ such that

$$h(t^*, x(t^*; t_0, \phi)) = \epsilon, \quad \text{and} \quad h(t, x(t; t_0, \phi)) < \epsilon, \quad t \in [t_0 - r, t^*). \quad (4.35)$$

Denote $x(s) = x(s; t_0, \phi)$, $s \in [t_0 - r, t^*]$.

If we assume that $h_0(t^*, x(t^*)) \leq \delta_2 < \rho$ then from the choice of δ_2 follows $h(t^*, x(t^*)) \leq \psi_2(h_0(t^*, x(t^*))) \leq \psi_2(\delta_2) < \epsilon$ that contradicts (4.35).

Therefore
$$h_0(t^*, x(t^*)) > \delta_2. \tag{4.36}$$

From inequality (4.36) and $h_0(t_0+s, \phi(s)) < \delta_6 \leq \delta_2$ for $s \in [-r, 0]$ it follows that there exists a point $t_0^* \in (t_0, t^*)$ such that $\delta_2 = h_0(t_0^*, x(t_0^*))$ and $(t, x(t)) \in S(h, \epsilon)$ for $t \in [t_0^*, t^*)$. From the choice of ϵ follows that

$$(t, x(t)) \in S(h, \rho) \bigcap S^C(h_0, \delta_2), \quad t \in [t_0^*, t^*]. \tag{4.37}$$

Let $r_1(t; t_0, u_0)$ be the maximal solution of differential equation (4.7) where $u_0 = sup_{s \in [-r,0]} V_1(t_0 + s, \phi(s))$. From inequality (4.32) follows $u_0 < \delta_3$ and according to inequality (4.29) we get

$$|r_1(t; t_0, u_0)| < \frac{\delta_1}{2} \text{ for } t \in [t_0, t^*]. \tag{4.38}$$

From condition 4 of Theorem 4.1.3 follows that the condition 1 of Lemma 4.1.1 is satisfied for $T = t^*$. According to Lemma 4.1.1 the inequality

$$V_1(s, x(s)) \leq r_1(s; t_0, u_0), \quad s \in [t_0, t^*] \tag{4.39}$$

holds.

Inequalities (4.38) and (4.39) imply

$$V_1(t_0^* + s, x(t_0^* + s)) < \frac{\delta_1}{2} \text{ for } s \in [-r, 0]. \tag{4.40}$$

Consider the function $V_2^{(\delta_2)}(t, x)$ that is defined in condition 5 of Theorem 4.1.3 and define the function $m : [t_0 - r, \infty) \times \mathbb{R}^n \to \mathbb{R}^n$ by equality

$$m(t, x) = V_1(t, x) + V_2^{(\delta_2)}(t, x). \tag{4.41}$$

From condition 5 of Theorem 4.1.3 follows that condition 1 of Lemma 4.1.1 is satisfied for the function $V(t, x) = m(t, x)$, $T = t^*$, and $t_0 = t_0^*$. According to Lemma 4.1.1 the inequality

$$m(t, x(t; t_0, \phi)) \leq r^*(t; t_0^*, w_0^*), \quad t \in [t_0^*, t^*] \tag{4.42}$$

holds, where $r^*(t; t_0^*, w_0^*)$ is the maximal solution of (4.25) through the point (t_0^*, w_0^*), $w_0^* = sup_{s \in [-r,0]} m(t_0^* + s, x(t_0^* + s; t_0, \phi))$.

From inequality (4.28) and condition (iii) of Theorem 4.1.3 it follows that

$$V_2^{(\delta_2)}(t_0^* + s, x(t_0^* + s)) < a(h_0(t_0^* + s, x(t_0^* + s))) \leq a(\delta_2) < \frac{\delta_1}{2},$$
$$s \in [-r, 0].$$

$$\tag{4.43}$$

From inequalities (4.40) and (4.43) we obtain

$$m(t_0^* + s, x(t_0^* + s)) < \delta_1 \quad \text{for } s \in [-r, 0]. \tag{4.44}$$

From inequality (4.44) follows that $|w_0^*| < \delta_1$ and therefore according to inequality (4.27)

$$r^*(t; t_0^*, w_0^*) < \beta(\epsilon), \quad t \geq t_0^*. \tag{4.45}$$

From inequalities (4.42) and (4.45) the choice of the point t^*, and condition (iii) of Theorem 4.1.3 we obtain

$$\begin{aligned} b(\epsilon) &> r^*(t^*; t_0^*, w_0^*) \geq m(t^*, x(t^*; t_0, \phi)) \\ &\geq V_2^{(\delta_2)}(t^*, x(t^*; t_0, \phi)) \geq b(h(t^*, x(t^*; t_0, \phi))) = b(\epsilon). \end{aligned}$$

The obtained contradiction proves the validity of inequality (4.34) for $t \geq t_0$.

Inequality (4.34) proves the equi-stability in terms of measures (h_0, h) of the considered system of differential equations with "maxima."

□

Theorem 4.1.4. *Let the conditions 1, 2, 3, 4, 5 and 7 of Theorem 4.1.3 be satisfied, and the zero solution of scalar differential equation (4.7) be uniformly stable.*

Then the system of differential equations with "maxima" (4.1) is uniformly stable in terms of measures (h_0, h).

The proof of Theorem 4.1.4 is similar to the proof of Theorem 4.1.3 but in this case δ_3, and therefore, δ_4 and δ_6 depend only on ϵ.

Remark 4.1.1. *Note that in the case $r = 0$ results in this section reduce to sufficient conditions for stability in terms of two measures for ordinary differential equations (see [Lakshmikantham and Liu 1993] and references cited therein).*

4.1.2   Stability of Zero Solution

Let $\rho >$ be a constant. Define sets:

$$\begin{aligned} \mathcal{S}(\rho) &= \{(t, x) \in \mathbb{R}_+ \times \mathbb{R}^n : \ \|x\| < \rho\}; \\ \mathcal{S}^C(\rho) &= \{(t, x) \in \mathbb{R}_+ \times \mathbb{R}^n : \ \|x\| \geq \rho\}. \end{aligned}$$

Definition 4.1.5. *The zero solution of the system of differential equations with "maxima" (4.1), (4.2) is said to be*

(S4.1.3) *equi-stable if for every $\epsilon > 0$ and for any $t_0 \geq 0$, there exists $\delta = \delta(t_0, \epsilon) > 0$ such that for any $\phi \in C([-r, 0], \mathbb{R}^n)$ inequality*

$$\max_{s \in [-r, 0]} \|\phi(s)\| < \delta \quad implies \quad \|x(t; t_0, \phi)\| < \epsilon$$

for $t \geq t_0$, where $x(t; t_0, \phi)$ is a solution of the initial value problem for the system of differential equations with "maxima"(4.1), (4.2);

(S4.1.4) *uniformly stable if δ in (S4.1.3) is independent on t_0;*

Definition 4.1.6. *The function $V(t, x) \in \Lambda$ is said to be weakly decrescent if there exists a constant $\delta > 0$ and a function $a \in CK$ such that $\|x\| < \delta$ implies that $V(t, x) \leq a(t, \|x\|)$.*

Definition 4.1.7. *Function $V(t, x) \in \Lambda$ is said to be decrescent if there exists a constant $\delta > 0$ and a function $a \in K$ such that $\|x\| < \delta$ implies that $V(t, x) \leq a(\|x\|)$.*

In the case when both measures, used in Subsection 4.1.1, are equal to the regular norm in $\mathbb{R}^n$ as a particular case of results in the Subsection 4.1.1 we obtain sufficient conditions for equi-stability and uniform stability of the zero solution of differential equations with "maxima". The proofs of the results are similar to the proofs of the corresponding Theorems in Subsection 4.1.1 and we omit them.

Theorem 4.1.5. *Let the following conditions be fulfilled:*

1. *The function $F \in C(\mathbb{R}_+ \times \mathbb{R}^n \times \mathbb{R}^n, \mathbb{R}^n)$, $F(t, 0, 0) \equiv 0$.*

2. *The function $g_1 \in C(\mathbb{R}_+ \times \mathbb{R}, \mathbb{R}_+)$, $g_1(t, 0) \equiv 0$.*

3. *There exists a function $V(t, x) \in \Lambda$ such that:*

   (i) *for any number $t \in [t_0, T]$ and any function $\psi \in C([t - r, t], \mathbb{R}^n)$ such that $V(t, \psi(t)) > V(t + s, \psi(t + s))$ for $s \in [-r, 0)$ the inequality*

   $$D_{(4.1)}V(t, \psi(t)) \leq g_1(t, V(t, \psi(t)))$$

   holds.

(ii) $b(\|x\|) \leq V(t,x) \leq a(\|x\|)$ for $(t,x) \in \mathcal{S}(\rho)$, where functions $a, b \in K$ and the constant $\rho > 0$.

4. The zero solution of the scalar differential equation (4.7) is equistable.

Then the zero solution of the differential equations with "maxima" (4.1) is equi-stable.

Theorem 4.1.6. *Let the conditions 1, 2, 3 of Theorem 4.1.5 be satisfied and the zero solution of the scalar differential equation (4.7) be uniformly stable.*

Then the zero solution of the differential equations with "maxima" (4.1) is uniformly stable.

As a partial case of Theorem 4.1.6 we get the following result:

Corollary 4.1.1. *Let the following conditions be fulfilled:*

1. *The function $F \in C(\mathbb{R}_+ \times \mathbb{R}^n \times \mathbb{R}^n, \mathbb{R}^n)$, $F(t,0,0) \equiv 0$.*

2. *There exists a function $V(t,x) \in \Lambda$ such that:*

   (i) *for any number $t \in [t_0, T]$ and any function $\psi \in C([t - r, t], \mathbb{R}^n)$ such that $V(t, \psi(t)) > V(t+s, \psi(t+s))$ for $s \in [-r, 0)$ the inequality*

   $$D_{(4.1)}V(t, \psi(t)) \leq 0$$

   holds.

   (ii) $b(\|x\|) \leq V(t,x) \leq a(\|x\|)$ *for* $(t,x) \in \mathcal{S}(\rho)$, *where functions $a, b \in K$ and the constant $\rho > 0$.*

Then the zero solution of the differential equations with "maxima" (4.1) is uniformly stable.

The proof the Corollary 4.1.1 follows from Theorem 4.1.6 for $g_1(t, u) \equiv 0$ and the fact that the zero solution of the scalar ordinary differential equation $u' = 0$ is uniformly stable.

In the case, when two different Lyapunov function are employed, the following two theorems are partial cases of Theorem 4.1.3 and Theorem 4.1.4 correspondingly.

Theorem 4.1.7. *Let the following conditions be fulfilled:*

1. *The conditions 1, 2, 6, 7 of Theorem 4.1.3 are satisfied.*

2. *There exists a function $V_1 \in \Lambda$ that is weakly decrescent and*

   (i) *for $t \geq 0$ and $\psi \in C([t-r,t], \mathbb{R}^n)$ such that $(t, \psi(t)) \in S(\rho)$ and $V_1(t, \psi(t)) > V_1(t+s, \psi(t+s))$ for $s \in [-r,0)$ the inequality*

   $$D_{(4.1)}V_1(t, \psi(t)) \leq g_1(t, V_1(t, \psi(t)))$$

   holds, where $\rho > 0$ is a constant;

3. *For any number $\mu > 0$ there exists a function $V_2^{(\mu)} \in \Lambda$ such that*

   (ii) *$b(\|x\|) \leq V_2^{(\mu)}(t, x) \leq a(\|x\|)$ for $(t, x) \in S(\rho) \cap S^C(\mu)$, where $a, b \in K$.*

   (iii) *for any number $t \geq 0$ and any function $\psi \in C([t-r,t], \mathbb{R}^n)$ such that $(t, \psi(t)) \in S(\rho) \cap S^C(\mu)$ and $V_1(t, \psi(0)) + V_2^{(\mu)}(t, \psi(t)) > V_1(t+s, \psi(t+s)) + V_2^{(\mu)}(t+s, \psi(t+s))$ for $s \in [-r,0)$ the inequality*

   $$D_{(4.1)}V_1(t, \psi(t)) + D_{(4.1)}V_2^{(\mu)}(t, \psi(t))$$
   $$\leq g_2\Big(t, V_1(t, \psi(t)) + V_2^{(\mu)}(t, \psi(t))\Big)$$

   holds.

Then the zero solution of the differential equations with "maxima" (4.1) is equi-stable.

Theorem 4.1.8. *Let the conditions 1, 2 and 3 of Theorem 4.1.3 and condition (i) of Theorem 4.1.6 be satisfied, where the function $V(t, x)$ be decrescent.*

If the zero solution of scalar differential equation (4.7) is uniformly stable, then the zero solution of the differential equations with "maxima" (4.1) is uniformly stable.

Now we will apply some of the obtained sufficient conditions to study stability properties of the solutions of differential equations with "maxima."

Example 4.1.2. *Consider the linear system of differential equations*

$$x' = -x + 2y$$
$$y' = -x - y. \tag{4.46}$$

The general solution of (4.46) is

$$x(t) = e^{-t}\sqrt{2}(c_1 \sin \sqrt{2}t - c_2 \cos \sqrt{2}t)$$
$$y(t) = e^{-t}(c_1 \cos \sqrt{2}t + c_2 \sin \sqrt{2}t).$$

The solution of the system (4.46) with initial condition $x(0) = 1$, $y(0) = 1$

$$x(t) = \sqrt{2}e^{-t}(\sin \sqrt{2}t + \frac{\sqrt{2}}{2} \cos \sqrt{2}t)$$

$$y(t) = e^{-t}(\cos \sqrt{2}t - \frac{\sqrt{2}}{2} \sin \sqrt{2}t)$$

approaches 0 as t increases without bound.

The solution of the system (4.46) with initial condition $x(0) = \frac{1}{2}$, $y(0) = \frac{1}{2}$

$$x(t) = \sqrt{2}e^{-t}(\frac{1}{2} \sin \sqrt{2}t + \frac{\sqrt{2}}{4} \cos \sqrt{2}t)$$

$$y(t) = e^{-t}(\frac{1}{2} \cos \sqrt{2}t - \frac{\sqrt{2}}{4} \sin \sqrt{2}t)$$

approaches 0 as t increases without bound.

Now let the system (4.46) be perturbed by the maximum function, i.e., consider the system of differential equations with "maxima"

$$x' = -x + 2y + C_1 \max_{s \in [-r,0]} x(s)$$
$$y' = -x - y + C_2 \max_{s \in [-r,0]} y(s). \tag{4.47}$$

We will apply Theorem 4.1.1 to investigate the stability of the solution of the system with maximum (4.47). Note that the solution of (4.47) could not be obtained in a closed form.

Consider the function $V : \mathbb{R}^2 \to \mathbb{R}_+$, $V(x,y) = \frac{1}{2}x^2 + y^2$ and the measures $h(t,x,y) = \sqrt{x^2 + y^2}$, $h_0(t,x,y) = |x| + |y|$. Using the inequality $\sqrt{x^2 + y^2} \le |x| + |y|$ it is easy to check the validity of the condition (ii) of Theorem 4.1.1 for $a(u) = u^2$ and $b(u) = \frac{1}{2}u^2$.

Let $t \ge 0$ be a number and $\psi \in C([t - r, t], \mathbb{R}^2)$, $\psi = (\psi_1, \psi_2)$ be such that

$$\frac{1}{2}\psi_1^2(t) + \psi_2^2(t) > \frac{1}{2}\psi_1^2(t+s) + \psi_2^2(t+s), \quad s \in [-r, 0). \tag{4.48}$$

From (4.48) follows that the derivative the function V among the trajectory of the system (4.47) is

$$D_{(4.47)}V(\psi_1(t), \psi_2(t))$$
$$= -\psi_1^2(t) - 2\psi_2^2(t) + c_1\psi_1(t) \max_{s\in[-r,0]} \psi_1(t+s) + 2c_2\psi_2(t) \max_{s\in[-r,0]} \psi_2(t+s).$$

Note that

$$\psi_1(t) \max_{s\in[-r,0]} \psi_1(t+s)$$

$$\leq |\psi_1(t)| \, | \max_{s\in[-r,0]} \psi_1(t+s)| = \sqrt{(\psi_1(t))^2} \sqrt{(\max_{s\in[-r,0]} \psi_1(t+s))^2}$$

$$\leq \sqrt{(\psi_1(t))^2 + 2(\psi_2(t))^2} \sqrt{(\max_{s\in[-r,0]} \psi_1(t+s))^2}$$

$$= \sqrt{(\psi_1(t))^2 + 2(\psi_2(t))^2} \sqrt{(\psi_1(\xi))^2}$$
$$\leq \sqrt{(\psi_1(t))^2 + 2(\psi_2(t))^2} \sqrt{(\psi_1(\xi))^2 + 2(\psi_1(\xi))^2} \leq (\psi_1(t))^2 + 2(\psi_2(t))^2$$

and

$$2\psi_2(t) \max_{s\in[-r,0]} \psi_2(t+s) \leq (\psi_1(t))^2 + 2(\psi_2(t))^2$$

where $\xi \in [t-r, t]$.

Hence

$$D_{(4.47)}V(\psi_1(t), \psi_2(t)) \leq$$
$$-\psi_1^2(t) - 2\psi_2^2(t) + c_1(\psi_1^2(t) + 2\psi_2^2(t)) + c_2(\psi_1^2(t) + 2\psi_2^2(t)) =$$
$$= (\psi_1^2(t) + 2\psi_2^2(t))(-1 + c_1 + c_2).$$

Since $\psi_1^2(t) + 2\psi_2^2(t) \geq 0$, then for $(-1 + c_1 + c_2) < 0$, or $c_1 + c_2 < 1$, we get

$$D_{(4.47)}V(\psi_1(t), \psi_2(t)) \leq 0.$$

Consider the scalar equation (4.7) for $g_1(t, u) \equiv 0$ which solution is $u(t; t_0, u_0) = u_0$. The zero solution of $u' = 0$ is uniformly stable. Therefore according to Theorem 4.1.2 the solution of (4.47) is uniformly stable in terms of measures (h_0, h), i.e., for $\epsilon > 0$ there exists $\delta = \delta(\epsilon) > 0$ such that the inequality $|\psi_1(s)| + |\psi_2(s)| < \delta, \quad s \in [-r, 0]$ implies $\sqrt{x^2(s) + y^2(s)} < \epsilon$ for $t \geq t_0$.

4.2 Integral Stability in Terms of Two Measures

In this section, we study the integral stability in terms of two different measures for differential equations involving the maximum of the
unknown function. Integral stability for ordinary differential equations
was introduced by I.Vrkoc ([Vrkoc 1959]) and later studied for various
types of differential equations by many authors (see for example, [Hris-
tova 2009a], [Hristova 2010a], and [Soliman and Abdalla 2008]). The
concept of integral stability occurs in connection with the stability under persistent perturbations when the perturbations are small enough
everywhere except on a small interval. The presence of maximum in the
equation requires initially well-defined and proved comparison result.
An appropriate definition for integral stability of differential equations
with "maxima" is given. Sufficient conditions for uniformly integral stability in terms of two measures are obtained. These results are derived
using Lyapunov functions, Razumikhin method and comparison results
for scalar differential equations.

In this section we will study stability properties of the system of
nonlinear differential equations with "maxima" (4.1) with initial condition (4.2), the solution of which is $x(t; t_0, \phi)$.

Consider the perturbed system of differential equations with "maxima"

$$x' = F(t, x(t), \max_{s \in [t-r,t]} x(s)) + G(t, x(t), \max_{s \in [t-r,t]} x(s)) \text{ for } t \geq t_0, \quad (4.49)$$

where $x \in \mathbb{R}^n$, $F, G : \mathbb{R}_+ \times \mathbb{R}^n \times \mathbb{R}^n \to \mathbb{R}^n$.

In our further investigations we will use Lyapunov functions from
the class Λ, introduced by Defintion 4.1.1 , and derivative of functions
from Λ, defined by equality (4.4).

Similarly we define a derivative of the function $V(t, x) \in \Lambda$ along
the trajectory of solution of the perturbed system (4.49) for $t \in \mathbb{R}_+$,
and $\phi \in C([t - r, t], \mathbb{R}^n)$ as follows

$$D_{(4.49)} V(t, \phi(t)) = \limsup_{\epsilon \to 0} \frac{1}{\epsilon} \left\{ V \left(t + \epsilon, \phi(t) + \epsilon \left(F(t, \phi(t), \max_{s \in [-r,0]} \phi(t + s)) \right. \right. \right.$$

$$\left. \left. \left. + G(t, \phi(t), \max_{s \in [-r,0]} \phi(t + s)) \right) \right) - V(t, \phi(t)) \right\}.$$

Consider the set K, defined by(4.3) and the set of measures Γ, defined by (4.5).

Let $\rho, t, T > 0$ be constants, $h \in \Gamma$. Consider the sets $S(h, \rho)$, $S^C(h, \rho)$, defined by equality (4.6) and the set

$$\Omega(t, h, \rho) = \{(x, y) \in \mathbb{R}^n \times \mathbb{R}^n : h(t, x) \leq \rho \text{ and } h(s, y) \leq \rho, \ s \in [t-r, t]\}.$$

We will use the following comparison scalar ordinary differential equations

$$u' = f(t, u), \tag{4.50}$$

and

$$u' = g(t, u), \tag{4.51}$$

and its perturbed scalar ordinary differential equation

$$w' = g(t, w) + q(t), \tag{4.52}$$

where $u, w \in \mathbb{R}$, $f, g : \mathbb{R}_+ \times \mathbb{R} \to \mathbb{R}$, $q : \mathbb{R}_+ \to \mathbb{R}$.

In our further investigations we will assume that solutions of the scalar differential equations (4.50), (4.51), and (4.52) exist on $[t_0, \infty)$ for any initial values.

Definition 4.2.1. ([Soliman and Abdalla 2008]) The ordinary differential equation (4.51) is said to be uniform-integrally stable *if for every* $\alpha > 0$ *and for any* $t_0 \geq 0$, *there exists* $\beta = \beta(\alpha) \in K$ *such that for any initial value* $|w_0| < \alpha$, *and for any perturbation* $q \in C(\mathbb{R}_+, \mathbb{R})$ *such that for every* $T > 0 : \int_{t_0}^{t_0+T} |q(s)| ds < \alpha$ *the inequality* $|w(t; t_0, w_0)| < \beta$ *holds for* $t \geq t_0$, *where* $w(t; t_0, w_0)$ *is a solution of (4.52) with initial condition* $w(t_0) = w_0$.

Based on the definition of integral stability for ordinary differential equations we will introduce integral stability in terms of two measures for differential equations with "maxima."

Definition 4.2.2. *Let* $h, h_0 \in \Gamma$. *System of differential equations with "maxima" (4.1) is said to be* uniform-integrally stable *in terms of measures* (h_0, h) *if for every* $\alpha > 0$ *and for any* $t_0 \geq 0$, *there exists* $\beta = \beta(\alpha) \in K$ *such that for any initial function* $\phi \in C([-r, 0], \mathbb{R}^n)$ *such that* $\max_{s \in [-r, 0]} h_0(t_0 + s, \phi(s)) < \alpha$, *and for any perturbation* $G \in C(\mathbb{R}_+ \times \mathbb{R}^n \times \mathbb{R}^n, \mathbb{R}^n)$ *which satisfies for every* $T > 0$ *the inequality* $\int_{t_0}^{t_0+T} \sup_{(x,y)\in\Omega(s,h,\beta)} ||G(s, x, y)|| ds < \alpha$, *it follows the validity of*

$h(t, y(t; t_0, \phi)) < \beta$ *for* $t \geq t_0$, *where* $y(t; t_0, \phi)$ *is the solution of the initial value problem for the perturbed system of differential equations with "maxima" (4.49), (4.2).*

We will give an example to illustrate the integral stability.

Example 4.2.1. *Consider the scalar differential equation* $x' = 0$, *where* $x \in \mathbb{R}$ *and its perturbed differential equation* $y' = 0 + ye^{-t}$. *The solution of the perturbed equation is* $y(t; t_0, y_0) = y_0 e^{e^{-t_0}} e^{-e^{-t}}$ *for* $t \geq t_0$. *Applying the inequalities* $e^{-e^{-t}} < 1$ *and* $e^{e^{-t_0}} < e$ *we obtain* $|y(t; t_0, y_0)| \leq |y_0|e$, *i.e., for any* $\alpha > 0$ *the inequality* $|y(t; t_0, y_0)| \leq \alpha$ *holds provided* $|y_0| < \beta = \frac{\alpha}{e}$.

On the other side, for any $\alpha > 0$, $T > 0$ and $\beta = \frac{\alpha}{3}$ the inequality

$$\int_{t_0}^{t_0+T} \sup_{w: |w| < \beta} |we^{-s}| ds = \beta e^{-t_0}(1 - e^{-T}) < \beta = \alpha$$

holds. Therefore, the differential equation $x' = 0$ is uniform-integrally stable.

□

In our further investigations we need some properties of both measures and the functions from the class Λ, defined by Definition 4.1.1.

We will use the properties h-decrescent and uniformly finer defined in Definition 4.1.3 and Definition 4.1.4 correspondingly.

We will obtain sufficient conditions for integral stability in terms of two measures for systems of differential equations with "maxima." We will apply two different types of Lyapunov functions from the class Λ and comparison results for scalar ordinary differential equations.

Theorem 4.2.1. *Let the following conditions be fulfilled:*

1. *The function* $F \in C[\mathbb{R}_+ \times \mathbb{R}^n \times \mathbb{R}^n, \mathbb{R}^n]$, $F(t, 0, 0) \equiv 0$.

2. *The functions* $h_0, h \in \Gamma$, h_0 *is uniformly finer than* h.

3. *There exists a function* $V_1 : [-r, \infty) \times \mathbb{R}^n \rightarrow \mathbb{R}_+$, $V_1 \in \Lambda$, *it is* h_0-decrescent and

   (i) *for any number* $t \geq 0$ *and any function* $\psi \in C([t - r, t], \mathbb{R}^n)$ *such that* $V_1(t, \psi(t)) > V_1(t + s, \psi(t + s))$ *for* $s \in [-r, 0)$ *and* $(t, \psi(t)) \in S(h, \rho)$ *the inequality*

   $$D_{(4.49)} V_1(t, \psi(t)) \leq f(t, V_1(t, \psi(t)))$$

holds, where $f \in C(\mathbb{R}_+ \times \mathbb{R}, \mathbb{R}_+]$, $f(t,0) \equiv 0$, $\rho > 0$ is a constant.

4. For any number $\mu > 0$ there exists a function $V_2^{(\mu)} : [-r, \infty) \times \mathbb{R}^n \to \mathbb{R}_+$, $V_2^{(\mu)} \in \Lambda$ such that

    (ii) $b(h(t,x)) \leq V_2^{(\mu)}(t,x) \leq a(h_0(t,x))$ for $(t,x) \in [-r, \infty) \times \mathbb{R}^n$,

         where $a, b \in K$ and $\lim_{u \to \infty} b(u) = \infty$.

    (iii)   for any number $t \geq 0$ and any function $\psi \in C([-r, 0], \mathbb{R}^n)$ such that $(t, \psi(t)) \in S(h, \rho) \bigcap S^C(h_0, \mu)$ and

$$V_1(t, \psi(t)) + V_2^{(\mu)}(t, \psi(t)) > V_1(t+s, \psi(t+s)) + V_2^{(\mu)}(t+s, \psi(t+s))$$

         for $s \in [-r, 0)$ the inequality

$$D_{(4.1)} V_1(t, \psi(t)) + D_{(4.1)} V_2^{(\mu)}(t, \psi(t))$$
$$\leq g\left(t, V_1(t, \psi(t)) + V_2^{(\mu)}(t, \psi(t))\right)$$

         holds, where $g \in C(\mathbb{R}_+ \times \mathbb{R}, \mathbb{R}_+]$, $g(t,0) \equiv 0$;

5. The zero solution of the scalar differential equation (4.50) is equi-stable.

6. The scalar differential equation (4.51) is uniform-integrally sta-ble.

Then the system of differential equations with "maxima" (4.1) is uniform-integrally stable in terms of measures (h_0, h).

Proof. Since the function $V_1(t, x)$ is h_0-decrescent, there exist a con-stant $\rho_1 \in (0, \rho)$ and a function $\psi_1 \in K$ such that for any point $(t, x) \in [-r, \infty) \times \mathbb{R}^n$ such that $h_0(t, x) < \rho_1$ the inequality

$$V_1(t, x) \leq \psi_1(h_0(t, x)) \tag{4.53}$$

holds.

     Since $h_0(t, x)$ is uniformly finer than $h(t, x)$, there exist $\rho_0 \in (0, \rho_1)$ and a function $\psi_2 \in K : \psi_2(\rho_0) < \rho_1$ such that $h_0(t, x) < \rho_0$ implies

$$h(t, x) \leq \psi_2(h_0(t, x)). \tag{4.54}$$

Let $t_0 \geq 0$ be a fixed point and $\alpha > 0$ be a number such that $\alpha < \rho_0$.

According to condition 4 of Theorem 4.2.1 there exists a function $V_2^{(\alpha)}(t, x)$ with Lipschitz constant M_2. Let M_1 be Lipschitz constant of the function $V_1(t, x)$.

Denote $(M_1 + M_2)\alpha = \alpha_1$. Without loss of generality we assume $\alpha_1 < b(\rho)$.

From condition 5 of Theorem 4.2.1 it follows there exists $\delta_1 = \delta_1(t_0, \alpha_1) > 0$ such that the inequality $|u_0| < \delta_1$ implies that

$$|u(t)| < \frac{\alpha_1}{2}, \quad t \geq t_0, \tag{4.55}$$

where $u(t)$ is a solution of (4.50) with the initial condition $u(t_0) = u_0$.

Since the function $\psi_1 \in K$, there exists $\delta_2 = \delta_2(\delta_1) > 0$, $\delta_2 < \rho_1$ such that for $|u| < \delta_2$ the inequality

$$\psi_1(u) < \delta_1 \tag{4.56}$$

holds.

From condition 6 of Theorem 4.2.1 it follows that there exists $\beta_1 = \beta_1(\alpha_1) \in K$, $b(\rho) > \beta_1 \geq \alpha_1$ such that, for every solution $w(t)$ of the perturbed equation (4.52) with the initial condition $w(t_0) = w_0$, the inequality

$$|w(t))| < \beta_1, \quad t \geq t_0, \tag{4.57}$$

holds, provided that $|w_0| < \alpha_1$ and for every $T > 0$: $\int_{t_0}^{t_0+T} |q(s)| ds < \alpha_1$.

Since the function $b \in K$, $\lim_{s\to\infty} b(s) = \infty$, and $\psi_2(\alpha) < \psi_2(\rho_0) < \rho_1 < \rho$, we choose $\beta = \beta(\beta_1) > 0$, $\rho > \beta > \alpha$, $\beta > \psi_2(\alpha)$ such that

$$b(\beta) \geq \beta_1. \tag{4.58}$$

Since the functions $a \in K$, $\psi_2 \in K$, and $\beta > \psi_2(\alpha)$, we can find $\delta_3 = \delta_3(\alpha_1, \beta) > 0$, $\alpha < \delta_3 < \min(\delta_2, \rho_0)$ such that the inequalities

$$a(\delta_3) < \frac{\alpha_1}{2}, \quad \psi_2(\delta_3) < \beta \tag{4.59}$$

hold.

Now let the initial function $\phi \in C([-r, 0], \mathbb{R}^n)$ and perturbation $G(t, x, y)$ of the right-hand side of the system of differential equations (4.49) be such that

$$\max_{s \in [-r,0]} h_0(t_0 + s, \phi(s)) < \alpha,$$

and for every $T > 0$:

$$\int_{t_0}^{t_0+T} \sup_{(x,y)\in\Omega(s,h,\beta)} \|G(s,x,y)\| ds < \alpha.$$

We will prove that

$$h(t, y(t)) < \beta, \quad t \geq t_0. \tag{4.60}$$

From (4.54) and the choice of β it follows that $h_0(t_0 + s, \phi(s)) < \alpha < \rho_0$ implies that $h(t_0 + s, \phi(s)) \leq \psi_2(h_0(t_0 + s, \phi(s))) < \psi_2(\alpha) < \beta$, i.e.,

$$h(t_0 + s, \phi(s)) < \beta \quad \text{for} \quad s \in [-r, 0].$$

Suppose inequality (4.60) is not true. Therefore, there exists a point $t^* > t_0$ such that

$$h(t^*, y(t^*)) = \beta, \quad h(t, y(t)) < \beta, \quad t \in [t_0 - r, t^*). \tag{4.61}$$

From inequality (4.61) and $\beta < \rho$ it follows the validity of the inclusions

$$(t, y(t)) \in S(h, \rho), \quad t \in [t_0, t^*], \quad \text{and} \quad (y(t), \max_{s\in[t-r,t]} y(s)) \in \Omega(t, h, \beta),$$

$$t \in [t_0, t^*]. \tag{4.62}$$

If we assume that $h_0(t^*, y(t^*)) \leq \delta_3$ then from the choice of δ_3 and inequality (4.54) it follows $h(t^*, y(t^*)) \leq \psi_2(h_0(t^*, y(t^*)) \leq \psi_2(\delta_3) < \beta$, which contradicts (4.61).

Therefore,

$$h_0(t^*, y(t^*)) > \delta_3, \quad h_0(t_0 + s, \phi(s)) < \alpha < \delta_3 \quad \text{for} \quad s \in [-r, 0]. \tag{4.63}$$

Then there exists a point $t_0^* \in (t_0, t^*)$ such that $\delta_3 = h_0(t_0^*, y(t_0^*))$ and $(t, y(t)) \in S(h, \beta) \cap S^c(h_0, \delta_3)$ for $t \in [t_0^*, t^*)$. Since $\beta < \rho$ and $\delta_3 > \alpha$ it follows that

$$(t, y(t)) \in S(h, \rho) \cap S^c(h_0, \alpha), \quad t \in [t_0^*, t^*]. \tag{4.64}$$

Let $r_1(t)$ be the maximal solution of scalar differential equation (4.50) with the initial condition $r_1(t_0) = u_0$ where $u_0 = \max_{s\in[-r,0]} V_1(t_0 + s, \phi(t_0 + s))$. From condition (i) of Theorem 4.2.1, according to Lemma 4.1.1, we obtain

$$V_1(t, y(t)) \leq r_1(t), \quad t \in [t_0, t^*]. \tag{4.65}$$

From inequality (4.56), we obtain

$$u_0 = V_1(\xi, \phi(\xi)) \leq \psi_1(h_0(\xi, \phi(\xi))) < \psi_1(\alpha) < \psi_1(\delta_2) < \delta_1, \quad (4.66)$$

where $\xi \in [t_0 - r, t_0]$.

From inequalities (4.55), (4.65), and (4.66) it follows that $V_1(t, y(t)) \leq r_1(t) < \frac{\alpha_1}{2}$ for $t \in [t_0, t^*]$, or

$$\max_{s \in [-r,0]} V_1(t_0^* + s, y(t_0^* + s)) < \frac{\alpha_1}{2}. \quad (4.67)$$

From inequality (4.59) and condition (ii) of Theorem 4.2.1 it follows that

$$V_2^{(\alpha)}(t_0^* + s, y(t_0^* + s)) < a(h_0(t_0^* + s, y(t_0^* + s))) = a(\delta_3) < \frac{\alpha_1}{2}, \quad s \in [-r, 0]. \quad (4.68)$$

Consider the function $V : \mathbb{R}_+ \times \mathbb{R}^n \to \mathbb{R}_+$, $V \in \Lambda$ defined by

$$V(t, x) = V_1(t, x) + V_2^{(\alpha)}(t, x). \quad (4.69)$$

From inequalities (4.67) and (4.68) it follows that

$$\max_{s \in [-r,0]} V(t_0^* + s, y(t_0^* + s)) < \alpha_1. \quad (4.70)$$

Let the point $t \in [t_0^*, t^*]$ and the function $\psi \in C([-r, 0], \mathbb{R}^n)$ be such that

$$(t, \psi(0)) \in S(h, \beta) \bigcap S^c(h_0, \alpha), \qquad (\psi(0), \max_{s \in [-r,0]} \psi(s)) \in \Omega(t, h, \beta),$$

and $V(t, \psi(0)) > V(t + s, \psi(s))$ for $s \in [-r, 0)$.

Using Lipschitz conditions for functions $V_1(t, x)$ and $V_2^{(\alpha)}(t, x)$, and condition (iii) of Theorem 4.2.1 we obtain

$$D_{(4.49)}V(t, \psi(t)) = D_{(4.49)}V_1(t, \psi(t)) + D_{(4.49)}V_2^{(\alpha)}(t, \psi(t))$$

$$\leq \limsup_{\epsilon \to 0} \frac{1}{\epsilon} \Big\{ \{V_1(t + \epsilon, \psi(t) + \epsilon F(t, \psi(t), \max_{s \in [-r,0]} \psi(t + s)))$$

$$- V_1(t, \psi(t))\}$$

$$+ \{V_2^{(\alpha)}(t + \epsilon, \psi(t) + \epsilon F(t, \psi(t), \max_{s \in [-r,0]} \psi(t + s))) - V_2^{(\alpha)}(t, \psi(t))\}\Big\}$$

$$+ \limsup_{\epsilon \to 0} \frac{1}{\epsilon} \Big\{ \{V_1(t + \epsilon, \psi(t)$$

$$(4.71)$$

$$+\epsilon[F(t,\psi(t),\max_{s\in[-r,0]}\psi(t+s))+G(t,\psi(t),\max_{s\in[-r,0]}\psi(t+s)])$$
$$-V_1(t+\epsilon,\psi(t)+\epsilon F(t,\psi(t),\max_{s\in[-r,0]}\psi(t+s)))\}$$
$$+\{V_2^{(\alpha)}(t+\epsilon,\psi(t)+\epsilon[F(t,\psi(t),\max_{s\in[-r,0]}\psi(t+s))$$
$$+G(t,\psi(t),\max_{s\in[-r,0]}\psi(t+s))])$$
$$-V_2^{(\alpha)}(t+\epsilon,\psi(t)+\epsilon F(t,\psi(t),\max_{s\in[-r,0]}\psi(t+s)))\}\}$$
$$\leq D_{(4.1)}V_1(t,\psi(t))+D_{(4.1)}V_2^{(\alpha)}(t,\psi(t))$$
$$+(M_1+M_2)\|G(t,\psi(t),\max_{s\in[-r,0]}\psi(t+s))\|$$
$$\leq g(t,V(t,\psi(t)))+(M_1+M_2)\sup_{(x,y)\in\Omega(t,h,\beta)}\|G(t,x,y)\|$$
$$=g(t,V(t,\psi(t)))+q(t).$$

Consider the scalar differential equation (4.52) where the perturbation on the right-hand side is given by $q(t) = (M_1 + M_2)\sup_{(x,y)\in\Omega(t,h,\beta)}\|G(t,x,y)\|$ for $t\in[t_0^*,t^*]$.

Let $r^*(t)$ be the maximal solution of (4.52) with the initial condition $r^*(t_0^*)=w_0^*$, where $w_0^*=\max_{s\in[-r,0]}V(t_0^*+s,y(t_0^*+s))$. According to Lemma 4.1.1 the inequality

$$V(t,y(t))\leq r^*(t),\quad t\in\Xi\bigcap[t_0^*,t^*] \tag{4.72}$$

holds, where $\Xi\subseteq[t_0^*,\infty)$ is the interval of existence of $r^*(t)$.

Choose a point $T^*>t^*$ such that

$$\int_{t_0^*}^{t^*}q(s)ds+\frac{1}{2}(T^*-t^*)q(t^*)<\alpha_1.$$

Now define the continuous function $q^*(t):[t_0^*,\infty)\to\mathbb{R}$ by

$$q^*(t)=\begin{cases} q(t) & \text{for } t\in[t_0^*,t^*] \\ \frac{q(t^*)}{t^*-T^*}(t-T^*) & \text{for } t\in[t^*,T^*] \\ 0 & \text{for } t\geq T^*. \end{cases} \tag{4.73}$$

From the choice of the perturbation $G(t,x,y)$ it follows that for every $T>0$ the inequality

$$\int_{t_0^*}^{t_0^*+T}q^*(s)ds\leq(M_1+M_2)\int_{t_0}^{t_0+T}\sup_{(x,y)\in\Omega(s,h,\beta)}\|G(s,x,y)\|ds<\alpha_1$$

holds.

Let $r^{**}(t)$ be the maximal solution of scalar differential equation (4.52) with the initial condition $r^{**}(t_0^*) = w_0^*$, where the perturbation of the right-hand side is defined above function $q^*(t)$. We note that $r^{**}(t) = r^*(t; t_0^*, w_0^*)$, $t \in [t_0^*, t^*)$.

From inequality (4.70) it follows that $|w_0^*| < \alpha_1$ and therefore inequality (4.57) holds, i.e.,

$$r^{**}(t) < \beta_1, \quad t \geq t_0^*. \tag{4.74}$$

From inequalities (4.72) and (4.74), the choice of the point t^*, and condition (iii) of Theorem 4.2.1 we obtain

$$b(\beta) \geq \beta_1 > r^{**}(t^*) = r^*(t^*) \geq V(t^*, y(t^*)) \geq V_2^{(\alpha)}(t^*, y(t^*))$$
$$\geq b(h(t^*, y(t^*))) = b(\beta).$$

The obtained contradiction proves the validity of inequality (4.60) for $t \geq t_0$.

Inequality (4.60) proves uniform-integral stability in terms of measures (h_0, h) of the considered system of differential equations with "maxima."

$\square$

In the case when the Lyapunov function is bounded both from above and from below, the following result is true:

Theorem 4.2.2. *Let the following conditions be fulfilled:*
1. *The function $F \in C[\mathbb{R}_+ \times \mathbb{R}^n \times \mathbb{R}^n, \mathbb{R}^n]$, $F(t, 0, 0) \equiv 0$.*
2. *The function $g \in C[\mathbb{R}_+ \times \mathbb{R}, \mathbb{R}_+]$, $g(t, 0) \equiv 0$.*
3. *The functions $h_0, h \in \Gamma$, h_0 are uniformly finer than h.*
4. *There exists a function $V : [-r, \infty) \times \mathbb{R}^n \to \mathbb{R}_+$, $V \in \Lambda$ such that*
   (i) $b(h(t, x)) \leq V(t, x) \leq a(h_0(t, x))$ *for* $(t, x) \in [0, \infty) \times \mathbb{R}^n$, *where $a, b \in K$ and $\lim_{u \to \infty} b(u) = \infty$;*
   (ii) *for any point $t \geq 0$ and for any function $\psi \in C([t - r, t], \mathbb{R}^n)$ such that $(t, \psi(t))) \in S(h, \rho)$ and $V(t, \psi(t)) > V(t + s, \psi(t + s))$ for $s \in [-r, 0)$ the inequality*

$$D_{(4.1)}V(t, \psi(t)) \leq g(t, V(t, \psi(t)))$$

holds, where $\rho > 0$ is a constant.
5. *The scalar differential equation (4.51) is uniform-integrally stable.*

    Then the system of differential equations with "maxima" (4.1) is uniform-integrally stable in terms of measures (h_0, h).

    The proof of Theorem 4.2.2 is similar to the one of Theorem 4.2.1 and we omit it.

Remark 4.2.1. *In the case $r = 0$ the obtained results reduce to results for integral stability in terms of two measures of ordinary differential equations studied in [Lakshmikantham and Liu 1993].*

    In the case $h(t, x) = h_0(t, x) \equiv \|x\|$ the obtained results reduce to results for integral stability of differential equations with "maxima."

    In the case $h(t, x) = h_0(t, x) \equiv \|x\|$ and $r = 0$ the obtained results reduce to results for integral stability of ordinary differential equations studied in [Soliman and Abdalla 2008].

4.3   Stability and Cone Valued Lyapunov Functions

One of the main problems of Lyapunov's second method is related to the construction of an appropriate Lyapunov function. Often, it is easier to construct a vector Lyapunov function rather than a scalar one. However, vector functions require comparison systems of differential equations. In order to involve scalar differential equations in the comparison method instead of comparison systems, we use a special type of stability that combines the ideas of two different measures and a dot product.

    In this section we will study the system of nonlinear differential equations with "maxima" (4.1) with initial condition (4.2), where $x \in \mathbb{R}^n$, $F : [0, \infty) \times \mathbb{R}^n \times \mathbb{R}^n \to \mathbb{R}^n$, $r > 0$ is a constant, $\phi \in C([-r, 0], \mathbb{R}^n)$ and $t_0 \in \mathbb{R}_+$ is a fixed point.

    Let $x, y \in \mathbb{R}^n$. Denote by $(x \bullet y)$ the dot product of both vectors x and y, i.e., $\sum_{i=1}^n x_i y_i$.

    Let $\mathcal{K} \subset \mathbb{R}^n$ be a cone. Consider the set

$$\mathcal{K}^* = \{\varphi \in \mathbb{R}^n : (\varphi \bullet x) \geq 0 \text{ for any } x \in \mathcal{K}\}. \qquad (4.75)$$

We assume that $\mathcal{K}^*$ is a cone.

Consider the set of measures

$$\mathcal{G}(\varphi_0) = \{h \in C([-r, \infty) \times \mathbb{R}^n, \mathcal{K}) : \inf_{x \in \mathbb{R}^n} (\varphi_0 \bullet h(t, x)) = 0$$

$$\text{for each } t \geq -r\}, \qquad (4.76)$$

where $\varphi_0 \in \mathcal{K}^*$.

Note, for example, $\mathcal{K} \equiv \mathbb{R}_+^n$ is a cone, φ_0 is the unit vector, and the function

$$h(t, x) = (e^{-t}|x_1|, e^{-t}|x_2|, \ldots, e^{-t}|x_n|), \quad x \in \mathbb{R}^n, \ x = (x_1, x_2, \ldots, x_n)$$

is from the set $\mathcal{G}(\varphi_0)$.

Let ρ be positive constant, $\varphi_0 \in \mathcal{K}^*$, $h \in \mathcal{G}(\varphi_0)$. Define set:

$$\mathcal{S}(h, \rho, \varphi_0) = \{(t, x) \in [0, \infty) \times \mathbb{R}^n : \ (\varphi_0 \bullet h(t, x)) < \rho\}.$$

We will introduce the definition of a new type of stability for differential equations with "maxima," which combines the ideas of stability in terms of two measures (see [Lakshmikantham and Liu 1993] and [Movchan 1960]) and a dot product.

Definition 4.3.1. *Let $\varphi_0 \in \mathcal{K}^*$, $h, h_0 \in \mathcal{G}(\varphi_0)$. The system of differential equations with "maxima" (4.1) is said to be*

(S4.3.1) d-stable in terms of two measures h_0 and h with a vector φ_0 if *for every $\epsilon > 0$ and $t_0 \geq 0$ there exists $\delta = \delta(t_0, \epsilon) > 0$ such that for any $\phi \in C([-r, 0], \mathbb{R}^n)$ inequality $(\varphi_0 \bullet h_0(t_0 + s, \phi(s))) < \delta$ for $s \in [-r, 0]$ implies $(\varphi_0 \bullet h(t, x(t; t_0, \phi))) < \epsilon$ for $t \geq t_0$, where $x(t; t_0, \phi)$ is a solution of differential equations with "maxima" (4.1) with initial condition (4.2);*

(S4.3.2) uniformly d-stable in terms of two measures h_0 and h with a *vector φ_0 if (S4.3.1) is satisfied, where δ is independent on t_0;*

Remark 4.3.1. *The vector φ_0, which is introduced in Definition 4.3.1, is a proxy for the weights of the solution's components.*

Remark 4.3.2. *In the partial case of one-dimensional cone $\mathcal{K} = \mathbb{R}_+$ the measures h and h_0 are scalar valued nonnegative functions. The above defined d-stability in terms of two measures reduces to stability in terms of two measures for differential equations with "maxima."*

In our further investigations we will use the following comparison scalar ordinary differential equations

$$u' = g(t, u), \tag{4.77}$$

where $u \in \mathbb{R}$, $g(t, 0) \equiv 0$.

We will use some properties of the functions from class $\mathcal{G}$.

Definition 4.3.2. *Let the vector $\varphi_0 \in \mathcal{K}^*$ and the functions $h, h_0 \in \mathcal{G}(\varphi_0)$. The function h_0 is uniformly φ_0-finer than the function h if there exist a constant $\sigma > 0$ and a function $p \in K$ such that for any point $(t, x) \in [-r, \infty) \times \mathbb{R}^n : (\varphi_0 \bullet h_0(t, x)) < \sigma$ the inequality $(\varphi_0 \bullet h(t, x)) \leq p((\varphi_0 \bullet h_0(t, x))$ holds.*

We will introduce the following class of functions:

Definition 4.3.3. *We will say that function $V(t, x) : \Omega \times \mathbb{R}^n \to \mathcal{K}$, $\Omega \subset \mathbb{R}_+$, $V = (V_1, V_2, \dots, V_n)$, belongs to the class $\mathcal{L}$ if:*

1. $V(t, x) \in C^1(\Omega \times \mathbb{R}^n, \mathcal{K})$;

2. There exist constants $M_i > 0$, $i = 1, 2, \dots, n$, such that $|V_i(t, x) - V_i(t, y)| \leq M_i \|x - y\|$ for any $t \in \Omega$, $x, y \in \mathbb{R}^n$.

Let function $V \in \mathcal{L}$, $V = (V_1, V_2, \dots, V_n)$, $t \in \mathbb{R}_+$, and $\phi \in C([t - r, t], \mathbb{R}^n)$. We define a derivative $\mathcal{D}_{(4.1)}V(t, x)$ of the function V among the system (4.1) by the equalities

$$\mathcal{D}_{(4.1)}V_i(t, \phi(t)) = \frac{\partial V_i(t, \phi(t))}{\partial t}$$
$$+ \sum_{j=1}^{n} \frac{\partial V_i(t, \phi(t))}{\partial x_j} F_j(t, \phi(t), sup_{s \in [-r, 0]}\phi(t + s)) \quad (4.78)$$

for $i = 1, 2, \dots, n$, where

$$\mathcal{D}_{(4.1)}V(t, x) = (\mathcal{D}_{(4.1)}V_1(t, x), \mathcal{D}_{(4.1)}V_2(t, x), \dots, \mathcal{D}_{(4.1)}V_n(t, x)).$$

In the further investigations we will use the following comparison result:

Lemma 4.3.1. *Let the following conditions be fulfilled:*

1. The condition 1 of Lemma 4.1.1 is satisfied.

2. Vector $\varphi_0 \in \mathcal{K}^*$ and function $V(t, x) : [t_0, T] \times \mathbb{R}^n \to \mathcal{K}$, $V \in \mathcal{L}$ are such that for any function $\psi \in C([-r, 0], \mathbb{R}^n)$ and any number $t \in [t_0, T]$ such that $(\varphi_0 \bullet V(t, \psi(0))) > (\varphi_0 \bullet V(t + s, \psi(s)))$ for $s \in [-r, 0)$ the inequality

$$\left(\varphi_0 \bullet \mathcal{D}_{(4.1)}V(t, \psi(0))\right) \leq g(t, (\varphi_0 \bullet V(t, \psi(0))))$$

holds, where $g \in C(\mathbb{R}_+ \times \mathbb{R}_+, \mathbb{R}_+), g(t, 0) \equiv 0$.

3. *Function* $x(t) = x(t; t_0, \varphi)$ *is a solution of (4.1) with initial condition* $x(t_0 + s) = \varphi(s)$, $s \in [-r, 0]$, *which is defined for* $t \in [t_0 - r, T]$ *where* $\varphi \in C([-r, 0], \mathbb{R}^n)$.

4. *Function* $u^*(t) = u^*(t; t_0, u_0)$ *is the maximal solution of (4.77) with initial condition* $u^*(t_0) = u_0$, *which is defined for* $t \in [t_0, T]$.

Then the inequality $\max_{s \in [-r,0]} (\varphi_0 \bullet V(t_0 + s, \varphi(s))) \le u_0$ *implies the validity of the inequality* $(\varphi_0 \bullet V(t, x(t))) \le u^*(t)$ *for* $t \in [t_0, T]$.

Proof. Let $v_n(t)$ be the maximal solution of the initial value problem

$$
\begin{aligned}
u' &= g(t, u) + \frac{1}{n}, \\
u(t_0) &= u_0 + \frac{1}{n},
\end{aligned}
\tag{4.79}
$$

where $\max_{s \in [-r,0]} (\varphi_0 \bullet V(t_0 + s, \varphi(s))) \le u_0$ and n is a natural number. Assume that $v_n(t)$ is defined for $t \in [t_0, T]$.

Define a function $m(t) \in C([t_0 - r, T], \mathbb{R}_+)$ by $m(t) = (\varphi_0 \bullet V(t, x(t)))$.

The rest of the proof is similar to the proof of Lemma 4.1.1. □

We will consider the cone $\mathcal{K} \subset \mathbb{R}^n$, $n > 1$ and we will obtain sufficient conditions for d-stability in terms of two measures of systems of differential equations with "maxima." We will employ cone-valued Lyapunov functions from class $\mathcal{L}$. The proofs are based on the Razumikhin method and the comparison method with scalar ordinary differential equations.

Theorem 4.3.1. *Let the following conditions be fulfilled:*

1. *The function* $F \in C(\mathbb{R}_+ \times \mathbb{R}^n \times \mathbb{R}^n, \mathbb{R}^n)$, $F(t, 0, 0) \equiv 0$.

2. *The vector* $\varphi_0 \in \mathcal{K}^*$ *and the functions* $h_0, h \in \mathcal{G}(\varphi_0)$, h_0 *is uniformly* φ_0-*finer than* h.

3. *There exists a function* $V(t, x) : \mathbb{R}_+ \times \mathbb{R}^n \to \mathcal{K}$, $V \in \mathcal{L}$ *such that*

   (i) $b((\varphi_0 \bullet h(t, x))) \le (\varphi_0 \bullet V(t, x)) \le a((\varphi_0 \bullet h_0(t, x)))$
   for $(t, x) \in \mathcal{S}(h, \rho, \varphi_0)$, *where* $a, b \in K$, *the set* K *is defined by (4.3)*;

(ii) *For any number $t \geq 0$ and any function $\psi \in C([t-r,t],\mathbb{R}^n)$ such that $(\varphi_0 \bullet V(t,\psi(t))) > (\varphi_0 \bullet V(t+s,\psi(t+s)))$ for $s \in [-r,0)$ and $(t,\psi(t)) \in \mathcal{S}(h,\rho,\varphi_0)$ the inequality*

$$\left(\varphi_0 \bullet \mathcal{D}_{(4.1)} V(t,\psi(t))\right) \leq g(t,(\varphi_0 \bullet V(t,\psi(t))))$$

holds, where $g \in C(\mathbb{R}_+ \times \mathbb{R}, \mathbb{R}_+)$, $g(t,0) \equiv 0$, $\rho > 0$ is a constant.

4. *For any initial function $\phi \in C([-r,0],\mathbb{R}^n)$ the solution of the initial value problem for the system of differential equations with "maxima" (4.1) and (4.2) exists on $[t_0 - r, \infty)$, $t_0 \geq 0$.*

5. *For any initial point $(t_0,u_0) \in \mathbb{R}_+ \times \mathbb{R}$ the solution of scalar differential equation (4.77) exists on $[t_0,\infty)$, $t_0 \geq 0$.*

6. *Zero solution of scalar differential equation (4.77) is equi-stable.*

Then system of differential equations with "maxima" (4.1) is d-stable in terms of two measures h_0 and h with a vector φ_0.

Proof. Let $\epsilon > 0$ be a fixed number, $\epsilon < \rho$, and $t_0 \geq 0$ be a fixed point.

From condition 2 of Theorem 4.3.1 it follows that there exist a constant $\sigma > 0$ and a function $p \in K$ such that for any point $(t,x) \in [-r,\infty) \times \mathbb{R}^n : (\varphi_0 \bullet h_0(t,x)) < \sigma$ the inequality

$$(\varphi_0 \bullet h(t,x)) < p((\varphi_0 \bullet h_0(t,x))) \qquad (4.80)$$

holds.

We can find $\delta_1 = \delta_1(\epsilon) > 0$, $\delta_1 < \rho$ such that the inequality

$$p(\delta_1) < \epsilon \qquad (4.81)$$

holds.

Since the zero solution of scalar differential equation (4.77) is equi-stable there exists $\delta_2 = \delta_2(t_0,\epsilon) > 0$ such that inequality $|u_0| < \delta_2$ implies

$$|u(t;t_0,u_0)| < b(\epsilon), \quad t \geq t_0, \qquad (4.82)$$

where $u(t;t_0,u_0)$ is the maximal solution of (4.77) with initial condition $u(t_0) = u_0$.

Since the function $a \in K$, which is defined in condition (i), we can find $\delta_3 = \delta_3(t_0, \epsilon) > 0$, $\delta_3 < \rho$ such that the inequality

$$a(\delta_3) < \delta_2 \tag{4.83}$$

holds.

Now let function $\phi \in C([-r, 0], \mathbb{R}^n)$ be such that

$$(\varphi_0 \bullet h_0(t_0 + s, \phi(s))) < \delta_4 \quad \text{for} \quad s \in [-r, 0], \tag{4.84}$$

where $\delta_4 = \min\{\delta_1, \delta_3, \sigma\}$, $\delta_4 = \delta_4(t_0, \epsilon) > 0$.

From condition 2, inequalities (4.81) and (4.84), and the choice of δ_4, it follows that for $s \in [-r, 0]$ the inequality

$$(\varphi_0 \bullet h(t_0 + s, x(t_0 + s; t_0, \phi))) < p((\varphi_0 \bullet h_0(t_0 + s, \phi(s))))$$
$$< p(\delta_4) \leq p(\delta_1) < \epsilon \tag{4.85}$$

holds.

We will prove that if inequality (4.84) is satisfied, then

$$(\varphi_0 \bullet h(t, x(t; t_0, \phi))) < \epsilon \quad \text{for} \quad t \geq t_0. \tag{4.86}$$

Suppose inequality (4.86) is not true. From inequality (4.85) it follows that there exists a point $t^* > t_0$ such that

$$(\varphi_0 \bullet h(t^*, x(t^*; t_0, \phi))) = \epsilon, \quad (\varphi_0 \bullet h(t, x(t; t_0, \phi))) < \epsilon, \quad t \in [t_0 - r, t^*). \tag{4.87}$$

Inequalities (4.85) and (4.87), $\epsilon < \rho$ and the inclusion $\mathcal{S}(h, \epsilon, \varphi_0) \subset \mathcal{S}(h, \rho, \varphi_0)$ prove that

$$(t, x(t; t_0, \phi)) \in \mathcal{S}(h, \rho, \varphi_0) \quad \text{for} \quad t \in [t_0 - r, t^*]. \tag{4.88}$$

From inequality (4.83), condition (i) and the inclusion (4.88) it follows that

$$max_{s \in [-r, 0]}(\varphi_0 \bullet V(t_0 + s, \varphi(s))) \leq a((\varphi_0 \bullet h_0(t_0 + s, x(t_0 + s; t_0, \phi))))$$
$$\leq a(\delta_3) < \delta_2. \tag{4.89}$$

Therefore according to inequalities (4.81) and (4.89) we get

$$u^*(t; t_0, u_0^*) < b(\epsilon), \quad t \geq t_0, \tag{4.90}$$

where $u^*(t; t_0, u_0^*)$ is the maximal solution of the scalar equation (4.77) with initial condition $u(t_0) = u_0^*$, $u_0^* = max_{s \in [-r, 0]}(\varphi_0 \bullet V(t_0 + s, \varphi(s)))$.

According to a suitable modification of Lemma 4.1.1 the inequality

$$(\varphi_0 \bullet V(t, x(t; t_0, \phi))) \leq u^*(t; t_0, u_0^*) \quad \text{for} \quad t \in [t_0, t^*] \tag{4.91}$$

holds.

From inequalities (4.90) and (4.91), the choice of the point t^*, and condition (i) of Theorem 4.3.1 we obtain

$$
\begin{aligned}
b(\epsilon) \;>\; & u^*(t^*; t_0^*, u_0^*) \geq (\varphi_0 \bullet V(t^*, x(t^*; t_0, \phi))) \\
\geq\; & b((\varphi_0 \bullet h(t^*, x(t^*; t_0, \phi)))) = b(\epsilon).
\end{aligned}
$$

The obtained contradiction proves the validity of inequality (4.86) for $t \geq t_0$.

□

Inequality (4.86) proves d-stability in terms of two measures of the considered system of differential equations with "maxima".

Theorem 4.3.2. *Let the following conditions be fulfilled:*

1. *Conditions 1, 2, 3, 4, 5 of Theorem 4.3.1 are satisfied.*

2. *Zero solution of scalar differential equation (4.77) is uniformly stable.*

Then the system of differential equations with "maxima" (4.1) is uniformly d-stable in terms of two measures h_0 and h with a vector φ_0.

The proof of Theorem 4.3.2 is similar to the one of Theorem 4.3.1 and we omit it.

Now we will illustrate the application of the above-defined stability in terms of two measures and the obtained sufficient conditions on an example.

Example 4.3.1. *Consider the system of differential equations with "maxima"*

$$x'(t) = -x(t) + 4y(t) + \frac{1}{2} \max_{s \in [t-r, t]} x(s) \tag{4.92}$$

$$y'(t) = -x(t) - y(t) + \frac{1}{2} \max_{s \in [t-r, t]} y(s), \quad t \geq t_0, \tag{4.93}$$

with initial conditions

$$x(t) = \phi_1(t - t_0), \quad y(t) = \phi_2(t - t_0) \quad \text{for } t \in [t_0 - r, t_0], \tag{4.94}$$

where $x, y \in \mathbb{R}$, $r > 0$ is a small fixed constant, $t_0 \geq 0$.

We will study the stability of the solution of (4.92) and (4.93) by applying different Lyapunov functions and two measures for the initial function and the solution.

Case 1. (vector Lyapunov function and a dot product).

Consider the cone $\mathcal{K} = \{(x, y) : x \geq 0, y \geq 0\} \subset \mathbb{R}^2$.

Let functions $h_0(t, x, y) = (|x|, |y|)$, $h(t, x, y) = (x^2, y^2)$ and vector $\varphi_0 = (1, 4)$.

Note that the vector $\varphi_0 \in \mathcal{K}^*$, the functions $h, h_0 \in \mathcal{G}(\varphi_0)$, and the function h_0 is uniformly φ_0-finer than the function h since there exist a constant $\delta = \frac{1}{4} > 0$ and a function $p \in K$, $p(u) \equiv u$ such that for any point $(t, x, y) \in [-r, \infty) \times \mathbb{R}^2 :$ $|x| + 4|y| < \delta$ the inequality $x^2 + 4y^2 \leq |x| + 4|y|$ holds.

Consider the set $\mathcal{S}(h, 1, \varphi_0) = \{(x, y) \in \mathbb{R}^2 : x^2 + 4y^2 < 1\}$.

Define the Lyapunov function $V : \mathbb{R}^2 \to \mathcal{K}$, $V = (V_1, V_2)$, where

$V_1(x, y) = \frac{1}{2}x^2$, $V_2(x, y) = \frac{1}{2}y^2$.

Then the condition (i) of Theorem 4.3.1 is fulfilled for functions $b(s) = \frac{1}{2}s \in K$ and $a(s) = \frac{1}{2}s^2 \in K$ since $\frac{1}{2}(x^2 + 4y^2) \leq \frac{1}{2}(|x| + 4|y|)^2$ for $(x, y) \in \mathcal{S}(h, 1, \varphi_0)$.

Let $t \in \mathbb{R}_+$ and the function $\psi \in C([t - r, t], \mathbb{R}^2)$, $\psi = (\psi_1, \psi_2)$ be such that the inequality

$$(\varphi_0 \bullet V(\psi_1(t), \psi_2(t))) = \frac{1}{2}\psi_1^2(t) + 2\psi_2^2(t)$$
$$> \frac{1}{2}\psi_1^2(t + s) + 2\psi_2^2(t + s) = (\varphi_0 \bullet V(\psi_1(t + s), \psi_2(t + s)))$$

$$(4.95)$$

holds for $s \in [-r, 0)$.

Then

$$\psi_1(t) \max_{s \in [t-r,t]} \psi_1(s) \leq |\psi_1(t)| \, | \max_{s \in [t-r,t]} \psi_1(s)|$$
$$= \sqrt{(\psi_1(t))^2} \, \sqrt{(\max_{s \in [t-r,t]} \psi_1(s))^2}$$
$$\leq \sqrt{2(\varphi_0 \bullet V(\psi_1(t), \psi_2(t)))} \, \sqrt{2(\varphi_0 \bullet V(\psi_1(t), \psi_2(t)))}$$
$$\leq 2(\varphi_0 \bullet V(\psi_1(t), \psi_2(t)))$$

and

$$\psi_2(t) \max_{s\in[t-r,t]} \psi_2(s) \leq |\psi_2(t)| \; | \max_{s\in[t-r,t]} \psi_2(s)|$$

$$= \sqrt{(\psi_2(t))^2} \; \sqrt{(\max_{s\in[t-r,t]} \psi_2(s))^2}$$

$$\leq \sqrt{\frac{1}{2}(\varphi_0 \bullet V(\psi_1(t),\psi_2(t)))} \sqrt{\frac{1}{2}(\varphi_0 \bullet V(\psi_1(s),\psi_2(s)))}$$

$$\leq \frac{1}{2}(\varphi_0 \bullet V(\psi_1(t),\psi_2(t))).$$

Therefore, if inequality (4.95) is fulfilled, then

$$\left(\varphi_0 \bullet \mathcal{D}_{(4.92),(4.93)} V(\psi_1(t),\psi_2(t))\right)$$

$$= -(\psi_1(t))^2 - 4(\psi_2(t))^2 + \frac{1}{2}\psi_1(t) \max_{s\in[t-r,t]} \psi_1(s) + 2\psi_2(t) \max_{s\in[t-r,t]} \psi_2(s)$$

$$\leq -(\psi_1(t))^2 - 4(\psi_2(t))^2 + \frac{1}{2}(\varphi_0 \bullet V(\psi_1(t),\psi_2(t)))$$

$$+ \frac{1}{2}(\varphi_0 \bullet V(\psi_1(t),\psi_2(t)))$$

or

$$\left(\varphi_0 \bullet \mathcal{D}_{(4.92),(4.93)} V(\psi_1(t),\psi_2(t))\right) \leq -(\varphi_0 \bullet V(\psi_1(t),\psi_2(t))) \leq 0. \tag{4.96}$$

Consider the scalar comparison equation $u' = 0$ which zero solution is uniformly stable and according to Theorem 4.3.2 the system of differential equations with "maxima" (4.92) and (4.93) is uniformly d- stable in terms of two measures, i.e., for every $\epsilon > 0$ there exists $\delta = \delta(\epsilon) > 0$ such that inequality $\max_{s\in[-r,0]}\left(|\phi_1(s)| + 4|\phi_2(s)|\right) < \delta$ implies

$$(x(t))^2 + 4(y(t))^2 < \epsilon \quad \text{for} \quad t \geq t_0, \tag{4.97}$$

where $x(t), y(t)$ is the solution of initial value problems (4.92)-(4.94).

Case 2. (scalar Lyapunov function).

Now we will apply a scalar Lyapunov function to study the stability of the system of differential equations with "maxima" (4.92) and (4.93)

Consider the scalar Lyapunov function $\tilde{V}(x,y) = \frac{1}{2}x^2 + 2y^2$ and both measures $h_0(t,x,y) = |x| + |y|$, $h(t,x,y) = x^2 + y^2$. As above, if (4.125) is satisfied then the derivative of Lyapunov function, defined by (4.4) satisfies the inequality $\mathcal{D}_{(4.92),(4.93)}\tilde{V}(\phi_1(t),\phi_2(t)) =$

$\phi_1(t)\Big(-\phi_1(t)+4\phi_2(t)+\frac{1}{2}\ \max_{s\in[t-r,t]}\phi_1(s)\Big)+4\phi_2(t)\Big(-\phi_1(t)-\phi_2(t)+$

$\frac{1}{2}\ \max_{s\in[t-r,t]}\phi_2(s)\Big)\le 0.$ According to Theorem 4.1.1 the solution of the initial value problems (4.92)-(4.94) is stable since the zero solution of the scalar equation $u'=0$ is stable, i.e.,

$$h(t,x(t),y(t)) = (x(t))^2 + (y(t))^2 < \epsilon \qquad (4.98)$$

provided that $|\phi_1(s)| + |\phi_2(s)| < \delta$.

Note that the estimate (4.97) is better than (4.98) and the application of a dot product gives us an opportunity to use various weights to components of the solution. This is very applicable in the case when some of the components of the solution play a more important role on the stability than others.

4.4 Practical Stability on a Cone

It is well known (see [Salle and Lefschetz 1961]) that stability and even asymptotic stabilities themselves are neither necessary nor sufficient to ensure practical stability. The desired state of a system may be mathematically unstable; however, the system may oscillate sufficiently close to the desired state, and its performance is deemed acceptable. The practical stability is neither weaker nor stronger than the usual stability; an equilibrium can be stable in the usual sense, but not practically stable, and vice versa. For example an aircraft may oscillate around a mathematically unstable path, yet its performance may be acceptable. Practical stability is, in a sense, a uniform boundedness of the solution relative to the initial conditions, but the bound must be sufficiently small.

In this section we apply cone-valued multidimentional Lypunov functions to study practical stability of differential equations with "maxima." Note that in the applications, such kinds of functions are comparatively easier for construction. To avoid applications of comparison systems of differential equations and to apply scalar differential equations, we introduce a scalar product on a cone and appropriate modifications of stability definitions.

Consider the initial value problem for the system of nonlinear differential equations with "maxima" (4.1), (4.2), where $x \in \mathbb{R}^n$, $F : [0,\infty) \times \mathbb{R}^n \times \mathbb{R}^n \to \mathbb{R}^n$, $F = (F_1, F_2, \ldots, F_n)$, $r > 0$ is a constant, $\phi \in C([-r,0], \mathbb{R}^n)$ and $t_0 \in \mathbb{R}_+$ is a fixed point.

Let $\mathcal{K} \subset \mathbb{R}^n$ be a cone. Consider the sets $\mathcal{K}^*$ and $\mathcal{G}(\varphi_0)$, defined by (4.75) and (4.76) correspondingly.

We assume that $\mathcal{K}^*$ is a cone.

Consider the set of functions K, defined by (4.3), the set $\mathcal{L}$ of measures, defined by (4.76) and the vector $\varphi_0 \in \mathcal{K}^*$.

Let ρ be positive constant, $\varphi_0 \in \mathcal{K}^*$, $h \in \mathcal{G}(\varphi_0)$. Define set:

$$\tilde{\mathcal{S}}(h, \rho, \varphi_0) = \{(t, x) \in [0, \infty) \times \mathbb{R}^n : \ (\varphi_0 \bullet h(t, x)) < \rho\}.$$

In our further investigations we will use the following comparison scalar ordinary differential equation

$$u' = g(t, u), \tag{4.99}$$

where $u \in \mathbb{R}$, $g(t, 0) \equiv 0$.

We will use the property uniformly φ_0-finer of functions from class $\mathcal{G}$ which is given in Definition 4.3.2.

We will consider the class of functions $\mathcal{L}$, introduced by Definition 4.3.3 and we will define a derivative $\mathcal{D}_{(4.1)}V(t, x)$ of the function $V \in \mathcal{L}$ along the system (4.1) by the equality (4.78).

4.4.1   Practical Stability

We will introduce the definition of a practical stability for differential equations with "maxima," based on the ideas of stability in terms of two measures (see [Lakshmikantham and Liu 1993]) and a dot product.

Definition 4.4.1. *Let $\varphi_0 \in \mathcal{K}^*$, $h, h_0 \in \mathcal{G}(\varphi_0)$, $\lambda, A = const : 0 < \lambda < A$ be given. The system of differential equations with "maxima" (4.1) is said to be*

(S4.4.1)   *d-practically stable with respect to (λ, A) in terms of measures h_0 and h with a vector φ_0 if there exists $t_0 \geq 0$ such that for any $\phi \in C([-r, 0], \mathbb{R}^n)$ inequality $\max_{s \in [-r, 0,]}(\varphi_0 \bullet h_0(t_0 + s, \varphi_0(s))) < \lambda$ implies $(\varphi_0 \bullet h(t, x(t; t_0, \phi))) < A$ for $t \geq t_0$, where $x(t; t_0, \phi)$ is a solution of differential equations with "maxima" (4.1) with initial condition (4.2);*

(S4.4.2)   *d-uniformly practically stable with respect to (λ, A) in terms of measures h_0 and h with a vector φ_0 if for any point $t_0 \geq 0$ and function $\phi \in C([-r, 0], \mathbb{R}^n)$ inequality $\max_{s \in [-r, 0]}(\varphi_0 \bullet h_0(t_0 + s, \phi(s)) < \lambda$ implies $(\varphi_0 \bullet h(t, x(t; t_0, \phi))) < A$ for $t \geq t_0$.*

Note that as a partial case of Definition 4.4.1, we obtain the definition for practical stability of differential equations with "maxima:"

Definition 4.4.2. *The system of differential equations with "maxima" (4.1) is said to be*

(S4.4.3) practically stable *with respect to* (λ, A) *if there exists a point* $t_0 \geq 0$ *such that for any* $\phi \in C([-r, 0], \mathbb{R}^n)$ *inequality* $\sup_{s \in [-r,0]} \|\phi(s)\| < \lambda$ *implies* $\|x(t; t_0, \phi)\| < A$ *for* $t \geq t_0$, *where* $x(t; t_0, \phi)$ *is a solution of the initial value problem for differential equations with "maxima" (4.1), (4.2);*

(S4.4.4) uniformly practically stable *with respect to* (λ, A) *if for any* $t_0 \geq 0$ *and* $\phi \in C([-r, 0], \mathbb{R}^n)$ *inequality* $\max_{s \in [-r,0]} \|\phi(s)\| < \lambda$ *implies* $\|x(t; t_0, \phi))\| < A$ *for* $t \geq t_0$.

Note that in the case $r = 0$, the above given Definition 4.4.2 reduces to a definition for practical stability of ordinary differential equations, given in the book by [Lakshmikantham et al. 1990].

We will obtain sufficient conditions for d-practical stability in terms of two measures of systems of differential equations with "maxima." We will employ Lyapunov functions from class $\mathcal{L}$. The proof is based on the Razumikhin method combined with the comparison method, employed scalar ordinary differential equations.

Theorem 4.4.1. *Let the following conditions be fulfilled:*

1. *The function* $F \in C(\mathbb{R}_+ \times \mathbb{R}^n \times \mathbb{R}^n, \mathbb{R}^n)$, $F(t, 0, 0) \equiv 0$.

2. *The vector* $\varphi_0 \in \mathcal{K}^*$, *the functions* $h_0, h \in \mathcal{G}(\varphi_0)$, *the positive constants* λ, A *are such that* $\lambda < A$.

3. *There exists a function* $V(t, x) : \mathbb{R}_+ \times \mathbb{R}^n \to \mathcal{K}$, $V(t, x) : \mathbb{R}_+ \times \mathbb{R}^n \to \mathcal{K}$, $V \in \mathcal{L}$ *such that*

   (i) $b((\varphi_0 \bullet h(t, x))) \leq (\varphi_0 \bullet V(t, x)) \leq a((\varphi_0 \bullet h_0(t, x)))$, $(t, x) \in \tilde{\mathcal{S}}(h, A, \varphi_0)$ *where* $a, b \in K$ *and* $a(\lambda) < b(A)$;

   (ii) *for any number* $t \geq 0$ *and any function* $\psi \in C([t - r, t], \mathbb{R}^n)$ *such that* $(\varphi_0 \bullet V(t, \psi(t))) \geq (\varphi_0 \bullet V(t + s, \psi(t + s)))$ *for* $s \in [-r, 0)$ *and* $(t, \psi(t))) \in \tilde{\mathcal{S}}(h, A, \varphi_0)$ *the inequality*

   $$\left(\varphi_0 \bullet \mathcal{D}_{(4.1)} V(t, \psi(t))\right) \leq g(t, (\varphi_0 \bullet V(t, \psi(t))))$$

   holds, where $g \in C(\mathbb{R}_+ \times \mathbb{R}, \mathbb{R}_+)$, $g(t, 0) \equiv 0$, $\rho > 0$ *is a constant.*

4. *For any initial function $\phi \in C([-r, 0], \mathbb{R}^n)$ and any initial point $t_0 \geq 0$ the solution of the initial value problem for the system of differential equations with "maxima" (4.1),(4.2) exists on $[t_0 - r, \infty)$, $t_0 \geq 0$.*

5. *For any initial point $(t_0, u_0) \in \mathbb{R}_+ \times \mathbb{R}$ the solution of scalar equation (4.77) exists on $[t_0, \infty)$, $t_0 \geq 0$.*

6. *The scalar differential equation (4.77) is practically stable with respect to $(a(\lambda), b(A))$.*

Then the system of differential equations with "maxima" (4.1) is d-practically stable with respect to (λ, A) in terms of measures h_0 and h with a vector φ_0.

Proof. From condition 5 it follows that there exists a point $t_0 \geq 0$ such that $|u_0| < a(\lambda)$ implies

$$|u(t; t_0, u_0)| < b(A) \quad \text{for} \quad t \geq t_0, \tag{4.100}$$

where $u(t; t_0, u_0)$ is a solution of scalar differential equation (4.77) with initial condition $u(t_0) = u_0$.

Choose a function $\phi \in C([-r, 0], \mathbb{R}^n)$ such that

$$\max_{s \in [-r, 0]} (\varphi_0 \bullet h_0(t_0 + s, \phi(s)) < \lambda \tag{4.101}$$

and let $x(t; t_0, \phi)$ be a solution of (4.1) with initial condition (4.2).

Let $u_0 = max_{s \in [-r, 0]}(\varphi_0 \bullet V(t_0 + s, \phi(s)))$. From a suitable modification of Lemma 4.1.1 follows the validity of the inequality

$$(\varphi_0 \bullet V(t, x(t; t_0, \phi))) \leq u^*(t; t_0, u_0) \quad \text{for} \quad t \geq t_0, \tag{4.102}$$

From condition (i) and inequality (4.101) we obtain

$$(\varphi_0 \bullet V(t_0 + s, \phi(s))) \leq a((\varphi_0 \bullet h_0(t_0 + s, \phi(s)))) < a(\lambda). \tag{4.103}$$

From inequalities (4.100) and (4.103) it follows that

$$(\varphi_0 \bullet V(t, x(t; t_0, \phi))) \leq u^*(t; t_0, u_0) < b(A) \quad \text{for} \quad t \geq t_0, \tag{4.104}$$

From inequality (4.104) and condition (i) we get

$$b((\varphi_0 \bullet h(t, x(t; t_0, \phi)))) \leq (\varphi_0 \bullet V(t, x(t; t_0, \phi)))$$
$$\leq u^*(t; t_0, u_0) < b(A) \quad \text{for} \quad t \geq t_0, \tag{4.105}$$

or

$$(\varphi_0 \bullet h(t, x(t; t_0, \phi))) < A \quad \text{for} \quad t \geq t_0. \qquad (4.106)$$

$\square$

Theorem 4.4.2. *Let the following conditions be fulfilled:*

1. *The conditions 1, 2, 3, 4, and 5 of Theorem 4.4.1 are satisfied.*

2. *The scalar differential equation (4.77) is uniformly practically stable with respect to $(a(\lambda), b(A))$.*

Then the system of differential equations with "maxima" (4.1) is uniformly d-practically stable with respect to (λ, A) in terms of measures h_0 and h with the vector φ_0.

Proof. From condition 2 of Theorem 4.4.2 it follows that for every point $t_0 \geq 0$ and $|u_0| < a(\lambda)$ the inequality

$$|u(t; t_0, u_0)| < b(A) \quad \text{for} \quad t \geq t_0, \qquad (4.107)$$

holds, where $u(t; t_0, u_0)$ is a solution of scalar differential equation (4.77) with initial condition $u(t_0) = u_0$.

From condition (i) we get that for every function $\phi \in C([-r, 0], \mathbb{R}^n)$ such that $\max_{s \in [-r, 0]} (\varphi_0 \bullet h_0(t_0 + s, \phi(s)) < \lambda$ the inequalities

$$b((\varphi_0 \bullet h(t_0 + s, \phi(t_0 + s)))) \leq a((\varphi_0 \bullet h_0(t_0 + s, \phi(t_0 + s)))) < a(\lambda) < b(A)$$

holds, for $s \in [-r, 0]$, or

$$(\varphi_0 \bullet h(t_0 + s, \phi(s))) < A \quad \text{on the interval} \quad [-r, 0]. \qquad (4.108)$$

We will prove that for every point $t_0 \geq 0$ and every function $\phi \in C([-r, 0], \mathbb{R}^n)$ such that

$$\max_{s \in [-r, 0]} (\varphi_0 \bullet h_0(t_0 + s, \phi(s))) < \lambda \qquad (4.109)$$

the inequality

$$(\varphi_0 \bullet h(t, x(t; t_0, \phi))) < A \quad \text{for} \quad t \geq t_0 \qquad (4.110)$$

holds, where $x(t; t_0, \phi)$ is a solution of (4.1) with initial condition (4.2).

Assume the claim is not true. Therefore, there exist a point $t_0 \geq 0$ and a function $\phi \in C([-r, 0], \mathbb{R}^n)$ such that the inequality (4.109) holds and inequality (4.110) is not true. According to the assumption and inequality (4.108) there exists a point $t^* \geq t_0$ such that

$$
\begin{aligned}
(\varphi_0 \bullet h(t, x(t; t_0, \phi))) &< A \quad \text{for} \quad t \in [t_0 - r, t^*) \\
(\varphi_0 \bullet h(t^*, x(t^*; t_0, \phi))) &= A \\
(\varphi_0 \bullet h(t, x(t; t_0, \phi))) &\geq A \quad \text{for} \quad t \in (t^*, t^* + \Delta],
\end{aligned}
\tag{4.111}
$$

where $\Delta > 0$ is a small enough number.

Let $u_0^* = max_{s \in [-r,0]}(\varphi_0 \bullet V(t_0 + s, \phi(s)))$. From a suitable modification of Lemma 4.1.1 and condition (ii) it follows the validity of the inequality

$$
(\varphi_0 \bullet V(t, x(t; t_0, \phi))) \leq u^*(t; t_0, u_0^*) \quad \text{for} \quad t \in [t_0, t^*],
\tag{4.112}
$$

where $u^*(t; t_0, u_0^*)$ is a solution of scalar differential equation (4.77) with initial condition $u(t_0) = u_0^*$.

From condition (i) we obtain for $s \in [-r, 0]$

$$
(\varphi_0 \bullet V(t_0 + s, \phi(s))) \leq a((\varphi_0 \bullet h_0(t_0 + s, \phi(s)))) < a(\lambda).
\tag{4.113}
$$

Inequality (4.113) proves that $|u_0^*| < a(\lambda)$ and therefore, according to inequality (4.107) we get

$$
u^*(t; t_0, u_0) < b(A) \quad \text{for} \quad t \in [t_0, t^*].
\tag{4.114}
$$

From inequality (4.114), the choice of the point t^*, and condition (i) we get

$$
\begin{aligned}
b(A) = b((\varphi_0 \bullet h(t^*, x(t^*; t_0, \phi)))) &\leq (\varphi_0 \bullet V(t^*, x(t^*; t_0, \phi))) \\
&\leq u^*(t^*; t_0, u_0) < b(A).
\end{aligned}
\tag{4.115}
$$

The obtained contradiction proves the validity of inequality (4.110). $\qquad\square$

In the case when both measures are equal to a regular norm in $\mathbb{R}^n$ and the vector φ_0 is the unit vector, we obtain the following result:

Theorem 4.4.3. *Let the following conditions be fulfilled:*

1. *The function $F \in C(\mathbb{R}_+ \times \mathbb{R}^n \times \mathbb{R}^n, \mathbb{R}^n)$, $F(t, 0, 0) \equiv 0$.*

2. *There exists a function $V(t, x) : \mathbb{R}_+ \times \mathbb{R}^n \to \mathcal{K}$, $V(t, x) : \mathbb{R}_+ \times \mathbb{R}^n \to \mathcal{K}$, $V \in \mathcal{L}$ such that*

(i) $b(\|x\|) \leq \sum_{i=1}^{n} V_i(t, x) \leq a(\|x\|)$ *for* $t \in \mathbb{R}_+$, $\|x\| < A$,
where $a, b \in K$ *and* $a(\lambda) < b(A)$;

(ii) *for any number* $t \geq 0$ *and any function* $\psi \in C([t - r, t], \mathbb{R}^n)$
such that if

$$\sum_{i=1}^{n} V_i(t, \psi(t)) \geq \sum_{i=1}^{n} V_i(t + s, \psi(t + s)) \quad \text{for } s \in [-r, 0)$$

$$\text{and } \|\psi(t)\| < A$$

then the inequality

$$\sum_{i=1}^{n} \mathcal{D}_{(4.1)} V_i(t, \psi(t)) \leq g(t, \sum_{i=1}^{n} V_i(t, \psi(t)))$$

holds, where $g \in C(\mathbb{R}_+ \times \mathbb{R}, \mathbb{R}_+)$, $g(t, 0) \equiv 0$, $\rho > 0$ *is a
constant.*

3. *For any initial function* $\phi \in C([-r, 0], \mathbb{R}^n)$ *the solution of the
initial value problem for the system of differential equations with
"maxima" (4.1), (4.2) exists on* $[t_0 - r, \infty)$, $t_0 \geq 0$.

4. *For any initial point* $(t_0, u_0) \in \mathbb{R}_+ \times \mathbb{R}$ *the solution of scalar
equation (4.77) exists on* $[t_0, \infty)$, $t_0 \geq 0$.

Then the (uniform) practical stability with respect to $(a(\lambda), b(A))$
*of scalar differential equation (4.77) implies (uniform) practical stabil-
ity with respect to* (λ, A) *of the system of differential equations with
"maxima" (4.1).*

The proof of Theorem 4.4.3 is similar to the proofs of Theorem 4.4.1
and Theorem 4.4.2 and we omit it.

4.4.2   Eventual Practical Stability

We will introduce the definition of a new type of eventual practical
stability for differential equations with "maxima", based on the ideas of
stability in terms of two measures (see [Lakshmikantham and Liu 1993])
and a dot product. The application of a dot product allows us to use
scalar comparison ordinary differential equations for investigation of
stability properties of the solutions. At the same time, the fixed vector,
involved in the definition, plays the role of a weight of components of
the solution.

Consider the system of differential equations with "maxima" (4.1) with initial condition (4.2).

We introduce the following set (**H4.4**) of conditions:

H 4.4.1. The function $F \in C[\mathbb{R}_+ \times \mathbb{R}^n \times \mathbb{R}^n, \mathbb{R}^n]$, $F(t, 0, 0) \equiv 0$.

H 4.4.2. For any initial function $\phi \in C([-r, 0], \mathbb{R}^n)$ the solution of the initial value problem for the system of differential equations with "maxima" (4.1),(4.2) exists on $[t_0 - r, \infty)$, $t_0 \geq 0$.

H 4.4.3. For any initial point $(t_0, u_0) \in \mathbb{R}_+ \times \mathbb{R}$ the solution of scalar differential equation (4.77) exists on $[t_0, \infty)$, $t_0 \geq 0$.

H 4.4.4. The vector $\varphi_0 \in \mathcal{K}^*$ and the functions $h_0, h \in \mathcal{G}(\varphi_0)$.

H 4.4.5. The functions $h_0, h \in \mathcal{G}(\varphi_0)$ are such that h is eventually φ_0-stronger than h_0.

H 4.4.6. The functions $h_0, h \in \mathcal{G}(\varphi_0)$ are such that, for $(t, x) \in \tilde{\mathcal{S}}(h_0, \rho_0, \varphi_0)$, the inequality $(\varphi_0 \bullet h(t, x)) \leq Q((\varphi_0 \bullet h_0(t, x)))$ holds, where $Q \in K$ such that $Q(s) \leq s$ and $\rho_0 > 0$ is a constant.

Definition 4.4.3. *Let $\varphi_0 \in \mathcal{K}^*$, $h, h_0 \in \mathcal{G}(\varphi_0)$. The system of differential equations with "maxima" (4.1) is said to be*

(S4.4.5) *d-eventually practically stable in terms of measures h_0 and h with a vector φ_0 if for any couple (λ, A) such that $0 < \lambda < A$ there exists $\tau(\lambda, A) > 0$, such that for some $t_0 \geq \tau(\lambda, A)$ and $\phi \in C([-r, 0], \mathbb{R}^n)$ such that $\max_{s \in [-r, 0]}(\varphi_0 \bullet h_0(t_0 + s, \phi(s)) < \lambda$ the inequality $(\varphi_0 \bullet h(t, x(t; t_0, \phi))) < A$ holds for $t \geq t_0$, where $x(t; t_0, \phi)$ is a solution of the initial value problem (4.1), (4.2);*

(S4.4.6) *uniformly d-eventually practically stable in terms of measures h_0 and h with a vector φ_0 if for any couple (λ, A) such that $0 < \lambda < A$ there exists $\tau(\lambda, A) > 0$, such that for any point $t_0 \geq \tau(\lambda, A)$ and function $\phi \in C([-r, 0], \mathbb{R}^n)$ such that $\max_{s \in [-r, 0]}(\varphi_0 \bullet h_0(t_0 + s, \phi(s)) < \lambda$ the inequality $(\varphi_0 \bullet h(t, x(t; t_0, \phi))) < A$ holds for $t \geq t_0 \geq \tau(\lambda, A)$.*

The vector φ_0, introduced in Definition 4.4.3, plays the role of a weight of components of the solution.

As a partial case of Definition 4.4.3 we obtain a definition for eventually practical stability of differential equations with "maxima:"

Definition 4.4.4. *The system of differential equations with "maxima" (4.1) is said to be*

(S4.4.7)   eventually practically stable *if for any* $(\lambda, A):\ \ 0 < \lambda < A$ *there exists* $\tau(\lambda, A) > 0$, *such that for some point* $t_0 \geq \tau(\lambda, A)$ *and a function* $\phi \in C([-r, 0], \mathbb{R}^n):\ \ \sup_{t \in [-r, 0]} \|\phi(t)\| < \lambda$ *the inequality* $\|x(t; t_0, \phi)\| < A$ *holds for* $t \geq t_0$, *where* $x(t; t_0, \phi)$ *is a solution of the initial value problem (4.1), (4.2);*

(S4.4.8)   uniformly eventually practically stable *if for any couple* $(\lambda, A):\ \ 0 < \lambda < A$ *there exists* $\tau(\lambda, A) > 0$, *such that for any* $t_0 \geq \tau(\lambda, A)$ *and any function* $\phi \in C([-r, 0], \mathbb{R}^n):$ $\sup_{t \in [-r, 0]} \|\phi(t)\| < \lambda$ *the inequality* $\|x(t; t_0, \phi)\| < A$ *holds for* $t \geq t_0 \geq \tau(\lambda, A)$.

We will obtain sufficient conditions for d-eventual practical stability in terms of two measures of systems of differential equations with "maxima." We will employ Lyapunov functions from class $\mathcal{L}$. The proof is based on the Razumikhin method combined with the comparison method and employed scalar ordinary differential equations.

Theorem 4.4.4. *Let the following conditions be fulfilled:*

1. *The conditions H 4.4.1–H 4.4.5 are satisfied.*

2. *There exists a function* $V(t, x): \mathbb{R}_+ \times \mathbb{R}^n \to \mathcal{K}$, *with* $V \in \mathcal{L}$ *such that*

   (i) $b((\varphi_0 \bullet h(t, x))) \leq (\varphi_0 \bullet V(t, x)) \leq a((\varphi_0 \bullet h_0(t, x)))$ *for* $(t, x) \in \tilde{\mathcal{S}}(h, \rho, \varphi_0)$, *where* $a, b \in K$;

   (ii)   *for any number* $t \geq 0$ *and any function* $\psi \in C([t - r, t], \mathbb{R}^n)$ *such that* $(\varphi_0 \bullet V(t, \psi(t))) > (\varphi_0 \bullet V(t + s, \psi(t + s)))$ *for* $s \in [-r, 0)$ *and* $(t, \psi(t)) \in \tilde{\mathcal{S}}(h, \rho, \varphi_0)$ *the inequality*

   $$\left(\varphi_0 \bullet \mathcal{D}_{(4.1)} V(t, \psi(t)) \right) \leq g(t, (\varphi_0 \bullet V(t, \psi(t))))$$

   holds, where $g \in C(\mathbb{R}_+ \times \mathbb{R}, \mathbb{R}_+)$, $g(t, 0) \equiv 0$, $\rho > 0$ *is a constant.*

3. *For any initial point* $(t_0, u_0) \in \mathbb{R}_+ \times \mathbb{R}$ *the solution of scalar equation (4.99) exists on* $[t_0, \infty)$, $t_0 \geq 0$.

4. *The scalar differential equation (4.99) is eventually practically stable.*

Then the system of differential equations with "maxima" (4.1) is d-eventually practically stable in terms of measures h_0 *and* h *with a vector* φ_0.

Proof. Let the couple (λ, A) such that $0 < \lambda < A$ be given.

Case 1. Let $A < \rho$. From condition 4, it follows that there exist $\tau(\lambda, A) > 0$ and a point $t_0 \geq \tau(\lambda, A)$ such that $|u_0| < a(\lambda)$ implies

$$|u(t; t_0, u_0)| < b(A) \qquad \text{for } t \geq t_0, \tag{4.116}$$

where $u(t; t_0, u_0)$ is a solution of the scalar differential equation (4.99) with initial condition $u(t_0) = u_0$.

Choose a function $\phi \in C([-r, 0], \mathbb{R}^n)$ such that $\sup_{s \in [-r,0]} (\varphi_0 \bullet h_0(t_0 + s, \phi(s))) < \lambda$, where t_0 is defined above. Let $x(t) = x(t; t_0, \phi)$ be a solution of differential equations with "maxima" (4.1) with the initial function ϕ.

From assumption H 4.4.5, it follows that the inequality

$$(\varphi_0 \bullet h(t, \phi(t - t_0))) < A \qquad \text{for } t \in [t_0 - r, t_0] \tag{4.117}$$

holds.

We claim that

$$(\varphi_0 \bullet h(t, x(t))) < A \qquad \text{for } t \geq t_0 \tag{4.118}$$

holds.

Assume the claim is not true. From the choice of the initial function ϕ and inequality (4.117), it follows there exists a point $t^* > t_0$ such that

$$\begin{aligned} (\varphi_0 \bullet h(t, x(t))) &< A, \qquad \text{for } t \in [t_0 - r, t^*), \\ (\varphi_0 \bullet h(t^*, x(t^*))) &= A. \end{aligned} \tag{4.119}$$

Since $A < \rho$, the inclusion $x(t; t_0, \phi) \in \tilde{\mathcal{S}}(h, \rho, \varphi_0)$ is valid for $t \in [t_0 - r, t^*]$.

Let $u_0^* = max_{s \in [-r,0]} (\varphi_0 \bullet V(t_0 + s, \phi(s)))$. From Lemma 4.3.1 and condition (ii), it follows the validity of the inequality

$$(\varphi_0 \bullet V(t, x(t))) \leq u^*(t; t_0, u_0^*) \qquad \text{for } t \in [t_0, t^*], \tag{4.120}$$

where $u^*(t; t_0, u_0^*)$ is a solution of scalar differential equation (4.99) with initial condition $u(t_0) = u_0^*$.

From condition (i) and the choice of the initial function ϕ, we obtain for $s \in [-r, 0]$

$$(\varphi_0 \bullet V(t_0 + s, \phi(s))) \leq a((\varphi_0 \bullet h_0(t_0 + s, \phi(s)))) < a(\lambda). \quad (4.121)$$

Inequality (4.121) proves that $|u_0^*| < a(\lambda)$, and therefore, according to inequalities (4.116) and (4.120), we get

$$(\varphi_0 \bullet V(t_0 + s, \phi(s))) \leq u^*(t; t_0, u_0) < b(A) \quad \text{for} \quad t \in [t_0, t^*]. \quad (4.122)$$

From inequality (4.122), the choice of t^*, and condition (i), we get

$$b(A) = b((\varphi_0 \bullet h(t^*, x(t^*)))) \leq (\varphi_0 \bullet V(t^*, x(t^*))) \leq u^*(t^*; t_0, u_0) < b(A).$$

This is a contradiction, which proves the validity of inequality (4.118).

Case 2. Let $A \geq \rho$. We repeat the proof of Case 1, but instead of the number a, we use the number ρ everywhere.

$\square$

Note the condition H 4.4.5 could be replaced by the condition H 4.4.6 in the sufficient condition for d-eventual practical stability in terms of two measures:

Theorem 4.4.5. *Let the conditions H 4.4.1-H 4.4.4, H4.4.6 and conditions 2, 3, 4 of Theorem 4.4.4 be fulfilled.*

Then the system of differential equations with "maxima" (4.1) is d-eventually practically stable in terms of measures h_0 and h with a vector φ_0.

The proof of Theorem 4.4.5 is similar to the one of Theorem 4.4.4. In this case we consider the constant $\rho_1 = \min\{\rho, \rho_0\}$, and from the choice of the initial function ϕ, it follows that $(t_0 + s, \phi(s)) \in \tilde{S}(h_0, \rho_1, \varphi_0)$ for $s \in [-r, 0]$, and then condition H 4.4.6 immediately shows the validity of inequality (4.117).

Theorem 4.4.6. *Let the following conditions be fulfilled:*

1. *The conditions 1, 2, 3 and 4 of Theorem 4.4.4 are satisfied.*

2. *The scalar differential equation (4.99) is uniformly eventually practically stable.*

Then the system of differential equations with "maxima" (4.1) is uniformly d-eventually practically stable in terms of measures h_0 and h with the vector φ_0.

The proof of Theorem 4.4.6 is similar to the proof of Theorem 4.4.4 and we omit it.

Note that condition H 4.4.5 could be replaced by condition H 4.4.6 in Theorem 4.4.6:

Theorem 4.4.7. *Let the conditions H 4.4.1-H 4.4.4, H 4.4.6, conditions 2, 3, 4 of Theorem 4.4.4, and condition 2 of Theorem 4.4.5 be fulfilled.*

Then the system of differential equations with "maxima" (4.1) is uniformly d-eventually practically stable in terms of measures h_0 and h with a vector φ_0.

Now we will illustrate the application of the above-obtained sufficient conditions on an example.

Example 4.4.1. *Consider the following system of differential equations with "maxima"*

$$x'(t) = -x(t)\Big(x^2(t) + y^2(t)\Big)\sin^2 t + e^{-t}\max_{s\in[t-r,t]} x(s),$$

$$y'(t) = -y(t)\Big(x^2(t) + y^2(t)\Big)\sin^2 t + e^{-t}\max_{s\in[t-r,t]} y(s), \quad t \geq t_0,$$

$$(4.123)$$

with initial conditions

$$x(t) = \phi_1(t - t_0), \quad y(t) = \phi_2(t - t_0) \qquad for \quad t \in [t_0 - r, t_0], \quad (4.124)$$

where $x, y \in \mathbb{R}$, $r > 0$ is small enough a constant, $t_0 \geq 0$.

Let $h_0(t, x, y) = (|x|, |y|)$, $h(t, x, y) = (x^2, y^2)$.

Consider $V : \mathbb{R}^2 \to \mathcal{K}$, $V = (V_1, V_2)$, $V_1(x, y) = \frac{1}{2}(x + 2y)^2$, $V_2(x, y) = \frac{1}{2}(x - y)^2$, where $\mathcal{K} = \{(x, y) : x \geq 0, y \geq 0\} \subset \mathbb{R}^2$ is a cone.

For the vector $\varphi_0 = (1, 2)$, then $(\varphi_0 \bullet h(t, x, y)) = x^2 + 2y^2$, $(\varphi_0 \bullet V(x, y)) = \frac{1}{2}(x + 2y)^2 + (x - y)^2 = \frac{3}{2}(x^2 + 2y^2)$ and $(\varphi_0 \bullet h_0(t, x, y)) = |x| + 2|y|$.

It is easy to check the validity of condition (i) of Theorem 4.4.4 for functions $a(s) = \frac{3}{2}s \in K$ and $b(s) = \frac{3}{2}s^2 \in K$. Let $t \in \mathbb{R}_+$ and $\psi \in C([t-r,t], \mathbb{R}^2)$, $\psi = (\psi_1, \psi_2)$ be such that

$$(\varphi_0 \bullet V(\psi_1(t), \psi_2(t))) = \frac{3}{2}(\psi_1^2(t) + 2\psi_2^2(t))$$

$$\geq \frac{3}{2}(\psi_1^2(t+s) + 2\psi_2^2(t+s)) = (\varphi_0 \bullet V(\psi_1(t+s), \psi_2(t+s)) \quad (4.125)$$

$$\text{for } s \in [-r, 0).$$

Then for $i = 1, 2$, we obtain

$$\psi_i(t) \max_{s \in [t-r,t]} \psi_i(s) \leq |\psi_i(t)| \, | \max_{s \in [t-r,t]} \psi_i(s)| = \sqrt{(\psi_i(t))^2} \sqrt{(\max_{s \in [t-r,t]} \psi_i(s))^2}$$

$$\leq \sqrt{\frac{2}{3}(\varphi_0 \bullet V(\psi_1(t), \psi_2(t)))} \sqrt{\frac{2}{3}(\varphi_0 \bullet V(\psi_1(t+s), \psi_2(t+s)))}$$

$$\leq \frac{2}{3}(\varphi_0 \bullet V(\psi_1(t), \psi_2(t))).$$

Therefore, if inequality (4.125) is fulfilled, we have

$$\left(\varphi_0 \bullet \mathcal{D}_{(4.92)} V(\psi_1(t), \psi_2(t))\right)$$

$$= 3e^{-t}\left(\psi_1(t) \max_{s \in [t-r,t]} \psi_1(s) + 2\psi_2(t) \max_{s \in [t-r,t]} \psi_2(s)\right)$$

$$\leq 6e^{-t}(\varphi_0 \bullet V(\psi_1(t), \psi_2(t))).$$

Now, consider the scalar comparison equation $u' = 6e^{-t}u$ with initial condition $u(t_0) = u_0$, whose solution is $u(t) = u_0 e^{6\left(e^{-t_0} - e^{-t}\right)}$ and $|u(t)| \leq |u_0| e^{6e^{-t_0}}$ for $t \geq t_0$. For any numbers $0 < \lambda < A$, we choose a number $\tau > \max\{0, \ln 6 - \ln(\ln(\frac{A}{\lambda}))\} > 0$. Note $\tau = \tau(\lambda, A) > 0$. It is easy to check that for $t_0 > \tau$ and $|u_0| < \lambda$ the inequality $|u(t)| < A$ holds, i.e., the scalar comparison equation is uniformly eventually practically stable, and therefore, according to Theorem 4.4.5 the system of differential equations with "maxima" (4.92) is uniformly d-eventually practically stable in terms of two measures, i.e., for any numbers $0 < \lambda < A$, there exists a number $\tau = \tau(\lambda, A) > 0$ such that, if $t_0 > \tau$ then the inequality $\sup_{s \in [-r,0]}(|\phi_1(s)| + 2|\phi_2(s)|) < \lambda$ implies $x^2(t; t_0, \phi) + 2y^2(t; t_0, \phi) < A$, for $t \geq t_0$.

Note that the choice of the vector φ_0 has a huge influence on the sufficient conditions. Let us, for example, consider the vector $\varphi_0 =$

$(1, 1)$. In this case

$$(\varphi_0 \bullet V(x, y)) = \frac{1}{2}(x + 2y)^2 + \frac{1}{2}(x - y)^2 = x^2 + xy + \frac{5}{2}y^2$$

and condition (i) of Theorem 4.4.4 is not satisfied for the above-defined function $V(x, y)$.

Now, let us consider the Lyapunov function $\tilde{V} : \mathbb{R}^2 \to \mathcal{K}$, $\tilde{V} = (\tilde{V}_1, \tilde{V}_2)$, defined by $\tilde{V}_1(x, y) = \frac{1}{2}(x + y)^2$ and $\tilde{V}_2(x, y) = \frac{1}{2}(x - y)^2$. In this case $(\varphi_0 \bullet \tilde{V}(x, y)) = x^2 + y^2$ and condition (i) of Theorem 4.4.4 is satisfied. But in this case condition (ii) is not satisfied.

Chapter 5

Oscillation Theory

The oscillation and nonoscillation of solutions of various types of differential equations have been the object of intensive studies in the last decades. The monographs of [Agarwal et al. 2000], [Bainov and Mishev 1991], and [Ladde et al. 1987] are devoted to the systematic investigation on this subject. However, the results for oscillation and nonoscillation are relatively scarce in literature, especially for differential equations with "maxima."

5.1 Differential Equations with "Maxima" versus Differential Equations with Delay

In this section, the oscillatory properties of various types of differential equations will be studied and the behavior of their solutions will be compared with the behavior of the corresponding delay differential equations. It will be demonstrated that the presence of the maximum function into the equation changes totally the behavior of the solution.

Initially we will give the basic definition in this chapter.

Definition 5.1.1. *The solution $x(t)$ of a scalar differential equation with "maxima" in the interval $J \subset \mathbb{R}_+$ is said to be:*

1. *A proper solution, if there exists a number $T_x \in J$ such that*

$$\sup\{|x(t)| : t \geq T_1\} > 0 \quad \text{for all} \quad T_1 \geq T_x.$$

2. *Nonoscillatory solution, if it is a proper solution and it is either positive or negative for $t \geq T_x$.*

3. *Oscillatory solution, if it is a proper solution and there is an infinite number of points on J at which the solution changes its sign.*

The differential equation with "maxima" is said to be *oscillatory* if all its proper solutions are oscillatory.

We will consider delay differential equations versus differential equations with "maxima."

Consider the differential equation with "maxima"

$$x'(t) + q(t) \max_{[t-r,t]} x(s) = 0, \tag{5.1}$$

and its corresponding delay differential equation

$$x'(t) + q(t)x(t - r) = 0, \tag{5.2}$$

where $x \in \mathbb{R}$, $q \in C(\mathbb{R}_+, \mathbb{R})$, $r > 0$.

Definition 5.1.2. *A solution of a differential equation is called from Z-type if it either nonpositive or nonnegative.*

Theorem 5.1.1. *If $q(t)$ is of one sign, then all solutions of equation (5.1) are nonoscillatory.*

Proof. When $q(t) \equiv 0$ or $r = 0$, the conclusion of Theorem 5.1.1 is obvious. Therefore, we assume that $q(t) \neq 0$ and $r > 0$.

If $q(t) \geq 0$, suppose that $x(t)$ is an oscillatory solution of Equation (5.1). Then $x(t)$ is not a Z-type solution and otherwise $x(t) \equiv 0$ eventually. Therefore, there exist t_1, t_2 and t_3 such that $x(t_1) = x(t_2) = x(t_3) = 0$ and $x(t) < 0$ for $t \in (t_1, t_2)$ and $x(t) > 0$ for $t \in (t_2, t_3)$. Thus, $x'(t_2) = -q(t_2) \max_{[t_2-r,t_2]} x(s) \leq 0$, which is a contradiction.

For $q(t) \leq 0$, Theorem 5.1.1 can be proved similarly.

$\square$

Remark 5.1.1. *If $q(t)$ has the same sign, by Theorem 5.1.1, the solutions of (5.1) are more nonoscillatory in nature than those of (5.2).*

When $q(t)$ is oscillatory, then differential equation with "maxima" (5.1) may have oscillatory solutions.

For example, consider the equation

$$x'(t) + \sin t \max_{[t-2\pi,t]} x(s) = 0, \tag{5.3}$$

where $x \in \mathbb{R}$.

The differential equation with "maxima" (5.3) has an oscillatory so-lution $x = \cos t$, but it also has a nonoscillatory solution $x = -2 - \cos t$. It is obviously different than the corresponding ordinary differential equation $x'(t) + (\sin t)x = 0$.

This example shows that the solution of (5.1) is different from the behavior of the solutions of the corresponding ordinary differen-tial equation.

Theorem 5.1.2. *If $q(t)$ is oscillatory, then (5.1) has at least one nonoscillatory solution.*

Proof. Assume that $q(t_n) = 0$ for $\{t_n\}_{n=1}^{\infty}$ and $\lim_{n \to \infty} t_n = \infty$ and $q(t) \geq 0$ for $t \in (t_1, t_2)$, $q(t) \leq 0$ for $t \in (t_2, t_3)$, $q(t) \geq 0$ for $t \in (t_3, t_4)$, ... We define a function $\phi(t)$ for $t \in [t_1 - r, t_1]$, which is nondecreasing and negative; then (5.1) has a solution $y(t) = \phi(t_1) \exp\left(-\int_{t_1-r}^{t} q(s)ds\right)$ for $t \in [t_1 - r, t_2]$. It is obvious that $y(t) < 0$ for $t \in [t_1 - r, t_2]$ and $\max_{[t-r,t]} y(s) = y(t)$. By the method of steps, we can obtain $y(t)$ for $t \geq t_1 - r$. In view of $q(t) \leq 0$ for $t \in (t_2, t_3)$, we know that $y(t) < 0$ for $t \in [t_1 - r, t_3]$. We note that $q(t) \geq 0$ for $t \in [t_3, t_4]$. If there exists $\xi \in (t_3, t_4)$ such that $\max_{[\xi-r,\xi]} y(s) = y(t_2)$, then we have $y'(t) = -q(t) \max_{[t-r,t]} y(s) = -q(t)y(t)$ for $\xi \leq t \leq t_4$. By induction we know that $y(t) < 0$ for $t \geq t_1 - r$. $\qquad\square$

Now, consider the differential equation with "maxima"

$$x'(t) + q_1(t) \max_{s \in [t-r,t]} x(s) + q_2(t)x(t-h) = 0 \qquad (5.4)$$

and its corresponding delay differential equations

$$x'(t) + q_1(t)x(t-r) + q_2(t)x(t-h) = 0 \qquad (5.5)$$

and

$$x'(t) + q_1(t)x(t) + q_2(t)x(t-h) = 0, \qquad (5.6)$$

where $x \in \mathbb{R}$, q_1, $q_2 \in C([t_0, \infty), \mathbb{R}_+)$, r, $h > 0$.

It is obvious that if $x(t)$ is an eventually positive solution of the differential equation with "maxima" (5.4), then it is a solution of (5.5).

If $x(t)$ is an eventually negative solution of the equation with "max-ima" (5.4) then it is a solution of (5.6).

By the comparison result, we obtain the following result.

Theorem 5.1.3. *If the delay differential equation (5.6) is oscillatory, then so is the differential equation with "maxima" (5.5).*

By comparing (5.5) and (5.6), we know that the solutions of (5.4) are more nonoscillatory in nature that those of equation (5.6). For example, it is well known that the equation

$$x'(t) + q_1(t)x(t - r) + q_2(t)x(t) = 0, \tag{5.7}$$

may have oscillatory solutions (see [Ladde et al. 1987]). But the equation

$$x'(t) + \big(q_1(t) + q_2(t)\big)x(t) = 0, \tag{5.8}$$

is nonoscillatory. By Theorem 5.1.3, the equation

$$x'(t) + q_1(t) \max_{s \in [t-r,t]} x(s) + q_2(t)x(t) = 0, \tag{5.9}$$

has nonoscillatory solutions.

Now, consider the forced differential equation with "maxima"

$$x'(t) + q(t) \max_{s \in [t-r,t]} x(s) = f(t) \tag{5.10}$$

and its corresponding delay differential equation

$$x'(t) + q(t)x(t - r) = f(t), \tag{5.11}$$

where $x \in \mathbb{R}$, $f \in C([t_0, \infty), \mathbb{R})$, q and r are the same as above.

Theorem 5.1.4. *Assume that $q(t) \geq 0$ and that there exists $P(t)$ such that $P'(t) = f(t)$. Let $P_+(t) = \big(|P(t)| + P(t)\big)/2$ and $P_-(t) = -\big(|P(t)| - P(t)\big)/2$ such that*

$$\begin{aligned}
\int_T^\infty q(t) \max_{s \in [t-r,t]} P_+(s)dt &= \infty, \\
\int_T^\infty q(t) \max_{s \in [t-r,t]} P_-(s)dt &= -\infty.
\end{aligned} \tag{5.12}$$

Then all solutions of differential equation with "maxima" (5.10) oscillate.

Since the proof is standard (see [Erbe et al. 1987]), we omit it.

Consider the differential equation with "maxima"

$$x'(t) + \max_{s \in [t-\pi,t]} x(s) = \cos t, \tag{5.13}$$

where $x \in \mathbb{R}$.

Theorem 5.1.4 does not hold for the differential equation with "maxima" (5.13) because $\max_{s \in [t-\pi,t]} P_-(s) \equiv 0$. In fact, $x = \sin t - t$ is a nonoscillatory solution of (5.13). But, by the known result (see [Erbe et al. 1987]), all solutions of the equation

$$x'(t) + x(t - \pi) = \cos t \qquad (5.14)$$

oscillate.

5.2 Oscillations of Differential Equations with "Maxima" and Delay

Now we will study the oscillatory properties for various types of first order differential equations with "maxima."

Consider the differential equations with "maxima"

$$\left[x(t) - p(t)x(t - h)\right]' + q(t) \max_{s \in [t-r,t]} x(s) = 0, \qquad (5.15)$$

and the corresponding differential equations with "minima"

$$\left[y(t) - p(t)y(t - h)\right]' + q(t) \min s \in_{[t-r,t]} y(s) = 0, \qquad (5.16)$$

where $x \in \mathbb{R}$, $h > 0$, $r \geq 0$ and p, $q \in C([t_0, \infty), \mathbb{R})$.

In our further investigations we will need the following result:

Lemma 5.2.1. *Let the following condition be fulfilled:*

(i) $p(t) \geq 0$ for $t \geq t_0$ and there exists a $T \geq t_0$ such that

$$p(T + jr) \leq 1, \qquad j = 0, 1, 2, \dots ; \qquad (5.17)$$

(ii) $q(t) \geq 0(\neq 0)$ for $t \geq t_0$;
(iii) $x(t)$ is an eventually positive solution of (5.15), or (5.16).
Then $y(t)$ is eventually positive, where

$$y(t) = x(t) - p(t)x(t - h). \qquad (5.18)$$

The proof is standard (see [Chuanxi et al. 1990]) and we omit it.

Theorem 5.2.1. *Let the conditions of Lemma 5.2.1 hold and either $p(t) > 0$ or $h > 0$ and $q(t) \geq 0(\neq 0)$ for $t \in [u - r, u]$ for all large u.*

Then equation (5.15) has eventually positive solutions if and only if

$$\left[x(t) - p(t)x(t - h)\right]' + q(t) \max_{s \in [t-r,t]} x(s) \leq 0 \qquad (5.19)$$

has eventually positive solutions and equation (5.16) has eventually positive solutions if and only if

$$\left[x(t) - p(t)x(t - h)\right]' + q(t) \min_{s \in [t-r,t]} x(s) \leq 0 \qquad (5.20)$$

also has eventually positive solutions.

The proof of Theorem 5.2.1 is similar to Theorem 1 in [Zhang et al. 1995].

Theorem 5.2.2. *Let the conditions (i) and (ii) of Lemma 5.2.1 hold and there exists an integer N such that*

$$\liminf_{t \to \infty} \int_{t-h}^{t} q(s) \max_{u \in [s-r,s]} \sum_{j=0}^{N-1} \prod_{i=0}^{j} p(u - ih) ds > \frac{1}{e}. \qquad (5.21)$$

Then each solution of differential equation with "maxima" (5.15) oscillates.

Proof. If $x(t)$ is an eventually positive solution of equation (5.15), then $y'(t) \leq 0$ and $y(t) = x(t) - p(t)x(t - h) > 0$ eventually. Then

$$
\begin{aligned}
x &= y(t) + p(t)x(t - h) \\
&= y(t) + p(t)y(t - h) + p(t)p(t - h)x(t - 2h) \\
&= \dots \\
&\geq y(t) + p(t)y(t - h) + \dots + \prod_{i=0}^{N-1} p(t - ih)y\big(t - (i+1)h\big) \\
&\geq \sum_{j=0}^{N-1} \prod_{i=0}^{j} p(t - ih)y(t - h).
\end{aligned}
$$

Hence,

$$\max_{s\in[t-r,t]} x(s) \geq \max_{s\in[t-r,t]} \sum_{j=0}^{N-1} \prod_{i=0}^{j} p(s-ih)y(s-h)$$

$$= \max_{s\in[t-r,t]} \sum_{j=0}^{N-1} \prod_{i=0}^{j} p(s-ih)y(t-r-h)$$

$$\geq \max_{s\in[t-r,t]} \sum_{j=0}^{N-1} \prod_{i=0}^{j} p(s-ih)y(t-h).$$

Substituting the last inequality into (5.15), we have

$$y'(t) + q(t) \max_{s\in[t-r,t]} \sum_{j=0}^{N-1} \prod_{i=0}^{j} p(s-ih)y(t-h) \leq 0, \qquad (5.22)$$

which contradicts the fact that, under condition (5.21), the inequality (5.22) has no eventually positive solution [Ladde et al. 1987].

If $Z(t)$ is an eventually negative solution of (5.15), then $x(t) = -Z(t)$ is an eventually positive solution of (5.16). Similarly, we have

$$y'(t) + q(t) \min_{[t-r,t]} \sum_{j=0}^{N-1} \prod_{i=0}^{j} p(s-ih)y(t-h) \leq 0.$$

That is,

$$[-y(t)]' + q(t) \max_{[t-r,t]} \sum_{j=0}^{N-1} \prod_{i=0}^{j} p(s-ih)\big[-y(t-h)\big] \geq 0.$$

This is also a contradiction for the same reason as the positive solution. The proof is complete.

□

In the partial case $p(t) \equiv 1$ the differential equations with "maxima" (5.15) reduces to the following differential equation with "maxima:"

$$\big[x(t) - x(t-h)\big]' + q(t) \max_{s\in[t-r,t]} x(s) = 0, \qquad (5.23)$$

where $x \in \mathbb{R}$.

For the equation (5.23) we obtain the following result:

Theorem 5.2.3. *Let $q(t) \geq 0$.*

Then (5.23) has nonoscillatory solutions if and only if

$$Z''(t) + \frac{1}{h}q(t)Z(t) = 0 \tag{5.24}$$

also has nonoscillatory solutions.

Proof. Assume that $x(t)$ is an eventually positive solution of (5.23). Let $y(t) = x(t) - x(t - h)$. Then $y(t) > 0$ and $y'(t) \leq 0$ eventually. Let T be a large number so that $x(t) > 0$, $y(t) > 0$ and $y'(t) \leq 0$ for $t \geq T - h$. Set $m = \min_{-h \leq t \leq T} x(t)$. When $N \leq t \leq N + h$ we have

$$x(t) = y(t) + x(t - h) \geq \frac{1}{h}\int_t^{t+h} y(s)ds + m.$$

By induction, for $T + kh \leq t \leq T + (k+1)h$,

$$x(t) \geq \frac{1}{h}\int_{t-kh}^{t+h} y(s)ds + m.$$

Hence,

$$x(t) \geq \frac{1}{h}\int_{T^*}^{t+h} y(s)ds + m, \quad t \geq T^* \geq T + h,$$

and

$$x(t) \geq \frac{1}{h}\int_{T^*+h}^{t} y(s)ds + m, \quad t \geq T^* + h.$$

Set

$$Z(t) = \frac{1}{h}\int_{T^*+h}^{t} y(s)ds + m.$$

Thus, we have

$$Z''(t) + \frac{1}{h}q(t)z(t) \leq 0, \tag{5.25}$$

which implies that (5.24) has an eventually positive solution.

If $x(t)$ is an eventually positive solution of the equation

$$[x(t) - x(t - h)]' + q(t) \min_{s \in [t-r,t]} x(s) = 0, \tag{5.26}$$

we can also prove that (5.24) has an eventually positive solution by the above method.

If (5.24) has an eventually positive solution $Z(t)$, then $Z''(t) \leq 0$ and $Z'(t) > 0$ eventually. Therefore, there exist T and $M > 0$ such that $Z(t) > M$ and $Z'(t) < M$ eventually.

Define a function $H : \mathbb{R} \to \mathbb{R}_+$ by

$$H(t) = \begin{cases} hZ'(t) & t \geq T, \\ (t - T + h)Z'(T) & T - h \leq t < T, \\ 0 & t < T - h. \end{cases}$$

Define

$$y(t) = \sum_{i=0}^{\infty} H(t - ih) > 0$$

and obtain

$$y(t) - y(t - h) = H(t) \quad \text{for } t \geq T.$$

That is,

$$y(t) - y(t - h) = hZ'(t).$$

Setting $\mu = \max_{T-h \leq t \leq T} y(t)$, we have

$$
\begin{aligned}
y(t) &= hZ'(t) + y(t - h) \leq \int_{t-h}^{t} Z'(s)ds + y(t - h) \\
&\leq \int_{t-2h}^{t} Z'(s)ds + y(t - 2h) \leq \int_{t-nh}^{t} Z'(s)ds + y(t - nh).
\end{aligned}
$$

Therefore, we get

$$y(t) \leq \int_{T}^{t} Z'(s)ds + \mu \leq Z(t) \quad \text{for } t \geq T$$

and

$$\max_{s \in [t-\mu,t]} y(s) \leq Z(t) \quad \text{and} \quad \min_{s \in [t-\mu,t]} y(s) \leq Z(t) \quad \text{for } t \geq T.$$

Then we obtain

$$\left[y(t) - y(t - h) \right]' + q(t) \max_{s \in [t-r,t]} y(s) \leq 0$$

and

$$\left[y(t) - y(t - h)\right]' + q(t) \min_{s \in [t-r,t]} y(s) \leq 0.$$

According to Theorem 5.2.1 the differential equation with "maxima" (5.23) has nonoscillatory solutions. The proof is complete.

□

Theorem 5.2.4. *Let* $p(t) \equiv p$, $p \neq -1$, $q(t) \geq 0$, *and*

$$\int_{t_0}^{\infty} q(s)ds = \infty. \tag{5.27}$$

Then any nonoscillatory solution $x(t)$ *of (5.15) satisfies* $\lim_{t \to \infty} x(t) = 0$.

Theorem 5.2.5. *Let* $p(t) \equiv -1$, $q(t) \geq 0$, $Q(t) = \min\{q(t), q(t - h)\}$ *and*

$$\int_{t_0}^{\infty} Q(t)dt = \infty.$$

Then any eventually positive solution $x(t)$ *of (5.15) satisfies* $\lim_{t \to \infty} x(t) = 0$.

The proofs of Theorem 5.2.4 and Theorem 5.2.5 are similar to the proofs of Theorem 2 and Theorem 1 in [Shabadikov and Yuldashev 1989].

Now we will give the existence and growth conditions of nonoscillatory solutions of equation (5.15). We begin with the following theorem.

Theorem 5.2.6. *([Zhang and Migda 2005]) Let* $p(t) \equiv p$, *where* $p \geq 0$. *Then equation (5.15) has an eventually positive solution.*

Proof. Choose a positive continuous function $H(t)$ such that

$$\int_{t_0}^{\infty} q(t)H(t)dt = \infty \tag{5.28}$$

and

$$\lim_{t \to \infty} \left\{ \frac{q(t)}{\exp\left(\int_{t_0}^{t} q(s)H(s)ds\right)} \right\} = 0. \tag{5.29}$$

Define a function v by

$$v(t) = \exp\left[\int_{t_0}^t \exp\left(\int_{t_0}^s q(u)H(u)du\right)ds\right]. \qquad (5.30)$$

Let BC be the Banach space of all bounded and continuous functions $y : [t_0, \infty) \to \mathbb{R}$ with the sup norm. Define a subset Ω of BC as follows:

$$\Omega = \left\{ y \in BC : 0 \le y(t) \le 1,\ t_0 \le t < \infty \right\}.$$

Clearly Ω is a bounded, closed and convex subset of BC. Now we define a mapping S on Ω as follows:

$$(Sy)(t) =$$
$$\begin{cases} p\dfrac{v(t-h)y(t-h)}{v(t)} + \dfrac{1}{v(t)}\displaystyle\int_T^t q(s)\max_{u\in[s-r,s]} v(u)y(u)ds + \dfrac{1}{2v(t)}, & t \ge T, \\[2ex] \dfrac{t}{T}(Sy)(T) + \left(1 - \dfrac{t}{T}\right), & t_0 \le t < T, \end{cases}$$

$$\qquad (5.31)$$

where T is sufficiently large so that $t - h \ge t_0$, $t - r \ge t_0$, $v(t) \ge 1$, and

$$p\frac{v(t-h)}{v(t)} + \frac{1}{v(t)}\int_T^t q(s)\max_{u\in[s-r,s]} v(u)ds \le \frac{1}{2} \qquad \text{for } t \ge T. \qquad (5.32)$$

In fact, from (5.28), (5.29) and (5.30) it is easy to see that

$$\frac{v(t-h)}{v(t)} \to 0 \quad \text{and} \quad \frac{\int_{t_0}^t q(s)\max_{u\in[s-r,s]} v(u)ds}{v(t)} \to 0 \text{ as } t \to \infty$$

which shows that (5.32) is true for large t. Thus we have $S\Omega \subset \Omega$. Let y_1 and y_2 be two functions in Ω. Then

$$\begin{aligned}
\left|(Sy_2)(t) - (Sy_1)(t)\right| &\le p\frac{v(t-h)}{v(t)}\left|y_2(t-h) - y_1(t-h)\right| \\
&\quad + \frac{1}{v(t)}\int_T^t q(s)\max_{u\in[s-r,s]} v(u)\left|y_2(u) - y_1(u)\right|ds \\
&\le \frac{1}{2}\|y_2 - y_1\|, \qquad\qquad\qquad\qquad\qquad t \ge T,
\end{aligned}$$

and

$$\left\|Sy_2 - Sy_1\right\| = \sup_{t \geq t_0}\left|(Sy_2)(t) - (Sy_1)(t)\right|$$

$$= \sup_{t \geq T}\left|(Sy_2)(t) - (Sy_1)(t)\right| \leq \frac{1}{2}\|y_2 - y_1\|, \qquad t \geq T$$

which shows that S is a contraction on Ω. Hence there is a function $y \in \Omega$ such that $Sy = y$. That is

$$y(t) = \begin{cases} p\frac{v(t-h)y(t-h)}{v(t)} + \frac{1}{v(t)}\int_T^t q(s)\max_{u \in [s-r,s]} v(u)y(u)ds + \frac{1}{2v(t)}, \\ \hspace{4cm} t \geq T \\ \frac{t}{T}y(T) + \left(1 - \frac{t}{T}\right), \qquad t_0 \leq t < T. \end{cases}$$

Obviously $y(t) > 0$ for $t \geq t_0$. Set $x(t) = v(t)y(t)$. Then

$$x(t) - px(t-h) = \int_T^t q(s)\max_{u \in [s-r,s]} x(u)ds + \frac{1}{2}, \quad t \geq T. \qquad (5.33)$$

Therefore $x(t)$ is a positive solution of (5.15) for $t \geq T$. The proof is complete.

□

Lemma 5.2.2. (*[Erbe et al. 1987]*). *Let* $x,\ z \in C([t_0, \infty), \mathbb{R})$ *satisfy*

$$z(t) = x(t) - px(t-h), \quad t \geq t_0 + \max\{0, h\},$$

where $p,\ h \in R$. *Assume that* x *is bounded on* $[t_0, \infty)$ *and* $\lim_{t \to \infty} z(t) = l$ *exists.*

Then the following statements hold:

(a) If $p = 1$, *then* $l = 0$;

(b) If $p \neq \pm 1$, *then* $\lim_{t \to \infty} x(t)$ *exists.*

Lemma 5.2.3. (*[Zhang and Migda 2005]*). *Let* $p(t) \equiv p,\ p \geq 0$ *and* $x(t)$ *be a positive solution of equation (5.15) such that* $x(t) - px(t-h) \geq 0$.

Then the solution $x(t)$ *satisfies*

$$(a)\ \lim_{t \to \infty} x(t) = L \neq 0,$$

or

$$(b)\ \lim_{t \to \infty} x(t) = \infty.$$

Proof. Let $x(t)$ be a positive solution of equation (5.15). Set $z(t) = x(t) - px(t-h)$. Then $z(t) \geq 0$ and $z'(t) \geq 0$ and is not identically zero. Hence $0 < \lim_{t \to \infty} z(t) = l \leq \infty$. If $l = \infty$, then $x(t) \geq z(t) \to \infty$ as $t \to \infty$, i.e., (*b*) holds.

If $l < \infty$ and $x(t)$ is bounded, Lemma 5.2.2 implies that $\lim_{t \to \infty} x(t)$ exists when $p \neq 1$. But $\lim_{t \to \infty} x(t) = 0$ is impossible. When $p = 1$, Lemma 5.2.2 implies $l = 0$. This is a contradiction.

If $l < \infty$ and $x(t)$ is unbounded, then there exists $\{t_n\}$ such that $x(t_n) = \max_{t \leq t_n} x(t) \to \infty$ as $n \to \infty$. For $p \in (0,1)$, $z(t_n) \geq x(t_n)(1-p) \to \infty$ as $n \to \infty$, which contradicts the boundedness of z. For $p = 1$, $x(t) \geq x(t-h) + l/2 \geq \ldots \geq x(t-nh) + nl/2 \to \infty$ as $n \to \infty$. For $p > 1$, we have

$$x(t) \geq px(t-h) \geq \ldots \geq p^n x(t-nh)$$

which implies that (*b*) holds. The proof is complete.

$\square$

Remark 5.2.1. *In Lemma 5.2.3, the condition $x(t) - px(t-h) \geq 0$ is necessary. For example, we consider the differential equation with "maxima"*

$$\left(x(t) - 2x(t-1)\right)' = (2e-1)e^{-t}\left\{ \max_{s \in [t-1,t]} \left(\varphi(s) + e^{-s}\right) \right\}^{-1} \max_{s \in [t-1,t]} x(s), \tag{5.34}$$

where $x \in \mathbb{R}$ and

$$\varphi(t) = \begin{cases} 2^n(t-n), & t \in [n, n+\frac{1}{2}] \\ 2^n(n+1-t), & t \in [n+\frac{1}{2}, n+1] \end{cases}, \quad n = 0, 1, \ldots.$$

The differential equation with "maxima" (5.34) has an eventually positive solution $x(t) = \varphi(t) + e^{-t}$. In fact, $x(t)$ satisfies $\limsup_{t \to \infty} x(t) = \infty$ and $\liminf_{t \to \infty} x(t) = 0$. As if $r = 0$, the condition $x(t) - px(t-h) \geq 0$ is also necessary.

Consider the delay differential equation

$$\left(x(t) - 2x(t-1)\right)' = (2e-1)e^{-t}\left(\varphi(s) + e^{-t}\right)^{-1}x(t), \tag{5.35}$$

where $x \in \mathbb{R}$.

The delay differential equation (5.35) has also a positive solution $x(t) = \varphi(t) + e^{-t}$. Thus, we see that Lemma 5.3.1 in [Erbe et al. 1987] is false.

By Theorem 5.2.6 and Lemma 5.2.3 we obtain immediately the following result.

Theorem 5.2.7. (*[Zhang and Migda 2005]). Let $p(t) \equiv p$, where p is a constant.*
Then the conclusions hold:

(a) *If $p = 1$, then the differential equation with "maxima" (5.15) has an unbounded solution $x(t)$ satisfying $\lim_{t\to\infty} x(t) = \infty$;*

(b) *If $p > 1$, then the differential equation with "maxima" (5.15) has an unbounded positive solution $x(t)$ which tends to infinity exponentially;*

(c) *If $0 \le p < 1$ and*

$$\int_{t_0}^{\infty} q(s)ds = \infty, \tag{5.36}$$

then the differential equation with "maxima" (5.15) has an unbounded positive solution.

Proof. By (5.33) the inequality

$$x(t) - px(t - h) > 0 \quad \text{for } t \ge T$$

holds.

The proofs of (a) and (b) are obvious.
In the case that $p > 1$ we have

$$x(t) \ge px(t - h) \ge \ldots \ge p^n x(t - nh),$$

or

$$x(t) \ge x(t_0)\exp\big(\mu(t - t_0)\big), \quad \text{for } t \ge t_0,$$

where $\mu = \frac{\ln p}{h} > 0$, which shows that (ii) is true. The proof is complete. $\square$

Remark 5.2.2. *For eventually negative solutions of equation (5.15), we can also obtain similar results. They are omitted.*

Now we will study asymptotic properties of nonoscillatory solutions of differential equation with "maxima"(5.15).

Theorem 5.2.8. ([Zhang and Migda 2005]). Let $p(t) \equiv p$, $0 \le p < 1$ and (5.36) holds.
   If $x(t)$ is a nonoscillatory solution of equation (5.15), then either

$$\lim_{t \to \infty} |x(t)| = \infty$$

or

$$\lim_{t \to \infty} x(t) = 0.$$

Proof. Let $x(t)$ be an eventually positive solution of (5.15) and set

$$z(t) = x(t) - px(t - h). \tag{5.37}$$

Then we have either $z(t) > 0$ or $z(t) < 0$. If $z(t) > 0$ holds, then there exists $c > 0$ such that $x(t) \ge z(t) > c$. Integrating (5.15) we have

$$z(t) = \int_T^t q(s) \max_{u \in [s-r,s]} x(u)ds + z(T) \ge c \int_T^t q(s)ds.$$

It is clear that $\lim_{t \to \infty} z(t) = \infty$. Thus, we have $\lim_{t \to \infty} x(t) = \infty$.
   If $z(t) < 0$ holds, then $p \ne 0$ and $\lim_{t \to \infty} z(t) = L \le 0$ is finite. Integrating (5.15) we have

$$L - z(t) = \int_t^\infty q(s) \max_{u \in [s-r,s]} x(u)ds.$$

Thus, we have $\liminf_{t \to \infty} x(t) = 0$. Note that $-px(t-h) < z(t)$, we have $\lim_{t \to \infty} z(t) = 0$. On the other hand, $x(t) < px(t-h) < x(t-h)$ implies that $x(t)$ is bounded above. By Lemma 4.1.2 we get that $\lim_{t \to \infty} x(t)$ exists. Let $\lim_{t \to \infty} x(t) = l$. In view of (5.37), we get that

$$0 = \lim_{t \to \infty} x(t) - p \lim_{t \to \infty} x(t - h) = l(1 - p)$$

which implies $l = 0$. The case where $x(t)$ is eventually negative is proved in a similar way. The proof is complete.   □

Corollary 5.2.1. Let $p(t) \equiv p$, $0 \le p < 1$ and (5.36) holds.
   If $x(t)$ is a bounded nonoscillatory solution of differential equation with "maxima" (5.15), then $\lim_{t \to \infty} x(t) = 0$.

Theorem 5.2.9. Let $p(t) \equiv p$, $p > 1$ and (5.36) holds.
   If $x(t)$ is a bounded nonoscillatory solution of differential equation with "maxima" (5.15), then $\lim_{t \to \infty} x(t) = 0$.

Proof. Let $x(t)$ be a bounded positive solution of (5.15) and $z(t)$ is defined by (5.37).

Then $z(t) > 0$ is impossible.

Indeed, then we have $\lim_{t\to\infty} x(t) = \infty$ which is a contradiction. Thus, we have $z(t) < 0$. In view of the proof of Theorem 5.2.8 we have $\lim_{t\to\infty} z(t) = 0$. Similarly, we have

$$0 = \lim_{t\to\infty} x(t) - p \lim_{t\to\infty} x(t-h) = l(1-p)$$

which implies $l = 0$. The case where $x(t)$ is eventually negative solution is similarly proved. The proof is complete.

$\square$

Theorem 5.2.10. *Let $p(t) \equiv p$, $p \geq 1$ and $0 < q \leq q(t)$.*

If $x(t)$ is an eventually positive solution of differential equation with "maxima" (5.15) then either $\lim_{t\to\infty} x(t) = \infty$ or $\lim_{t\to\infty} x(t) = 0$.

Proof. If $z(t) > 0$, we have similarly $\lim_{t\to\infty} x(t) = \infty$. If $z(t) < 0$ holds, then $\lim_{t\to\infty} z(t) = L \leq 0$ is finite. Suppose that $x(t)$ does not tend to zero as $t \to \infty$. Then $c = \limsup_{t\to\infty} x(t) > 0$ (if $x(t)$ is unbounded, set c to be an arbitrarily positive constant). There exists a sequence $\{t_n\}_1^\infty$ such that $t_{n+1} - t_n > r$ and $x(t_n) > c/2$ for each $n \in N$. Thus the inequality

$$\max_{s\in[t-r,t]} x(s) > \frac{c}{2}, \quad t \in [t_n, t_n+r], \quad n \geq N,$$

holds, and

$$\int_{t_n}^{t_n+r} q(s) \max_{u\in[s-r,s]} x(u)ds \geq \frac{cqr}{2}.$$

From the definition of the sequence $\{t_n\}$ and integrating (5.15), we have

$$\begin{aligned}
L - z(t_0) &= \int_{t_0}^{\infty} q(s) \max_{u\in[s-r,s]} x(u)ds \\
&\geq \sum_{n=1}^{\infty} \int_{t_n}^{t_n+h} q(s) \max_{u\in[s-r,s]} x(u)ds \\
&\geq \sum_{n=1}^{\infty} \frac{cqr}{2} = \infty.
\end{aligned}$$

This is a contradiction. The proof is complete. l

$\square$

Remark 5.2.3. *Theorem 5.2.10 is not valid if $x(t)$ is an eventually negative solution of equation (5.15) as the following example shows.*

Example 5.2.1. *Consider the differential equation with "maxima"*

$$\left(x(t) - 2x(t-1)\right)' = (2e-1)e^{-t}\left\{\max_{s\in[t-1,t]}\left(\varphi(s)+e^{-s}\right)\right\}^{-1}\max_{s\in[t-1,t]}x(s),$$
$$(5.38)$$

where $x \in \mathbb{R}$ and

$$\varphi(t) = \begin{cases} 2^n(t-n), & t \in [n, n+\frac{1}{2}] \\ 2^n(n+1-t), & t \in [n+\frac{1}{2}, n+1] \end{cases}, \quad n = 0, 1, \ldots.$$

It is easy to check that

$$(2e-1)e^{-t}\left\{\max_{s\in[t-1,t]}\left(\varphi(s)+e^{-s}\right)\right\}^{-1} \geq 2e - 1 > 0.$$

Therefore, the assumptions of Theorem 5.2.10 are satisfied. A straightforward verification yields that the function $x(t) = -\varphi(t) - e^{-t}$ is an eventually negative solution of (5.38). Moreover, since $x(n) = -e^{-n}$ and $x(n+\frac{1}{2}) = -2^{n-1} - e^{-n-\frac{1}{2}}$, $\limsup_{t\to\infty} x(t) = 0$ and $\liminf_{t\to\infty} x(t) = -\infty$.

Corollary 5.2.2. *Let $p(t) \equiv p$, $p = 1$ and $0 < q \leq q(t)$.*
Then any bounded eventually positive solution $x(t)$ of equation (5.15) satisfies $\lim_{t\to\infty} x(t) = 0$.

Remark 5.2.4. *Corollary 5.2.2 is not valid if $x(t)$ is a bounded eventually negative solution of equation (5.15).*

Example 5.2.2. *Consider the differential equation with "maxima"*

$$\left(x(t) - x(t-1)\right)' = (e-1)e^{-t}\left\{\min_{s\in[t-1,t]}\left(\psi(s)+e^{-s}\right)\right\}^{-1}\max_{s\in[t-1,t]}x(s),$$
$$(5.39)$$

where $x \in \mathbb{R}$ and $\psi(t)$ is a 1-periodic function defined by the equality

$$\psi(t) = \begin{cases} t & t \in [0, \frac{1}{2}] \\ 1-t, & t \in [\frac{1}{2}, 1] \end{cases}.$$

It has an eventually negative solution $x(t) = -\psi(t) - e^{-t}$ such that

$$\limsup_{t\to\infty} x(t) = 0 \quad \text{and} \quad \liminf_{t\to\infty} x(t) = -\frac{1}{2}.$$

Example 5.2.3. *Consider the differential equation with "maxima"*

$$\big(x(t) - p(t)x(t - \tau)\big)' = q \max_{s \in [t-r,t]} x(s), \qquad (5.40)$$

where $x \in \mathbb{R}$ and $q = \frac{pe^\tau - 1}{e^\tau}$.

If $e^{-\tau} < p < 1$, then all conditions of Theorem 5.2.8 and Corollary 5.2.1 are satisfied for equation (5.40).

If $p > 1$, then all conditions of Theorem 5.2.9 are satisfied for equation (5.40).

If $p = 1$, then all conditions of Corollary 5.2.2 hold for equation (5.40).

In fact, equation (5.40) has a positive solution $x(t) = e^{-t}$.

Now we will discuss first the oscillation of the differential equation with "maxima" (5.15) for $p < 0$.

Theorem 5.2.11. *Let $p(t) \equiv p$, $p < 0$, $p \neq -1$ and the condition (5.36) holds.*

Then every bounded solution of equation (5.15) is oscillatory.

Proof. Let $x(t) > 0$ be a bounded positive solution of (5.15). Set $z(t) = x(t) - px(t - h)$, then $z'(t) \geq 0$. Thus $\lim_{t \to \infty} z(t) = l$ exists and $l > 0$. For $p \neq -1$, by Lemma 5.2.3, there exists $\lim_{t \to \infty} x(t)$. From (5.15) we have

$$\int_{t_0}^{\infty} q(t) \max_{s \in [t-r,t]} x(s)\,dt < \infty.$$

In view of (5.36), this implies that $\lim_{t \to \infty} x(t) = 0$ and so by (5.37) $l = (1 - p)0 = 0$. This contradicts $l > 0$. The proof in the case of eventually negative solution is similar and will be omitted.

$\square$

In the case when $p \geq 0$, we have the following results.

Theorem 5.2.12. *Let $p(t) \equiv p$, $p \geq 0$, $p \neq 1$ and the condition (5.36) holds.*

Then every bounded solution of equation (5.15) is either oscillatory or approaches zero as $t \to \infty$.

In view of Theorem 5.2.8 or Theorem 5.2.9 we know immediately that Theorem 5.2.12 is valid.

Theorem 5.2.13. *Let $p(t) \equiv p$, $p > 0$, $p \neq 1$, the condition (5.36) holds, and the inequality*

$$z'(t) + \frac{1}{p}q(t)z(t+h-r) \geq 0$$

has no eventually positive solution and the inequality

$$z'(t) + \frac{1}{p}q(t)z(t+h) \leq 0$$

has no eventually negative solution.

   Then every bounded solution of equation (5.15) is oscillatory.

   Indeed, from the proof of Theorem 5.2.9 we see that if $x(t)$ is a bounded positive solution of (5.15), then $z(t) < 0$ eventually. Note that $x(t0) > -z(t+h)/p$. Substituting it in (5.15) we obtain

$$z'(t) + \frac{1}{p}q(t)z(t+h-r) \geq 0.$$

This is a contradiction. For a bounded negative solution, we can obtain similarly a contradiction.

Theorem 5.2.14. *Let $p(t) \equiv p$, $p = 1$, the condition (5.36) holds and the inequality*

$$y''(t) \geq \frac{1}{h}q(t+h)y(t+h-r)$$

has no any eventually positive solution and the inequality

$$y''(t-h) \leq \frac{1}{h}q(t)y(t)$$

has no any eventually negative solution.

   Then every bounded solution of equation (5.15) oscillates.

Proof. Let $x(t)$ be a bounded eventually positive solution of (5.15) and set $z(t) = x(t-h) - x(t)$. Then $z(t) > 0$ and $z'(t) \leq 0$ eventually. Thus, we have also

$$
\begin{aligned}
x(t) &= x(t+h) + z(t+h) \\
&\geq x(t+h) + \frac{1}{h}\int_{t+h}^{t+2h} z(s)\,ds \\
&\geq \frac{1}{h}\int_{t+h}^{\infty} z(s)\,ds.
\end{aligned}
$$

Set

$$y(t) = \frac{1}{h} \int_{t+h}^{\infty} z(s)ds. \qquad (5.41)$$

Substituting it into (5.15) we obtain

$$y''(t) \geq \frac{1}{h}q(t+h)y(t+h-r)$$

which is a contradiction.

If $x(t)$ is a bounded eventually negative solution of (5.15), we have $z(t) < 0$ and $z'(t) \geq 0$ eventually. Thus, we have also

$$x(t) = x(t+h) + z(t+h) \leq \frac{1}{h} \int_{t+h}^{\infty} z(s)ds.$$

$y(t)$ is defined by (5.41), and substituting it into (5.15) we have

$$y''(t-h) \leq \frac{1}{h}q(t)y(t).$$

The proof is complete.

$\square$

5.3   Oscillations of Forced n-th Order Differential Equations with "Maxima"

The main purpose of this section is to establish some oscillatory properties of solutions of n-th order forced differential equation with "maxima."

Consider the following forced differential equation with "maxima"

$$x^{(n)}(t) + a(t)f\left(\max_{s \in I(t)} x(s)\right) = Q(t), \quad t \geq \alpha, \qquad (5.42)$$

where $n \geq 2$ is an integer, $x \in \mathbb{R}$, $f, a : \mathbb{R} \to \mathbb{R}$, $I(t) = [\sigma(t), \tau(t)]$, $\sigma(t) \leq \tau(t) \leq t$, $\alpha \geq 0$ is a constant.

Remark 5.3.1. *If there exists a point $\xi \geq \alpha$ such that $\sigma(\xi) = \tau(\xi)$ then we have $\max_{s \in I(\xi)} x(s) = x(\tau(\xi))$.*

Let $J = [\alpha, +\infty) \subseteq \mathbb{R}_+ = [0, +\infty)$ and $\mathbb{R}_0 = (-\infty, 0) \cup (0, +\infty)$. We introduce the following set **(H5.3)** of conditions:

H5.3.1. The functions $\sigma, \tau \in C(J, \mathbb{R})$, $\sigma(t) \leq \tau(t) \leq t$, $\lim\limits_{t \to +\infty} \sigma(t) = +\infty$ and

$$sup_{t \geq \alpha}\left(t - \sigma(t)\right) = h < +\infty. \tag{5.43}$$

H5.3.2. $a \in C(J, (0, +\infty))$, $Q \in C(J, \mathbb{R})$.

H5.3.3. $f \in C(\mathbb{R}, \mathbb{R})$ and $xf(x) > 0$ for $x \neq 0$.

H5.3.4. There exists an oscillatory function $\rho(t)$ such that:

$$\rho^{(n)}(t) = Q(t), \quad \text{for } t \geq \alpha, \tag{5.44}$$

$$\lim_{t \to +\infty} \rho^{(i)}(t) = 0, \quad \text{for } i = 0, 1, \dots n - 1. \tag{5.45}$$

Remark 5.3.2. *If condition H5.3.1 is satisfied then for any $t \geq \alpha$ the inequality $h = sup_{t \geq \alpha}\left(t - \sigma(t)\right) \geq t - \sigma(t) \geq \alpha - \sigma(t)$ holds, i.e., $\sigma(t) \geq \alpha - h$, i.e., the inclusion $I(t) \subset [\alpha - h, \infty)$ is valid for $t \geq \alpha$.*

Let $T \in J$ be a given number.

Definition 5.3.1. *The function $x(t)$ is called to be a solution of differential equation with "maxima" (5.42) on the interval $[T, +\infty)$, $T \geq \alpha$ if $x(t)$ is defined for $t \geq T - h$ and it satisfies (5.42) for $t \geq T$, where the constant h is defined by (5.43).*

In further investigations we will use the following result:

Lemma 5.3.1. *[Kiguradze 1964] Let $u(t)$ be a positive n-times differentiable function on the interval $[t_0, +\infty)$ and let $u^{(n)}(t)$ be a nonpositive function and it is not identically equal to zero on any sub-interval $[t_1, +\infty) \subset [t_0, +\infty)$.*
Then there exist a number $t_u \geq t_0$ and an integer k, $k \in \{0, 1, \dots, n - 1\}$, such that:

1. $n + k$ is odd.

2. $u^{(i)}(t) > 0$ for $t \geq t_u$, $i = 0, 1, \ldots, k - 1$.

3. $(-1)^{i+k} u^{(i)}(t) > 0$ for $t \geq t_u$, $i = k, \ldots, n$.

4. $(t - t_u)|u^{(k-i)}(t)| \leq (1 + i)|u^{(k-i-1)}(t)|$, where $t \geq t_u$, $i = 0, 1, \ldots, k - 1$.

We will study the oscillatory properties of the solutions of differential equations with "maxima."

Theorem 5.3.1. *Let the following conditions be fulfilled:*

1. *Conditions H5.3 are satisfied.*

2. *The equality $f(x) = g(x)h(x)$ holds for $x \in R_0$, where $g \in C(R_0, (0, +\infty))$ is a nondecreasing function on $(-\infty, 0)$ and a nonincreasing function on $(0, +\infty)$.*

3. *The function $h \in C(R_0, \mathbb{R})$ and*

$$\frac{h(x)}{x} \geq \gamma \quad for \quad x \neq 0. \tag{5.46}$$

4. *There exists a constant $\beta \geq \alpha$ such that the inequality*

$$\limsup_{t \to +\infty} \left[\frac{1}{t} \int_{\beta}^{t} a(u)g\left(c(\tau(u))^{n-1}\right) (\tau(u))^n \, du \right.$$

$$\left. + t^n \int_{t}^{+\infty} a(u)g\left(c(\tau(u))^{n-1}\right) \frac{(\tau(u))^n}{u^{n+1}} \, du \right] > \frac{(n-1)!}{\gamma} \tag{5.47}$$

holds for $c \geq 1$.

Then the following conclusions hold:

1. *If n is an even integer, then equation (5.42) is oscillatory.*

2. *If n is an odd integer, then every solution of equation (5.42) is either oscillatory or*

$$\lim_{t \to +\infty} x^{(i)}(t) = 0, \quad i = 0, 1, \ldots, n - 1. \tag{5.48}$$

Proof. Consider the following two cases:

Case 1. Let n be an odd integer. Assume $x(t)$ is a nonoscillatory solution of differential equation with "maxima" (5.42) in the interval $[a, +\infty)$.

Case 1.1. Let $x(t)$ be a proper positive solution, i.e., there exists $T_x \geq a$ such that $x(t) > 0$ for $t \geq T_x$. Let $t_0 = T_x + h$, where the constant h is defined by (5.43). Consider the function $y(t) = x(t) - \rho(t)$ for $t \geq t_0$, where $\rho(t)$ is the function defined in condition H5.3.4. From (5.42) and (5.44), condition H5.3.3 and the inclusion $I(t) \subset [T_x, \infty)$ for $t \geq t_0$, we have

$$y^{(n)}(t) = -a(t)f\left(\max_{s \in I(t)} x(s)\right) < 0, \quad t \geq t_0. \qquad (5.49)$$

Then the derivatives $y^{(i)}(t)$, $0 \leq i \leq n-1$ are monotone and one-signed functions for all $t \geq t_1$, where $t_1 \geq t_0$ is a sufficiently large number. From the inequality $x(t) > 0$ for $t \geq t_1$, equality (5.45) for $i = 0$, and the oscillatory properties of the function $\rho(t)$ it follows that there exists a number $t_2 \geq t_1$ such that $y(t) > 0$ for $t \geq t_2$. Therefore, the conditions of Lemma 5.3.1 are fulfilled and hence there exists an integer $k \in \{0, 1, \ldots, n-1\}$ and $T_k \geq t_2$ such that $n+k$ is an odd number and for $t \geq T_k$ the inequalities

$y^{(i)}(t) > 0$ for $0 \leq i \leq k-1$,

$(-1)^{i+k}y^{(i)}(t) > 0$ for $k \leq i \leq n$,

$(t - T_k)|y^{(k-i)}(t)| \leq (1+i)|y^{(k-i-1)}(t)|$ for $i = 0, 1, \ldots, k-1$

$$(5.50)$$

hold.

Since the derivative $y^{(n-1)}(t)$ is a positive and nonincreasing function for $t \geq T_k$, there exists $c_0 > 0$ such that

$$y^{(n-1)}(t) \leq c_0, \quad t \geq T_k.$$

Integrate n-times the above inequality on the interval (T_k, t) and obtain

$$y(t) \leq \frac{c_0}{(n-1)!}(t - T_k)^{n-1} + \cdots + y(T_k), \quad t \geq T_k.$$

Choose numbers t_3 and c_1 such that $t_3 \geq T_k$, $c_1 \geq c_0$ and

$$y(t) \leq c_1 t^{n-1} \quad \text{for} \quad t \geq t_3. \qquad (5.51)$$

Now we apply Taylor formula, use the equation (5.49) and obtain

$$
y^{(k)}(t) = \sum_{j=0}^{n-k-1} \frac{(-1)^j}{j!} y^{(k+j)}(\tau)(\tau - t)^j
$$

$$
+ \frac{(-1)^{n-k}}{(n-k-1)!} \int_t^\tau (s-t)^{n-k-1} a(s) f\left(\max_{u\in I(s)} x(u)\right) ds, \quad T_k \le t \le \tau.
$$

$$(5.52)$$

Using the second inequality of (5.50), we obtain the inequality

$$
y^{(k)}(t) \ge \int_t^\tau \frac{(s-t)^{n-k-1}}{(n-k-1)!} a(s) f\left(\max_{u\in I(s)} x(u)\right) ds, \quad T_k \le t \le \tau,
$$

or

$$
y^{(k)}(t) \ge \int_t^{+\infty} \frac{(s-t)^{n-k-1}}{(n-k-1)!} a(s) f\left(\max_{u\in I(s)} x(u)\right) ds, \quad t \ge T_k. \quad (5.53)
$$

Consider the following two cases:

Case 1.1.1. Let $k > 0$. Then $y(t)$ is a positive increasing function. Integrating (5.53) from T_k to t, we get

$$
y^{(k-1)}(t) \ge y^{(k-1)}(T_k)
$$

$$
+ \int_{T_k}^t \left[\int_s^{+\infty} \frac{(u-s)^{n-k-1}}{(n-k-1)!} a(u) f\left(\max_{v\in I(u)} x(v)\right) du \right] ds
$$

$$
= y^{(k-1)}(T_k) + \int_{T_k}^t \left[\int_{T_k}^u \frac{(u-s)^{n-k-1}}{(n-k-1)!} ds \right] a(u) f\left(\max_{v\in I(u)} x(v)\right) du
$$

$$
+ \int_t^{+\infty} \left[\int_{T_k}^t \frac{(u-s)^{n-k-1}}{(n-k-1)!} ds \right] a(u) f\left(\max_{v\in I(u)} x(v)\right) du \quad \text{for } t \ge T_k.
$$

$$(5.54)$$

Now, using the inequalities

$$
\int_{T_k}^u \frac{(u-s)^{n-k-1}}{(n-k-1)!} ds = \frac{(u-T_k)^{n-k}}{(n-k)!}, \quad T_k \le u \le t,
$$

$$\int_{T_k}^{t} \frac{(u-s)^{n-k-1}}{(n-k-1)!} ds = \frac{(u-T_k)^{n-k} - (u-t)^{n-k}}{(n-k)!} \geq \frac{(t-T_k)^{n-k}}{(n-k)!},$$

$$T_k \leq t \leq u$$

we obtain

$$y^{(k-1)}(t) \geq \int_{T_k}^{t} \frac{(u-T_k)^{n-k}}{(n-k)!} a(u) f\left(\max_{v \in I(u)} x(v) \right) du$$

$$+ (t-T_k)^{n-k} \int_{t}^{+\infty} \frac{1}{(n-k)!} a(u) f\left(\max_{v \in I(u)} x(v) \right) du \quad \text{for} \quad t \geq T_k.$$

$$(5.55)$$

Since $y(t)$ is positive and increasing for $t \geq T_k$ and $\max_{s \in I(t)} y(s) = y(\tau(t))$, it follows from (5.45) for $i = 0$ that there exists $c_2 > 1$ such that

$$\max_{s \in I(t)} x(s) \leq c_2 \max_{s \in I(t)} y(s) = c_2 y(\tau(t)) \quad \text{for} \quad t \geq t_3.$$

From condition H5.3.1 it follows that for $t \geq t_3 + h$ the inequalities $t - \tau(t) \leq t - \sigma(t) \leq h$ hold, i.e., $\tau(t) \geq t_3$. Therefore, from (5.51) we get

$$\max_{s \in I(t)} x(s) \leq c(\tau(t))^{n-1}, \quad t \geq t_3 + h, \quad (5.56)$$

where $c = c_1 c_2 > 1$.

From (5.47) it follows that there exists $\lambda \in (0,1)$ such that

$$\lambda \limsup_{t \to +\infty} \left[\frac{1}{t} \int_{\beta}^{t} a(u) g\left(c(\tau(u))^{n-1} \right) (\tau(u))^n du \right.$$

$$\left. + t^n \int_{t}^{+\infty} a(u) g\left(c(\tau(u))^{n-1} \right) \frac{(\tau(u))^n}{u^{n+1}} du \right] > \frac{(n-1)!}{\gamma}. \quad (5.57)$$

Set $\mu = \lambda^{\frac{1}{n+1}}$. Since $\mu \in (0,1)$, $\lim_{t \to +\infty} \rho(t) = 0$ and $y(t)$ is positive and increasing for $t \geq T_k$, then there exists $T_\mu : \quad T_\mu \geq t_3 + h \geq T_k$ such that

$$t - T_k \geq \mu t, \quad \tau(t) - T_k \geq \mu \tau(t) \quad \text{for} \quad t \geq T_\mu, \quad (5.58)$$

and

$$x(t) \geq \mu y(t) \quad \text{for} \quad t \geq T_\mu. \tag{5.59}$$

From condition 2 and (5.56) we have

$$f\left(\max_{s \in I(u)} x(s)\right) \geq g\left(c(\tau(u))^{n-1}\right) h\left(\max_{s \in I(u)} x(s)\right) \quad \text{for} \quad u \geq T_\mu. \tag{5.60}$$

Since $y(t)$ is increasing for $t \geq T_k$, then $\max_{s \in I(u)} y(s) = y(\tau(u))$ and it follows from (5.46), (5.59) and (5.60) that

$$
\begin{aligned}
f\left(\max_{s \in I(u)} x(s)\right) &\geq \gamma g\left(c(\tau(u))^{n-1}\right) \max_{s \in I(u)} x(s) \\
&\geq \gamma \mu g\left(c(\tau(u))^{n-1}\right) \max_{s \in I(u)} y(s) \\
&= \gamma \mu g\left(c(\tau(u))^{n-1}\right) y(\tau(u)), \quad u \geq T_\mu.
\end{aligned} \tag{5.61}
$$

From the last inequality of (5.50) for $t = \tau(u)$ we obtain

$$y(\tau(u)) \geq \frac{(\tau(u) - T_k)^{k-1}}{k!} y^{(k-1)}(\tau(u)), \tag{5.62}$$

where $u \geq T_\mu$ and $\tau(u) \geq t_3 \geq T_k$.

From the last inequality in (5.50) for $i = 0$ we obtain $(t - T_k)y^{(k)}(t) \leq y^{(k-1)}(t)$ for $t \geq T_k$ and therefore, the function $\frac{y^{(k-1)}(t)}{t - T_k}$ is nonincreasing since its derivative is nonpositive. From the monotonicity of the function $\frac{y^{(k-1)}(t)}{t - T_k}$ and the inequality $\tau(u) \leq u \leq t$ we obtain for $t \geq u \geq T_\mu \geq T_k$ the following inequality

$$y^{(k-1)}(\tau(u)) = \frac{\tau(u) - T_k}{\tau(u) - T_k} y^{(k-1)}(\tau(u)) \geq \frac{\tau(u) - T_k}{t - T_k} y^{(k-1)}(t). \tag{5.63}$$

Since the function $\frac{y^{(k-1)}(t)}{t - T_k}$ is nonincreasing, the function $y^{(k-1)}(t)$ is nondecreasing and $\tau(u) \leq u$ we get for $u \geq t \geq T_\mu$ the following inequality

$$y^{(k-1)}(\tau(u)) \geq \frac{\tau(u) - T_k}{u - T_k} y^{(k-1)}(u) \geq \frac{\tau(u) - T_k}{u - T_k} y^{(k-1)}(t). \tag{5.64}$$

Taking into account (5.55), the inequalities (5.58), (5.61), (5.62), (5.63), (5.64) and $(n-k)!k! \leq (n-1)!$ for $1 \leq k \leq n-1$, we obtain

$$y^{(k-1)}(t) \geq \int_{T_\mu}^{t} \frac{(u-T_k)^{n-k}}{(n-k)!k!} a(u)\gamma\mu^{k+1} g\left(c(\tau(u))^{n-1}\right) \frac{(\tau(u))^k}{t-T_k} y^{(k-1)}(t)du$$

$$+ (t-T_k)^{n-k} \int_{t}^{+\infty} \frac{1}{(n-k)!k!} a(u)\gamma\mu g\left(c(\tau(u))^{n-1}\right) \frac{(\tau(u))^k}{u-T_k} y^{(k-1)}(t)du,$$

$$\text{for} \quad t \geq T_\mu \geq T_k. \quad (5.65)$$

or

$$\frac{(n-1)!}{\gamma} \geq \frac{1}{t} \int_{T_\mu}^{t} a(u)\mu^{k+1} g\left(c(\tau(u))^{n-1}\right) (u-T_k)^{n-k}(\tau(u))^k du$$

$$+ \mu^{n+1}t^{n-k} \int_{t}^{+\infty} a(u)g\left(c(\tau(u))^{n-1}\right) \frac{u^{n-k}(\tau(u))^k}{u^{n-k}(u-T_k)}du,$$

$$\text{for} \quad t \geq T_\mu \geq T_k. \quad (5.66)$$

Since $u \geq \tau(u)$ and $u - T_k \geq \mu u \geq \mu(\tau(u))$ we get

$$\frac{(n-1)!}{\gamma} \geq \frac{\mu^{n+1}}{t} \int_{T_\mu}^{t} a(u)g\left(c(\tau(u))^{n-1}\right) (\tau(u))^n du$$

$$+ \mu^{n+1}t^{n} \int_{t}^{+\infty} a(u)g\left(c(\tau(u))^{n-1}\right) \frac{(\tau(u))^n}{t^k u^{n-k}(u-T_k)}du,$$

$$\text{for} \quad t \geq T_\mu \geq T_k. \quad (5.67)$$

Apply the inequality $u \geq t$ to the second integral of (5.67) and obtain for any $T \geq T_\mu$

$$\frac{(n-1)!}{\gamma} \geq \mu^{n+1} \left[\frac{1}{t} \int_{T}^{t} a(u)g\left(c(\tau(u))^{n-1}\right) (\tau(u))^n du \right.$$

$$\left. + t^n \int_{t}^{+\infty} a(u)g\left(c(\tau(u))^{n-1}\right) \frac{(\tau(u))^n}{u^{n+1}}du \right], \quad t \geq T.$$

Since $\mu^{n+1} = \lambda$ then

$$\frac{(n-1)!}{\gamma} \geq \lambda \left[\frac{1}{t} \int\limits_T^t a(u)g\left(c(\tau(u))^{n-1}\right)(\tau(u))^n du \right.$$

$$\left. + t^n \int\limits_t^{+\infty} a(u)g\left(c(\tau(u))^{n-1}\right)\frac{(\tau(u))^n}{u^{n+1}}du \right], \quad t \geq T. \quad (5.68)$$

Keeping in mind the relation

$$\lim_{t\to+\infty} \frac{1}{t} \int\limits_T^\beta a(u)g\left(c(\tau(u))^{n-1}\right)u^n du = 0,$$

we conclude from (5.68) that

$$\frac{(n-1)!}{\gamma} \geq \lambda \limsup_{t\to+\infty} \left[\frac{1}{t} \int\limits_\beta^t a(u)g\left(c(\tau(u))^{n-1}\right)(\tau(u))^n du \right.$$

$$\left. + t^n \int\limits_t^{+\infty} a(u)g\left(c(\tau(u))^{n-1}\right)\frac{(\tau(u))^{n^{n+1}}}{u}du \right].$$

The obtained inequality contradicts (5.57).

Case 1.1.2. Let $k = 0$. Then n is an odd number and it follows from condition 4 that

$$\int\limits_\beta^{+\infty} u^{n-1}a(u)g\left(c(\tau(u))^{n-1}\right)du = +\infty. \quad (5.69)$$

From equality (5.52) for $k = 1$ and inequality (5.50) we get

$$y(t) \geq \int\limits_t^{+\infty} \frac{(u-t)^{n-1}}{(n-1)!}a(u)f\left(\max_{v\in I(u)} x(v)\right)du, \quad t \geq T_k. \quad (5.70)$$

We shall prove that

$$\lim_{t\to+\infty} y(t) = 0. \quad (5.71)$$

Assume the opposite: there exists d : $0 < d < \infty$ such that $\lim_{t \to +\infty} y(t) = d$. Then there exists a constant $d_1 > 0$ and $T_* \geq T_\mu$ such that $y(\tau(u)) \geq d_1$ for $t \geq T_*$. From (5.61) and (5.70) it follows the inequality

$$y(t) \geq d_1 \gamma \mu \int_t^{+\infty} \frac{(u-t)^{n-1}}{(n-1)!} a(u) g\left(c(\tau(u))^{n-1}\right) du, \quad t \geq T_*,$$

which contradicts (5.69); therefore, (5.71) is proved.

Taking into account that $(-1)^i y^i(t)$, $(i = 0, 1, \ldots, n-1)$ are positive and decreasing functions for $t \geq T_k$, one can prove that

$$\lim_{t \to +\infty} y^i(t) = 0, \quad i = 1, \ldots, n-1.$$

Case 1.2. Let $x(t)$ be a negative proper solution, i.e., there exists $T_x \geq \alpha$ such that $x(t) < 0$ for $t \geq T_x$. As in Case 1.1., we obtain a contradiction.

Case 2. Let n be an even integer. Then the proof is similar to the one of Case 1. In this case, since $n + k$ is an odd integer it follows that $k > 0$ and again we obtain a contradiction.

Therefore, $x(t)$ is oscillatory.

$\square$

As a consequence of Theorem 5.3.1, we obtain the following result.

Theorem 5.3.2. *Let conditions (H5.3) hold,*

$$\frac{f(x)}{x} \geq \gamma > 0, \quad for \quad x \neq 0 \tag{5.72}$$

and there exists a number $\beta \geq \alpha$ such that

$$\limsup_{t \to +\infty} \left[\frac{1}{t} \int_\beta^t (\tau(s))^n a(s) ds + t^n \int_t^{+\infty} \frac{(\tau(s))^{n-1}}{s} a(s) ds \right] > \frac{(n-1)!}{\gamma}. \tag{5.73}$$

Then the following conclusions hold:

1. *If n is an even number, then equation (5.42) is oscillatory.*

2. *If n is an odd number, then every solution $x(t)$ of equation (5.42) is either oscillatory or*

$$\lim_{t\to+\infty} x^{(i)}(t) = 0, \quad i = 0, 1, \ldots, n-1.$$

In the case when the upper bound $\tau(t)$ of the retarded interval of maximum is a monotonic function, we obtain the following result:

Theorem 5.3.3. *Let the following conditions be fulfilled:*

1. *Conditions (H5.3) are satisfied and the function $\tau(t)$ is nondecreasing on J.*

2. *The equality $f(x) = g(x)h(x)$ for $x \in \mathbb{R}_0$ holds, where the function $g \in C(\mathbb{R}_0, (0, +\infty))$ is nondecreasing on $(-\infty, 0)$ and nonincreasing on $(0, +\infty)$ and the function $h \in C(\mathbb{R}_0, \mathbb{R})$ is nondecreasing on $\mathbb{R}_0$ and*

$$h(xy) \geq Kh(x)h(y), \quad h(-xy) \leq Kh(-x)h(y) \quad for \ x > 0, y > 0, \tag{5.74}$$

where $K > 0$ is a constant and

$$\lim_{x\to\pm\infty} \frac{x}{h(x)} = 0. \tag{5.75}$$

3. *The inequality*

$$\limsup_{t\to+\infty} (\tau(t))^{n-1} \int_t^{+\infty} a(u)g\left(c(\tau(u))^{n-1}\right) du > 0 \tag{5.76}$$

holds for every $c \geq 1$.

Then the following conclusions hold:

1. *If n is an even number, then equation (5.42) is oscillatory.*

2. *If n is an odd number and there exists a number $\beta \geq \alpha$ such that*

$$(\tau(t))^{n-1} \int_\beta^{+\infty} a(s)g\left(c(\tau(s))^{n-1}\right) ds = +\infty, \tag{5.77}$$

then every solution $x(t)$ of equation (5.42) is either oscillatory or

$$\lim_{t\to+\infty} x^{(i)}(t) = 0, \quad i = 0, 1, \ldots, n-1.$$

Proof. Consider the following two cases:

Case 1. Let n be an odd integer. Assume $x(t)$ is a nonoscillatory solution of differential equation with "maxima" (5.42) in the interval $[\alpha, +\infty)$.

Case 1.1. Let $x(t)$ be a proper positive solution, i.e., there exists $T_x \geq \alpha$ such that $x(t) > 0$ for $t \geq T_x$. Let $t_0 = T_x + h$, where the constant h is defined by (5.43). Consider the function $y(t) = x(t) - \rho(t)$ for $t \geq t_0$ and proceeding as in the proof of Theorem 4.1.1, we conclude that there exist $k \in \{0, 1, \ldots, n-1\}$ with $n+k$ odd and $T_k \geq t_0$ such that (5.50) holds and the inequality (5.53) is fulfilled.

Case 1.1.1. Let $k > 0$. Then $y(t)$ and $y^{(k-1)}(t)$ are nondecreasing for $t \geq T_k$ and inequalities (5.62) and (5.58)-(5.60) hold for some $\mu \in (0, 1)$ and $T_\mu \geq T_k$. Taking into account (5.58), (5.59) and (5.62) we obtain

$$\max_{s \in I(u)} x(s) \geq \frac{\mu^k}{k!} \left(\tau(u)\right)^{k-1} y^{(k-1)}(\tau(u)), \quad u \geq T_\mu. \tag{5.78}$$

From inequalities (5.56) and (5.78), and condition 2 of Theorem 5.3.3, we get

$$f\left(\max_{s \in I(u)} x(s)\right) \geq g\left(c(\tau(t))^{n-1}\right) h\left(\frac{\mu^k}{k!}\left(\tau(u)\right)^{k-1} y^{(k-1)}(\tau(u))\right)$$
$$\geq K^2 g\left(c(\tau(t))^{n-1}\right) h\left(\frac{\mu^k}{k!}\right) h\left((\tau(u))^{k-1}\right) h\left(y^{(k-1)}(\tau(u))\right), \quad u \geq T_\mu. \tag{5.79}$$

As in the proof of Theorem 5.3.1 from (5.55) it follows that

$$y^{(k-1)}(t) \geq \int_{T_k}^{t} \frac{(u - T_k)^{n-k}}{(n-k)!} a(u) f\left(\max_{v \in I(u)} x(v)\right) du$$
$$+ (t - T_k)^{n-k} \int_{t}^{+\infty} \frac{1}{(n-k)!} a(u) f\left(\max_{v \in I(u)} x(v)\right) du, \tag{5.80}$$
$$\geq (t - T_k)^{n-k} \int_{t}^{+\infty} \frac{1}{(n-k)!} a(u) f\left(\max_{v \in I(u)} x(v)\right) du \quad \text{for } t \geq T_\mu.$$

From inequalities (5.79) and (5.80), we obtain

$$y^{(k-1)}(t) \geq \frac{K^2}{(n-k)!} h\left(\frac{\mu^k}{k!}\right)(t-T_k)^{n-k}$$

$$\times \int_t^{+\infty} a(u)g\left(c(\tau(u))^{n-1}\right) h\left((\tau(u))^{k-1}\right)$$

$$\times h\left(y^{(k-1)}(\tau(u))\right) du$$

$$\geq M\frac{(t-T_k)^{n-1}}{t^{k-1}} \int_t^{+\infty} a(u)g\left(c(\tau(u))^{n-1}\right) h\left((\tau(u))^{k-1}\right)$$

$$\times h\left(y^{(k-1)}(\tau(u))\right) du \quad (5.81)$$

for $t \geq T_\mu$, where $M = \dfrac{K^2}{(n-k)!} h\left(\dfrac{\mu^k}{k!}\right)$.

Choose a number $T > T_\mu$ such that $t \geq T$ implies $\tau(t) \geq T_\mu$. Then from inequality (5.81) we get

$$y^{(k-1)}(\tau(t)) \geq M\frac{(\tau(t)-T_k)^{n-1}}{\tau(t)^{k-1}} \int_{\tau(t)}^{+\infty} a(u)g\left(c(\tau(u))^{n-1}\right) h\left((\tau(u))^{k-1}\right)$$

$$\times h\left(y^{(k-1)}(\tau(u))\right) du$$

$$\geq M_1\frac{(\tau(t))^{n-1}}{\tau(t)^{k-1}} \int_t^{+\infty} a(u)g\left(c(\tau(u))^{n-1}\right) h\left((\tau(u))^{k-1}\right)$$

$$\times h\left(y^{(k-1)}(\tau(u))\right) du \quad \text{for } t \geq T,$$

$$(5.82)$$

where $M_1 = \mu^{n-1}M = \dfrac{\mu^{n-1}K^2}{(n-k)!} h\left(\dfrac{\mu^k}{k!}\right)$.

Since $\tau(u)$, $y^{(k-1)}(u)$, and $h(u)$ are nondecreasing functions for $u \geq t$, we obtain the inequalities

$$h((\tau(u))^{k-1}) \geq h((\tau(t))^{k-1})$$

and

$$h\left(y^{(k-1)}(\tau(u))\right) \geq h\left(y^{(k-1)}(\tau(t))\right), \quad u \geq t \geq T.$$

Therefore inequality (5.82) implies

$$\frac{y^{(k-1)}(\tau(t))}{M_1 h\left(y^{(k-1)}(\tau(t))\right)} \frac{(\tau(t))^{k-1}}{h\left((\tau(t))^{k-1}\right)} \geq$$

$$(\tau(t))^{n-1} \int_t^{+\infty} a(u)g\left(c(\tau(u))^{n-1}\right) du \quad \text{for} \quad t \geq T. \tag{5.83}$$

According to [Foster and Crimer 1980], Theorem 2, $\lim_{t\to+\infty} y^{(k-1)}(t) = +\infty$ holds.

Then from inequality (5.75), we get

$$\lim_{t\to+\infty} \frac{y^{(k-1)}(\tau(t))}{h(y^{(k-1)}(\tau(t)))} = 0 \quad \text{and} \quad \lim_{t\to+\infty} \frac{(\tau(t))^{k-1}}{h((\tau(t))^{k-1})} = 0. \tag{5.84}$$

From equality (5.84) and inequality (5.83) we obtain

$$0 \geq \limsup_{t\to+\infty}(\tau(t))^{n-1} \int_t^{+\infty} a(u)g\left(c(\tau(u))^{n-1}\right) du. \tag{5.85}$$

The inequality (5.85) contradicts (5.76).

Case 1.1.2. Let $k = 0$. Then $y(t)$ is a nonincreasing function. From (5.80) for $k = 1$, it follows that

$$y(t) \geq (t - T_k)^{n-1} \int_t^{+\infty} \frac{1}{(n-1)!} a(u)f\left(\max_{v \in I(u)} x(v)\right) du \quad \text{for} \quad t \geq T_\mu. \tag{5.86}$$

We shall prove that $\lim_{t\to+\infty} y(t) = 0$. Assume the opposite: there exists $d : 0 < d < \infty$ such that $\lim_{t\to+\infty} y(t) = d$. Then there exists a constant $d_1 > 0$ and $T_* \geq T_\mu$ such that $y(\tau(u)) \geq d_1$ for $t \geq T_*$. Then from (5.79) it follows the inequality

$$y(t) \geq (t - T_k)^{n-1} \int_t^{+\infty} \frac{1}{(n-1)!} a(u)g\left(c(\tau(u))^{n-1}\right) h\left(\mu y(\tau(u))\right) du$$

$$\geq \frac{\mu^{n-1} h(\mu d)}{(n-1)!}(\tau(t))^{n-1} \int_t^{+\infty} a(u)g\left(c(\tau(u))^{n-1}\right) du,$$

which contradicts (5.77). Thus $\lim\limits_{t\to+\infty} y(t) = 0$.

The rest of the proof is the same as the proof of Theorem 5.3.1.

□

Now we will illustrate the above-obtained sufficient conditions as an example.

Example 5.3.1. *Consider the following scalar second order differential equation with "maxima"*

$$x''(t) + a \left(\max_{s\in[t-1,t]} x(s) \right) = e^{-t}, \quad t \geq a, \tag{5.87}$$

where a = const > 0.

In this case the functions $\sigma(t) = t-1$, $\tau(t) = t$ satisfy the condition H5.3.1, the function $f(x) = x$ satisfies the condition H5.3.3 and the function $\rho(t) = e^{-t}$ satisfies the condition H5.3.4. The function $g = \frac{1}{x^2}$, $g \in C(\mathbb{R}_0, (0, +\infty))$ is nondecreasing on $(-\infty, 0)$ and nonincreasing on $(0, +\infty)$, the function $h = x^3$, $h \in C(\mathbb{R}_0, \mathbb{R})$ is nondecreasing on the interval $\mathbb{R}_0$ and it satisfies the inequality (5.74) with a constant $K = 1 > 0$ and $\lim\limits_{x\to\pm\infty} \dfrac{x}{h(x)} = \lim\limits_{x\to\pm\infty} \dfrac{1}{x^2} = 0$.

In this case condition (5.76) reduces to

$$\limsup_{t\to+\infty} t \int_{t}^{+\infty} a\frac{c}{u^2} du = act(\frac{1}{t}) = ac > 0.$$

Then according to Theorem 5.3.3, equation (5.87) is oscillatory.

5.4 Oscillations and Almost Oscillations of n-th Order Differential Equations with "Maxima"

Consider the scalar n-th order differential equation with "maxima"

$$L_n x(t) + \delta F\left(t, \max_{s\in I(t)} x(s)\right) = 0 \quad \text{for} \ \ t \geq a, \tag{5.88}$$

where $x \in \mathbb{R}$, $n \geq 2$ is an integer, $\delta = \pm 1$, $\alpha \geq 0$, $I(t) = \left[\sigma(t), \tau(t)\right]$,
$r_k : \mathbb{R} \to \mathbb{R}, (k = 1, \ldots, n)$,

$$L_0 x(t) = x(t), \qquad L_k x(t) = r_k(t)\left(L_{k-1} x(t)\right)', \qquad (k = 1, \ldots, n).$$

Let $J = [\alpha, +\infty) \subseteq \mathbb{R}_+ = [0, +\infty)$.
Introduce the following set (**H5.4**) of conditions:

H5.4.1. $r_i \in C(J, (0, +\infty))$, $i = 1, \ldots, n-1$, $r_n \equiv 1$ and

$$\int^{\infty} \frac{ds}{r_i(s)} = +\infty, \quad i = 1, \ldots, n-1.$$

H5.4.2. $F \in C(J \times \mathbb{R}, \mathbb{R})$ and there exist functions $q \in C(J, \mathbb{R}_+)$ and $f \in C(\mathbb{R}, \mathbb{R})$ such that $xf(x) > 0$ for $x \neq 0$, $f(x)$ is nondecreasing on $\mathbb{R}$ and

$$F(t, x) \operatorname{sgn} x \geq q(t) f(x) \operatorname{sgn} x, \quad t \in J, \ x \in \mathbb{R}.$$

H5.4.3. $\sigma, \tau \in C(J, \mathbb{R})$, $\sigma(t) \leq \tau(t) \leq t$, $\lim_{t \to +\infty} \sigma(t) = +\infty$ and

$$\sup_{t \geq \alpha}\left(t - \sigma(t)\right) = h < +\infty. \tag{5.89}$$

The domain $D(L_n)$ of L_n is defined to be the set of all functions $x :$ $[t_0, +\infty) \to \mathbb{R}$ such that $L_k x(t)$, $k = 1, \ldots, n$ exist and are continuous on some interval $[t_0, +\infty) \subseteq J$.
Let $T \in J$ be a given number.

Definition 5.4.1. *The function $x \in D(L_n)$ is called to be a solution of differential equation with "maxima" (5.88) on the interval $[T, +\infty)$, $T \geq \alpha$ if $x(t)$ is defined for $t \geq T - h$ and it satisfies (5.88) for $t \geq T$, where the constant h is defined by (5.89).*

Definition 5.4.2. *The solution $x(t)$ of differential equation with "maxima" (5.88) in the interval $[T, +\infty)$ is said to be:*

1. A proper solution if there exists a number $T_x \geq T$ such that

$$\sup\{|x(t)| : t \geq T_1\} > 0 \quad \text{for all} \quad T_1 \geq T_x.$$

2. *Nonoscillatory solution, if it is a proper solution and it is either positive or negative for $t \geq T_x$.*

3. *Oscillatory solution, if it is a proper solution and there is an infinite number of points on $[T, +\infty)$ at which the solution changes its sign.*

We denote the set of all proper solutions, all oscillatory solutions and all nonoscillatory solutions of (5.88) by S, O and N, respectively. It is clear that $S = O \cup N$. According to conditions (H5.4) the set N has a decomposition such that (see [Trench 1975]):

$$\begin{aligned}
N &= N_1 \cup N_3 \cup \ldots \cup N_{n-1} && \text{if } \delta = 1 \text{ and } n \text{ is even,} \\
N &= N_0 \cup N_2 \cup \ldots \cup N_{n-1} && \text{if } \delta = 1 \text{ and } n \text{ is odd,} \\
N &= N_0 \cup N_2 \cup \ldots \cup N_n && \text{if } \delta = -1 \text{ and } n \text{ is even,} \\
N &= N_1 \cup N_3 \cup \ldots \cup N_n && \text{if } \delta = -1 \text{ and } n \text{ is odd,}
\end{aligned}$$

where $N_k \subset N$: consisting of all x satisfying

$$\begin{aligned}
N_k = \{x \in \mathbb{R} : \ &xL_ix > 0, \ i = 0, 1, \ldots, k \\
&\text{and } (-1)^{i-k}xL_ix > 0, \ i = k, \ldots, n \text{ on } [T_x, +\infty)\}.
\end{aligned} \tag{5.90}$$

Definition 5.4.3. *Differential equation with "maxima" (5.88) is said to be almost oscillatory, if:*

(i) for $\delta = 1$ and n even, equation (5.88) is oscillatory;

(ii) for $\delta = 1$ and n odd, every proper solution x of equation (5.88) is either oscillatory or strongly decreasing, i.e.,

$$\lim_{t \to +\infty} |L_ix(t)| = 0 \text{ monotonically, } i = 0, \ldots, n-1; \tag{5.91}$$

(iii) for $\delta = -1$ and n even, every proper solution x of equation (5.88) is oscillatory, strongly decreasing or strongly increasing, i.e.,

$$\lim_{t \to +\infty} |L_ix(t)| = +\infty \text{ monotonically, } i = 0, \ldots, n-1; \tag{5.92}$$

(iv) for $\delta = -1$ and n odd, every proper solution x of equation (5.88) is either oscillatory or strongly increasing.

We use the following notations:

$$J_0 = 1,$$

$$J_j(t, s; p_j, \ldots, p_1) = \int_s^t \frac{1}{p_j(u)} J_{j-1}(u, s; p_{j-1}, \ldots, p_1) du,$$

$$j = 1, 2, \ldots,$$

where $p_j \in C(J, (0, +\infty))$, $j = 1, 2, \ldots$.
It is easy to verify that for $j = 1, 2, \ldots, n-1$

$$J_j(t, s; p_j, \ldots, p_1) = (-1)^j J_j(s, t; p_1, \ldots, p_j),$$

$$J_j(t, s; p_j, \ldots, p_1) = \int_s^t \frac{1}{p_1(u)} J_{j-1}(t, u; p_j, \ldots, p_2) du. \qquad (5.93)$$

$$H_k(\sigma(t), T) = \int_T^{\sigma(t)} \frac{1}{r_k(s)} J_{k-1}(\sigma(t), s; r_1, \ldots, r_{k-1})$$
$$J_{n-k-1}(t, s; r_{n-1}, \ldots, r_{k+1}) ds$$

and

$$H_k[\sigma(t)] = H_k(\sigma(t), \alpha)$$

for $T \geq \alpha$, $k = 1, \ldots, n-1$.
It is easy to verify that for any fixed $T \geq \alpha$ there exist positive constants c_1 and c_2 such that

$$c_1 H_k[\sigma(t)] \leq H_k(\sigma(t), T) \leq c_2 H_k[\sigma(t)] \qquad (5.94)$$

for all sufficiently large t.
We need the following lemma:

Lemma 5.4.1. *If $x \in D(L_n)$ then for t, $s \in J$ and $0 \leq i < \nu \leq n$ the following equalities*

$$L_i x(t) = \sum_{j=i}^{\nu-1} J_{j-i}(t, s; r_{i+1}, \ldots, r_j) L_j x(s)$$
$$+ \int_s^t J_{\nu-i-1}(t, u; r_{i+1}, \ldots, r_{\nu-1}) \frac{L_\nu x(t)}{r_\nu(u)} du, \qquad (5.95)$$

$$L_i x(t) = \sum_{j=i}^{\nu-1} (-1)^{j-i} J_{j-i}(s,t;\ r_j,\ldots,\ r_{i+1}) L_j x(s)$$

$$+ (-1)^{\nu-i} \int_t^s J_{\nu-i-1}(u,t;\ r_{\nu-1},\ldots,\ r_{i+1}) \frac{L_\nu x(u)}{r_\nu(u)} du \qquad (5.96)$$

hold.

This lemma is a generalization of Taylor's formula with remainder encountered in calculus.

Consider the differential inequalities with "maxima"

$$\delta L_n x(t) + F\left(t,\ \max_{s\in I(t)} x(s)\right) \leq 0 \qquad (5.97)$$

and

$$\delta L_n x(t) + F\left(t,\ \max_{s\in I(t)} x(s)\right) \geq 0. \qquad (5.98)$$

We introduce the notations:

$$N^+ = \{x:\ x \text{ is a positive solution } (5.97) \text{ on } [T_x,+\infty) \text{ for some } T_x\},$$
$$N^- = \{x:\ x \text{ is a negative solution } (5.98) \text{ on } [T_x,+\infty) \text{ for some } T_x\},$$

and

$$N_k^\pm = \{x \in N^\pm :\ x \text{ satisfies } (5.90) \text{ for some } T_x\}.$$

$N^\pm$ has a decomposition such that:

$$\begin{aligned}
N^\pm &= N_1^\pm \cup N_3^\pm \cup \ldots \cup N_{n-1}^\pm && \text{if } \delta = 1 \text{ and } n \text{ is even,} \\
N^\pm &= N_0^\pm \cup N_2^\pm \cup \ldots \cup N_{n-1}^\pm && \text{if } \delta = 1 \text{ and } n \text{ is odd,} \\
N^\pm &= N_0^\pm \cup N_2^\pm \cup \ldots \cup N_n^\pm && \text{if } \delta = -1 \text{ and } n \text{ is even,} \\
N^\pm &= N_1^\pm \cup N_3^\pm \cup \ldots \cup N_n^\pm && \text{if } \delta = -1 \text{ and } n \text{ is odd.}
\end{aligned}$$

Theorem 5.4.1. Let conditions (H5.4) hold, $(-1)^{n-k}\delta = -1$ for $1 \leq k \leq n-1$, and

$$\int^{+\infty} H_k[\sigma(t)] q(t) dt = +\infty. \qquad (5.99)$$

Then the following equalities

$$N_k^+ = \emptyset, \text{ if } \int^{+\infty} \frac{dx}{f(x)} < +\infty, \qquad (5.100)$$

$$N_k^- = \emptyset, \text{ if } \int^{-\infty} \frac{dx}{f(x)} < +\infty \qquad (5.101)$$

hold.

Proof. We will prove the validity of (5.100). Assume that N_k^+ has an element x: $x(t) > 0$, $t \geq t_0 \geq \alpha$. Then there exists $t_k \geq t_0$ such that

$$L_i x(t) > 0, \quad i = 0, \ldots, k, \quad t \geq t_k, \tag{5.102}$$

$$(-1)^{i-k} L_i x(t) > 0, \quad i = k, \ldots, n, \quad t \geq t_k.$$

From condition H5.4.3, it follows that there exist $t_1 \geq t_k$ such that $\sigma(t) \geq t_k$ and $max_{s \in I(t)} x(s) > 0$ for $t \geq t_1$.

Choose $t_2 \geq t_1$ so large that

$$\sigma(t) > t_1, \quad t \geq t_2. \tag{5.103}$$

We fix $t_3 (t_3 \geq t_2)$ arbitrarily and choose $T \geq t_3$ so that

$$T \geq \sigma(t), \quad t_1 \leq t \leq t_3. \tag{5.104}$$

From equality (5.101) and inequality (5.102) we obtain

$$
\begin{aligned}
L_k x(t) &= \sum_{j=k}^{n-1} (-1)^{j-k} J_{j-k}(T, t; \; r_j, \ldots, r_{k+1}) L_j x(T) \\
&\quad + (-1)^{n-k} \int_t^T J_{n-k-1}(u, t; \; r_{n-1}, \ldots, r_{k+1}) L_n x(u) \, du \\
&\geq \int_t^T J_{n-k-1}(u, t; \; r_{n-1}, \ldots, r_{k+1})(-1)^{n-k} L_n x(u) \, du, \\
&\qquad\qquad\qquad\qquad\qquad t_1 \leq t \leq T.
\end{aligned}
$$

Since

$$(-1)^{n-k} L_n x(u) = -\delta L_n x(u) \geq q(u) f \left(\max_{s \in I(u)} x(s) \right), \quad t_1 \leq u$$

we have

$$L_k x(t) \geq \int_t^T J_{n-k-1}(u, t; \; r_{n-1}, \ldots, r_{k+1}) q(u) f \left(\max_{s \in I(u)} x(s) \right) du,$$
$$t_1 \leq t \leq T.$$

For $k \geq 1$ and $t \geq t_k$ the function $x(t)$ is increasing and hence the inequality $max_{s \in I(u)} x(s) \geq x(\sigma(u))$ holds for $u \geq t_1$. Then

$$L_k x(t) \geq \int_t^T I_{n-k-1}(u, t; \; r_{n-1}, \ldots, r_{k+1}) q(u) f \left(x(\sigma(u)) \right) du,$$
$$t_1 \leq t \leq T. \tag{5.105}$$

Let $k \geq 2$. Then from equality (5.100) and inequality (5.102), we have

$$
\begin{aligned}
x'(v) &= \frac{1}{r_1(v)} L_1 x(v) \\
&= \frac{1}{r_1(v)} \left\{ \sum_{j=1}^{k-1} J_{j-1}\big(v, t_1; \, r_2, \ldots, \, r_j\big) L_j x(t_1) \right. \\
&\quad \left. + \int_{t_1}^{v} J_{k-2}\big(v, t; \, r_2, \ldots, \, r_{k-1}\big) \frac{L_k x(t)}{r_k(t)} dt \right\} \\
&\geq \frac{1}{r_1(v)} \int_{t_1}^{v} J_{k-2}\big(v, t; \, r_2, \ldots, \, r_{k-1}\big) \frac{L_k x(t)}{r_k(t)} dt, \quad v \geq t_1. \qquad (5.106)
\end{aligned}
$$

From (5.105) and (5.106), after integrating on $[t_1, T]$ we obtain

$$
\int_{t_1}^{T} \frac{x'(v)}{f(x(v))} dv \geq \int_{t_1}^{T} \int_{t_1}^{v} \int_{t}^{T} \psi(v, t, u) du \, dt \, dv
$$

where

$$
\begin{aligned}
\psi(v, t, u) &= \frac{q(u) f\big(x(\sigma(u))\big)}{r_1(v) r_k(t) f(x(v))} J_{k-2}\big(v, t; \, r_2, \ldots, \, r_{k-1}\big) \\
&\quad J_{n-k-1}\big(u, t; \, r_{n-1}, \ldots, \, r_{k+1}\big).
\end{aligned}
$$

Interchanging the order of integration and taking into account (5.103) and (5.104) we get

$$
\begin{aligned}
\int_{x(t_1)}^{x(T)} \frac{dx}{f(x)} \\
\geq \int_{t_1}^{T} \int_{t_1}^{v} \int_{t}^{T} \psi(v, t, u) du \, dt \, dv &= \int_{t_1}^{T} \int_{t}^{T} \int_{t}^{T} \psi(v, t, u) du \, dv \, dt \\
= \int_{t_1}^{T} \int_{t}^{T} \int_{t}^{T} \psi(v, t, u) dv \, du \, dt &= \int_{t_1}^{T} \int_{t_1}^{u} \int_{t}^{T} \psi(v, t, u) dv \, dt \, du \\
\geq \int_{t_2}^{t_3} \int_{t_1}^{\sigma(u)} \int_{t}^{\sigma(u)} \psi(v, t, u) dv \, dt \, du.
\end{aligned}
$$

Taking into account the inequality

$$
\frac{f\big(x(\sigma(u))\big)}{f(x(v))} \geq 1, \quad t_1 \leq v \leq \sigma(u),
$$

which is a consequence of the increasing nature of f and x and using (5.93) and (5.94) we obtain

$$\int_{x(t_1)}^{x(T)} \frac{dx}{f(x)} \geq \int_{t_2}^{t_3} q(u) \int_{t_1}^{\sigma(u)} \frac{J_{n-k-1}\left(u,t;\; r_{n-1},\dots,\; r_{k+1}\right)}{r_k(t)}$$

$$\times \int_{t}^{\sigma(u)} \frac{J_{k-2}\left(v,t;\; r_2,\dots,\; r_{k-1}\right)}{r_1(v)}\, dv\, dt\, du$$

$$= \int_{t_2}^{t_3} q(u) \int_{t_1}^{\sigma_*(u)} \frac{1}{r_k(t)} J_{n-k-1}\left(u,t;\; r_{n-1},\dots,\; r_{k+1}\right)$$

$$\times J_{k-1}\left(\sigma(u),t;\; r_1,\dots,\; r_{k-1}\right) dt\, du$$

$$\geq c_1 \int_{t_2}^{t_3} q(u) H\left[\sigma(u)\right] du.$$

Letting $t_3 \to +\infty$ in the above inequality and using (5.100) we conclude that

$$c_1 \int_{t_2}^{t_3} q(u) H\left[\sigma(u)\right] du \leq \int_{x(t_1)}^{+\infty} \frac{dx}{f(x)} < +\infty.$$

This contradicts the assumption (5.99).

Let $k = 1$. Then from (5.105) we have

$$x'(t) = \frac{L_1 x(t)}{r_1(t)}$$

$$\geq \frac{1}{r_1(t)} \int_{t}^{T} J_{n-2}\left(u,t;\; r_{n-1},\dots,\; r_2\right) q(u) f\left(x(\sigma(u))\right) du, \quad t_1 \leq t \leq T.$$

Dividing the above inequality by $f(x(t))$ and integrating from t_1 to T, we get

$$\int_{t_1}^{T} \frac{x'(t)}{f(x(t))}\, dt \geq \int_{t_1}^{T} \int_{t}^{T} \tilde{\psi}(t,u) du\, dt$$

where $\tilde{\psi}(t,u) = \frac{q(u)}{r_1(t)} J_{n-2}\left(u,t;\; r_{n-1},\dots,\; r_2\right) \frac{f\left(x(\sigma(u))\right)}{f(x(t))}$.

It follows that

$$\int_{x(t_1)}^{x(T)} \frac{dx}{f(x)} \geq \int_{t_1}^{T} \int_{t}^{T} \tilde{\psi}(t,u) du dt = \int_{t_1}^{T} \int_{t_1}^{u} \tilde{\psi}(t,u) dt du$$

$$\geq \int_{t_2}^{t_3} \int_{t_1}^{\sigma(u)} \tilde{\psi}(t,u) dt\, du.$$

Using the inequality

$$\frac{f\big(x(\sigma(u))\big)}{f\big(x(t)\big)} \geq 1, \quad t_1 \leq t \leq \sigma(u)$$

we obtain

$$\int_{x(t_1)}^{x(T)} \frac{dx}{f(x)} \geq \int_{t_2}^{t_3} q(u) \int_{t_1}^{\sigma(u)} \frac{1}{r_1(t)} J_{n-2}\big(u,t; r_{n-1},\ldots, r_2\big)dt\, du$$

$$\geq c_1 \int_{t_2}^{t_3} q(u) H_1\big[\sigma(u)\big]du,$$

which gives in the limit as $t_3 \to +\infty$

$$c_1 \int_{t_2}^{+\infty} q(u) H_1\big[\sigma(u)\big]du \leq \int_{x(t_1)}^{+\infty} \frac{dx}{f(x)} < +\infty.$$

This contradicts (5.97).

If $x \in N_k^-$ is a negative solution of inequality (5.98), the proof is similar.

$\square$

Theorem 5.4.2. *Assume that conditions (H5.4) hold, $(-1)^{n-k}\delta = -1$ for $1 \leq k \leq n-1$, and*

$$\int^{\pm\infty} \frac{dx}{f(x)} < +\infty, \tag{5.107}$$

$$\int^{+\infty} H_k\big[\sigma(t)\big]q(t)dt = +\infty. \tag{5.108}$$

Then $N_k = \emptyset$ for $1 \leq k \leq n-1$, where N_k are defined by (5.90).

Proof. From Theorem 5.4.1 follows that $N_k^+ = \emptyset$ and $N_k^- = \emptyset$ for the inequalities

$$\delta L_n x(t) + q(t)f\big(\max_{s\in I(t)} x(s)\big) \leq 0 \tag{5.109}$$

and

$$\delta L_n x(t) + q(t)f\big(\max_{s\in I(t)} x(s)\big) \geq 0. \tag{5.110}$$

Assume that differential equation with "maxima" (5.88) has a positive solution $x \in N_k$, i.e., $x(t) > 0$ and $\max_{s\in I(t)} x(s) > 0$ for $t \geq t_0 > \alpha$. By condition H5.4.2 there exists $t_1 \geq t_0$ such that

$$F\big(t, \max_{s\in I(t)} x(s)\big) \geq q(t)f\big(\max_{s\in I(t)} x(s)\big), \quad t \geq t_1.$$

Therefore

$$0 = \delta L_n x(t) + F\left(t, \max_{s \in I(t)} x(s)\right)$$

$$\geq \delta L_n x(t) + q(t) f\left(\max_{s \in I(t)} x(s)\right), \quad t \geq t_1.$$

This means that inequality (5.109) has a positive solution $x \in N_k^+$, which leads to a contradiction.

If differential equation with "maxima" (5.88) has a negative solution $x \in N_k$, then

$$F\left(t, \max_{s \in I(t)} x(s)\right) \leq q(t) f\left(\max_{s \in I(t)} x(s)\right)$$

and analogously we obtain that the inequality (5.110) has a negative solution $x \in N_k^-$, which leads to a contradiction.

$\square$

Theorem 5.4.3. *Assume that conditions (H5.4) hold, $(-1)^n \delta = -1$ and*

$$\int^{+\infty} J_{n-1}\left(t, \alpha; \ r_{n-1}, \ldots, \ r_1\right) q(t) dt = +\infty. \tag{5.111}$$

Then every nonoscillatory solution x of equation (5.88), which is from N_0, satisfies (5.91).

Proof. Let $x \in N_0$ be a positive solution of equation (5.88) such that $x(t) > 0$, $\max_{s \in I(t)} x(s) > 0$ for $t \geq t_0 \geq \alpha$, $j = 1, \ldots, \ m$.
Then there exists $t_1 \geq t_0$ such that

$$(-1)^j L_j x(t) > 0, \quad j = 0, \ 1, \ldots, \ n, \quad t \geq t_0. \tag{5.112}$$

Since $x'(t) < 0$ the function $x(t)$ is decreasing and there exists the limit

$$\lim_{t \to +\infty} x(t) = c \geq 0 \quad \text{and} \quad x(t) \geq c, \quad t \geq t_1.$$

By equality (5.101) we have for $s \geq t_1$

$$x(t_1) = \sum_{j=0}^{n-1} (-1)^j J_j\left(s, t_1; \ r_j, \ldots, \ r_1\right) L_j x(s)$$

$$+ (-1)^n \int_{t_1}^{s} J_{n-1}\left(u, t_1; \ r_{n-1}, \ldots, \ r_1\right) L_n x(u) du.$$

Using (5.88) and (5.112) and condition H5.4.2, we get

$$x(t_1) \geq \int_{t_1}^{s} J_{n-1}(u, t_1; \ r_{n-1}, \ldots, \ r_1) q(u) f\left(\max_{s \in I(u)} x(s)\right) du$$

$$\geq \int_{t_1}^{s} J_{n-1}(u, t_1; \ r_{n-1}, \ldots, \ r_1) q(u) du \ f(c).$$

If $c > 0$, then letting $s \to +\infty$ and using (5.111) we get $x(t_1) = +\infty$. The obtained contradiction proves $c = 0$. Now using (5.112) we can show that x satisfies (5.91).

If $x \in N_0$ is a negative solution of equation (5.88), the proof is similar.

$\square$

We set

$$A(t, s) = J_{n-1}(t, s; \ r_1, \ldots, \ r_{n-1}), \quad A[t] = A(t, \alpha) \text{ for } t \geq s \geq \alpha.$$

Theorem 5.4.4. *Assume that conditions (H5.4) hold, $\delta = -1$ and*

$$\int^{+\infty} q(t) \left| f\left(cA\left[\sigma_j(t)\right]\right) \right| dt = +\infty \tag{5.113}$$

for all $c \neq 0$.

Then every nonoscillatory solution x of equation (5.88) belonging to N_n satisfies (5.92).

Proof. Assume that $x \in N_n$ is a positive solution of equation (5.88): $x(t) > 0$, $\max_{s \in I(t)} x(s) > 0$ for $t \geq t_0 \geq \alpha$. Then there exists $t_1 \geq t_0$ such that

$$L_i x(t) > 0, \quad t \geq t_1, \quad i = 0, 1, \ldots, n. \tag{5.114}$$

On the other hand, by L'Hôpital's rule

$$\lim_{t \to +\infty} \frac{x(t)}{J(t, t_1)} = \lim_{t \to +\infty} L_{n-1} x(t) > 0.$$

Since $\lim_{t \to +\infty} \sigma(t) = +\infty$ there exist a constant $c > 0$ and a $t_2 \geq t_1$ such that

$$\max_{s \in I(t)} x(s) \geq x(\sigma(t)) \geq cA(t, t_1), \quad t \geq t_2. \tag{5.115}$$

Integrating equation (5.88) (with $\delta = -1$) from t_2 to t and using (5.115) and condition H5.4.2 we have

$$
\begin{aligned}
L_{n-1}x(t) &= L_{n-1}x(t_2) + \int_{t_2}^{t} F\big(\max_{\xi \in I(s)} x(\xi)\big)ds \\
&\geq \int_{t_2}^{t} q(s)f\big(\max_{\xi \in I(s)} x(\xi)\big)ds \\
&\geq \int_{t_2}^{t} q(s)f\big(cA(s,t_1)\big)ds.
\end{aligned}
$$

Using (5.113) we get $\lim_{t \to +\infty} L_{n-1}x(t) = +\infty$. By (5.114) it follows that $L_j x(t)$ are strongly increasing and $\lim_{t \to +\infty} L_i x(t) = +\infty$, $i = 0, 1, \ldots, n-1$.

If $x \in N_n$ is a negative solution of equation (5.88), the proof is similar.

$\square$

Theorem 5.4.5. *Assume that conditions (H5.4) hold. Sufficient conditions for (5.88) to be almost oscillatory are:*

(i) *when $\delta = 1$ and n is even and (5.107) and (5.108) for $k = 1, 3, \ldots, n-1$ hold;*

(ii) *when $\delta = 1$ and n is odd, (5.107) and (5.108) hold for $k = 2, 4, \ldots, n-1$ and (5.111) is satisfied;*

(iii) *when $\delta = -1$ and n is even, (5.107) and (5.108) hold for $k = 2, 4, \ldots, n-2$ and (5.111) and (5.113) are satisfied;*

(iv) *when $\delta = -1$ and n is odd, (5.107) and (5.108) hold for $k = 1, 3, \ldots, n-2$ and (5.111) is satisfied.*

Proof. (i) By Theorem 5.4.2 we have $N = N_1 \cup N_3 \cup \ldots \cup N_{n-1} = \emptyset$. Hence $S = O$.

(ii) Theorem 5.4.2 implies that $S = O \cup N_0$. Assume that $x \in N_0$ is a positive solution of equation (5.88): $x(t) > 0$, $\max_{s \in I(t)} x(s) > 0$ for $t \geq t_0 \geq \alpha$, $j = 1, \ldots, m$. Then (5.112) holds for $t_1 \geq t_0$. Thus x is decreasing and $x(t)$ has nonnegative limit as $t \to +\infty$.

By Theorem 5.4.3 this limit must be 0 under the condition (5.111) and x satisfies (5.91).

If $x \in N_0$ is a negative solution of equation (5.88) the proof is similar.

(*iii*) From Theorem 5.4.2 it follows that $S = O \cup N_0 \cup N_n$.

Assume that $x \in N_n$ is a positive solution of equation (5.88), i.e., $x(t) > 0$, $\max_{s \in I(t)} x(s) > 0$ for $t \geq t_0 \geq \alpha$, $j = 1, \ldots, m$. Then (5.114) holds for $t \geq t_1 \geq t_0$, so that x is increasing and tends to a finite or infinite limit as $t \to +\infty$. By Theorem 5.4.4 this limit must be infinite under the condition (5.113) and $x(t)$ satisfies (5.92). If $x \in N_n$ is a negative solution of equation (5.88) the proof is similar.

If $x \in N_0$ then we prove as in (*ii*) that x satisfies (5.91).

(*iv*) We have $S = O \cup N_n$ by Theorem 5.4.2. Exactly as above we can show that a solution belonging to N_n is strongly increasing.

□

5.5   Oscillations of Differential Inequalities with "Maxima"

We consider the n-th order differential inequalities with "maxima"

$$(-1)^n L_n x(t) \operatorname{sgn} x(t) \geq \sum_{j=1}^{m} p_j(t) f_j\big(M_t^j x\big) \qquad (5.116)$$

and

$$(-1)^n L_n x(t) \operatorname{sgn} x(t) \geq p_0(t) \prod_{j=1}^{m} F_j\big(M_t^j x\big) \qquad (5.117)$$

where $n \geq 1$, $M_t^j x = \max_{s \in J_j(t)} x(s)$, $J_j(t) = [\sigma_j(t), \tau_j(t)]$, $j = 1, \ldots, m$, $\sigma_j(t) \leq \tau_j(t) \leq t$, $j = 1, \ldots, m$, $t \in J = [\alpha, +\infty) \subseteq \mathbb{R}_+ = [0, +\infty)$ and $L_0 x(t) = x(t)$, $L_k x(t) = r_k(t)\big(L_{k-1}x(t)\big)'$, $k = 1, \ldots, n$.

The domain $D(L_n)$ of L_n is defined to be the set of all functions $x :$ $[t_x, +\infty) \to \mathbb{R}$ such that $L_k x(t)$, $k = 1, \ldots, n$ exist and are continuous on an interval $[t_x, +\infty) \subseteq J$.

Definition 5.5.1. *The function $x \in D(L_n)$ is called a* proper *solution of inequality (5.116) or (5.117) if it satisfies (5.116) or (5.117) for all sufficiently large t and $\sup_{t \geq T} |x(t)| > 0$ for $T \geq t_x$.*

We assume that inequality (5.116) or (5.117) does possess proper solutions.

Definition 5.5.2. *A proper solution of inequality (5.116) or (5.117) is called* oscillatory *if it has arbitrarily large zeros.*

Otherwise it is called nonoscillatory.

Let $J = [\alpha, +\infty) \subseteq \mathbb{R}_+ = [0, +\infty)$.
Introduce the following set **(H5.5)** of conditions:

$(H5.5.1)$ $r_i \in C(J, (0, +\infty))$, $i = 1, \ldots, n$, $r_n \equiv 1$ and $\int^\infty \frac{ds}{r_i(s)} = +\infty$,
$i = 1, \ldots, n-1$.

$(H5.5.2)$ $p_j \in C(J, (0, +\infty))$, $j = 0, \ldots, m$.

$(H5.5.3)$ f_j, $F_j \in C(\mathbb{R}, \mathbb{R})$, $j = 1, \ldots, m$, $x f_j(x) > 0$ for $x \neq 0$,
$j = 1, \ldots, m$, $x F_j(x) > 0$ for $x \neq 0$, $j = 1, \ldots, m$; f_j and F_j
are nondecreasing, $j = 1, \ldots, m$.

$(H5.5.4)$ Functions σ_j, $\tau_j \in C(J, \mathbb{R})$, $j = 1, \ldots, m$ and

$$\lim_{t \to +\infty} \sigma_j(t) = +\infty, \ j = 1, \ldots, m,$$
$$\sigma_j(t) \leq \tau_j(t), \ j = 1, \ldots, m, \ t \in J.$$

$(H5.5.5)$ There exists one-to-one function $\tau : J \to \mathbb{R}$ such that $\tau_j(t) \leq$
$\tau(t) \leq t$, $j = 1, \ldots, m$, $t \in J$.

In the paper we use the following notations:

$$f(x) = \max_{1 \leq j \leq m} f_j(x), \quad F(x) = \prod_{j=1}^m F_j(x), \quad I_0 = 1,$$

$$I_j(t, s; p_j, \ldots, p_1) = \int_s^t \frac{1}{p_j(u)} I_{j-1}(u, s; p_{j-1}, \ldots, p_1) du, \ j = 1, 2, \ldots$$

where $p_j \in C(J, (0, +\infty))$, $j = 1, 2, \ldots$.
It is easy to verify that

$$I_j(t, s; p_j, \ldots, p_1) = (-1)^j I_j(s, t; p_1, \ldots, p_j),$$
$$I_j(t, s; p_j, \ldots, p_1) = \int_s^t \frac{1}{p_1(u)} I_{j-1}(t, u; p_j, \ldots, p_2) du.$$

We need the following lemma which generalized the well-known
lemma of T.T. Kiguradze [Kiguradze 1964].

Lemma 5.5.1. *Suppose condition H5.5.1 holds and the functions $L_n x$
and $x \in D(L_n)$ are of constant sign and not identically zero for $t \geq$
$t_* \geq \alpha$.*

Then there exist $t_k \geq t_*$ and an integer k, $0 \leq k \leq n$ with $n+k$ even for $x(t)L_n x(t)$ nonnegative or $n+k$ odd for $x(t)L_n x(t)$ nonpositive and such that for every $t \geq t_k$

$$x(t)L_i x(t) > 0, \qquad\qquad i = 0,\ 1, \ldots,\ k$$

$$(-1)^{i-k} x(t) L_i x(t) > 0, \quad i = k,\ k+1, \ldots,\ n.$$

Lemma 5.5.2. *If $x \in D(L_n)$ then for t, $s \in J$ and $0 \leq i < \nu \leq n$:*

$$^{\cdot}L_i x(t) \;\; = \;\; \sum_{j=i}^{\nu-1} (-1)^{j-i} I_{j-i}\big(s,t;\ r_j,\ldots,\ r_{i+1}\big) L_j x(s)$$

$$+ (-1)^{\nu-i} \int_t^s I_{\nu-i-1}\big(u,t;\ r_{\nu-1},\ldots,\ r_{i+1}\big) \frac{L_\nu x(u)}{r_\nu(u)} du.$$

This lemma is a generalization of Taylor's formula with a remainder encountered in calculus.

Theorem 5.5.1. *Assume that conditions (H5.5) hold and*

$$\lim_{t\to+\infty} \sup \int_{\tau(t)}^t I_{n-1}\big(u,\tau(t);\ r_{n-1},\ldots,\ r_1\big) \sum_{j=1}^m p_j(u)du > \lim_{x\to 0} \sup \frac{|x|}{f(x)}.$$

$$(5.118)$$

Then if $n \geq 2$ all proper bounded solutions of (5.116) are oscillatory, while if $n = 1$ all proper solutions of (5.116) are oscillatory.

Proof. Suppose there exists a bounded nonoscillatory solution of inequality (5.116).

Without loss of generality we may suppose that $x(t)$ is eventually positive: $x(t) > 0$, $t \geq t_0 \geq \alpha$. It follows from conditions $H5.5.2$, $H5.5.3$ and inequality (5.116) that $(-1)^n L_n x(t) \geq 0$, $t \geq t_1 \geq t_0$. By the boundedness of $x(t)$ and Lemma 5.5.1 it follows that there exists $t_2 \geq t_1$ such that

$$(-1)^i L_i x(t) > 0, \quad t \geq t_2, \quad i = 1,\ldots,\ n. \qquad (5.119)$$

From Lemma 5.5.2 with $\nu = n$ and $i = 0$ we have

$$x(t) = \sum_{j=0}^{n-1} (-1)^j I_j\big(s,t;\ r_j,\ldots,\ r_1\big) L_j x(s)$$

$$(5.120)$$

$$+ (-1)^n \int_t^s I_{n-1}\big(u,t;\ r_{n-1},\ldots,\ r_1\big) L_n x(u)du.$$

Using (5.119) and (5.116) from (5.120) we get

$$x(s) \geq x(t) + \int_s^t I_{n-1}(u, s;\ r_{n-1}, \ldots,\ r_1) \sum_{j=1}^m p_j(u) f_j(M_u^j x) du, \quad t \geq s.$$
$$(5.121)$$

By (5.119), $x'(t) \leq 0$ for $t \geq t_2$ so that $x(t)$ is decreasing for $t \geq t_2$. Then $M_t^j x \geq x(\tau_j(t)) \geq x(\tau(t))$ for $t \geq t_2$, $j = 1, \ldots, m$.

From (5.121) and the monotonicity of x, f_j, $j = 1, \ldots, m$ it follows that

$$x(s) \geq x(t) + \int_s^t I_{n-1}(u, s;\ r_{n-1}, \ldots,\ r_1) \sum_{j=1}^m p_j(u) f_j\Big(x(\tau(u))\Big) du$$

$$\geq x(t) + f\Big(x(\tau(t))\Big) \int_s^t I_{n-1}(u, s;\ r_{n-1}, \ldots,\ r_1) \sum_{j=1}^m p_j(u) du, \quad t \geq t_2.$$

Therefore

$$x(\tau(t)) \geq x(t) + f\Big(x(\tau(t))\Big) \int_{\tau(t)}^t I_{n-1}(u, \tau(t);\ r_{n-1}, \ldots,\ r_1) \sum_{j=1}^m p_j(u) du$$
$$(5.122)$$

for $t \geq t_3$, where t_3 is so large that $\tau(t) \geq t_2$ for $t \geq t_3$.

Since $x(t)$ is decreasing and positive for $t \geq t_2$ there exists $\lim_{t \to +\infty} x(t) = c \geq 0$. It follows from (5.118) and (5.122) that $c = 0$.

From (5.122) we find

$$\frac{x(\tau(t))}{f(x(\tau(t)))} \geq \int_{\tau(t)}^t I_{n-1}(u, \tau(t);\ r_{n-1}, \ldots,\ r_1) \sum_{j=1}^m p_j(u) du, \quad t \geq t_3.$$
$$(5.123)$$

Taking the limit superior as $t \to +\infty$ of both sides of (5.123), we obtain a contradiction to the hypothesis (5.118).

For $n = 1$ all solutions of (5.116) are oscillatory because every nonoscillatory solution of (5.116) is necessarily bounded. $\qquad \square$

In the same way we can prove the following theorem.

Theorem 5.5.2. *Assume that conditions (H5.5) hold and*

$$\lim_{t \to +\infty} \sup \int_{\tau(t)}^t I_{n-1}(u, \tau(t);\ r_{n-1}, \ldots,\ r_1) p_0(u) du > \lim_{x \to 0} \sup \frac{|x|}{F(x)}.$$
$$(5.124)$$

*Then if $n \geq 2$ all bounded solutions of inequality (5.117) are oscil-
latory and if $n = 1$ all solutions of (5.117) are oscillatory.*

Theorem 5.5.3. *Assume that conditions (H5.5) hold and*

$$\int_{+0}^{+a} \frac{dx}{f(x)} < +\infty, \quad \int_{-a}^{-0} \frac{dx}{f(x)} < +\infty \quad \text{for some } a > 0, \quad (5.125)$$

and

$$\int^{\infty} \frac{1}{r_1(s)} \left(\int_{s}^{\tau^{-1}(s)} I_{n-2}(u, s; \; r_{n-1}, \ldots, \; r_2) \sum_{j=1}^{m} p_j(u) du \right) ds = +\infty.$$

$$(5.126)$$

Then all bounded solutions of inequality (5.116) are oscillatory.

Proof. Let $x(t)$ be a bounded nonoscillatory solution of inequality
(5.116). Without loss of generality we assume that $x(t) > 0$. It
follows from conditions $H5.5.2$, $H5.5.3$ and inequality (5.116) that
$(-1)^n L_n x(t) \geq 0$, $t \geq t_1$ for some t_1. By the boundedness of $x(t)$
and Lemma 5.5.1, it follows that (5.119) holds for $t \geq t_2$ and t_2 is
sufficiently large. Applying Lemma 5.5.2 with $i = 1$, $\nu = n$ we have

$$L_1 x(t) \; = \; \sum_{j=1}^{n-1} (-1)^{j-1} I_{j-1}(s, t; \; r_j, \ldots, \; r_2) L_j x(s)$$

$$+ (-1)^{n-1} \int_{t}^{s} I_{n-2}(u, t; \; r_{n-1}, \ldots, \; r_2) L_n x(u) du.$$

Then

$$-r_1(t) x'(t) \; = \; \sum_{j=1}^{n-1} (-1)^j I_{j-1}(s, t; \; r_j, \ldots, \; r_2) L_j x(s)$$

$$+ (-1)^n \int_{t}^{s} I_{n-2}(u, t; \; r_{n-1}, \ldots, \; r_2) L_n x(u) du$$

and using (5.119) and (5.116) we get

$$-x'(s) \geq \frac{1}{r_1(s)} \int_{s}^{t} I_{n-2}(u, s; \; r_{n-1}, \ldots, \; r_2) \sum_{j=1}^{m} p_j(u) f_j(M_u^j x) du.$$

$$(5.127)$$

Since $x(t)$ is decreasing for $t \geq t_2$ and $M_t^j x \geq x(\tau_j(t)) \geq x(\tau(t))$, $j = 1, \ldots, m$, putting $t = \tau^{-1}(s)$ in (5.127) and taking the monotonicity of x, f_j, $j = 1, \ldots, m$ into account we obtain

$$-x'(s) \geq \frac{f(x(s))}{r_1(s)} \int_s^{\tau^{-1}(s)} I_{n-2}(u, s; \ r_{n-1}, \ldots, \ r_2) \sum_{j=1}^m p_j(u) du.$$

(5.128)

Dividing both sides of (5.128) by $f(x(s))$ and then integrating from t_3 to t where t_3 is so that $\tau(t) > t_2$ for $t \geq t_3$ we obtain

$$\int_{x(t)}^{x(t_3)} \frac{dx}{f(x)} \geq \int_{t_3}^t \frac{1}{r_1(s)} \left(\int_s^{\tau^{-1}(s)} I_{n-2}(u, s; \ r_{n-1}, \ldots, \ r_2) \sum_{j=1}^m p_j(u) du \right) ds.$$

(5.129)

It follows from (5.119) that $\lim_{t \to +\infty} x(t) = c \geq 0$. Therefore, by (5.125) the left-hand side of (5.129) remains bounded, while on account of (5.126) the right-hand side becomes unbounded as $t \to +\infty$. This is a contradiction.

□

Similarly, we obtain the following theorem.

Theorem 5.5.4. *Assume that conditions (H5.5) hold and*

$$\int_{+0}^{+a} \frac{dx}{F(x)} < +\infty, \quad \int_{-a}^{-0} \frac{dx}{F(x)} < +\infty \quad \text{for some } a > 0, \qquad (5.130)$$

and

$$\int^\infty \frac{1}{r_1(s)} \left(\int_s^{\tau^{-1}(s)} I_{n-2}(u, s; \ r_{n-1}, \ldots, \ r_2) p_0(u) du \right) ds = +\infty.$$

(5.131)

Then all bounded solutions of inequality (5.117) are oscillatory.

Chapter 6

Asymptotic Methods

Since the set of nonlinear problems, whose solutions could be presented as well-known functions, is too narrow, one needs to exploit various approximate methods. There are different analytic approximate methods applied to various types of differential equations.

In this chapter the monotone-iterative technique and method of quasilinearization are applied to initial value problems and periodic boundary value problems for differential equations with "maxima." Both considered methods combine the method of lower and upper solutions by an appropriate monotone method. The algorithm for constructing successive approximations is very simple and the conditions for the right side of the equations are natural. It makes the applications of both methods very successful to different types of differential equations (see the monograph [Ladde et al. 1985], and the references cited therein).

The monotone-iterative technique is studied for differential equations with "maxima" in [Bainov and Hristova 1995] and for impulsive differential equations with supremum in [Cai 2003], [Hristova and Roberts 2001], [Hristova and Bainov 1993], [Hristova and Bainov 1991], and [Qi and Chen 2008].

6.1 Monotone-Iterative Technique for Initial Value Problems

We will begin considerations with the initial value problem. In this section, we will study the application of the monotone-iterative technique to both a scalar case and a multidimensional case of differential equations with "maxima." We will define different types of lower and upper solutions of the studied system. We will give an algorithm for constructing two sequences of successive approximations of the solution of the considered problem.

Consider the following initial value problem for the system of differential equations with "maxima"

$$x' = f(t, x(t), \max_{s \in [t-h,t]} x(s)), \quad \text{for} \quad t \in [0, T], \tag{6.1}$$

$$x(t) = \varphi(t), \quad t \in [-h, 0], \tag{6.2}$$

where $x \in \mathbb{R}^n$, $f : [0, T] \times \mathbb{R}^n \times \mathbb{R}^n \to \mathbb{R}^n$, $T, h = const > 0$.

We will use notations that are analogous to those used in [Ladde et al. 1985] for systems of ordinary differential equations. These notations play an important role in the definitions of different types of lower and upper solutions for the systems of differential equations with "maxima."

Definition 6.1.1. Let $x = (x_1, x_2, \ldots, x_n)$, $y = (y_1, y_2, \ldots, y_n)$. We shall say that $x \leq (\geq) y$ if for each integer $j : 1 \leq j \leq n$ the inequalities $x_j \leq (\geq) y_j$ hold.

To each natural number $j : 1 \leq j \leq n$ we assign two nonnegative integers p_j and q_j such that $p_j + q_j = n - 1$ and for the points $x, y, z \in \mathbb{R}^n$ we introduce the notation

$$\left(z_j, [x]_{p_j}, [y]_{q_j}\right) = \begin{cases} \underbrace{(x_1, x_2, \ldots, x_{j-1}, z_j, x_{j+1}, \ldots, x_{p_j+1}}_{p_j}, \underbrace{y_{p_j+2}, \ldots, y_n)}_{q_j}, \\ \qquad \qquad \qquad \qquad \text{for } p_j > j, \\ \underbrace{(x_1, x_2, \ldots, x_{p_j}}_{p_j}, \underbrace{y_{p_j+1}, \ldots, y_{j-1}, z_j, y_{j+1}, \ldots, y_n)}_{q_j+1}, \\ \qquad \qquad \qquad \qquad \text{for } p_j < j, \\ \underbrace{(x_1, x_2, \ldots, x_{p_j}}_{p_j}, \underbrace{z_{j+1}, y_{p_j+1}, \ldots, y_{j-1}, , y_{j+1}, \ldots, y_n)}_{q_j}, \\ \qquad \qquad \qquad \qquad \text{for } p_j = j. \end{cases}$$

For example, let $n = 3$. Choose $p_1 = 2, q_1 = 0, p_2 = 1, q_2 = 1$ and $p_3 = 1, q_3 = 1$. Then $(x_1, [z]_{p_1}, [y]_{q_1}) = (x_1, z_2, z_3)$, $(x_2, [z]_{p_2}, [y]_{q_2}) = (z_1, x_2, y_3)$, $(x_3, [z]_{p_3}, [y]_{q_3}) = (z_1, y_2, x_3)$.

According to the above-introduced notation the initial value problems (6.1) and (6.2) can be rewritten in the form

$$x_j' = f_j\left(t, x_j(t), [x(t)]_{p_j}, [x(t)]_{q_j},\right.$$

$$\left.\max_{s\in[t-h,t]} x_j(s), [\max_{s\in[t-h,t]} x(s)]_{p_j}, [\max_{s\in[t-h,t]} x(s)]_{q_j}\right), \quad t \in [0, T],$$

(6.3)

$$x_j(t) = \varphi_j(t), \quad t \in [-h, 0], \quad j = 1, 2, \ldots, n. \tag{6.4}$$

Definition 6.1.2. *The pair of functions* $v, w \in C([-h, T], \mathbb{R}^n) \cup C^1([0, T], \mathbb{R}^n)$, $v = (v_1, v_2, \ldots, v_n)$, $w = (w_1, w_2, \ldots, w_n)$ *is called a pair of lower and upper quasisolutions of the initial value problem for the system of differential equations with "maxima" (6.1), (6.2) if*

$$v_j' \le f_j\left(t, v_j(t), [v(t)]_{p_j}, [w(t)]_{q_j},\right.$$

$$\left.\max_{s\in[t-h,t]} v_j(s), [\max_{s\in[t-h,t]} v(s)]_{p_j}, [\max_{s\in[t-h,t]} w(s)]_{q_j}\right), \tag{6.5}$$

$$w_j' \ge f_j\left(t, w_j(t), [w(t)]_{p_j}, [v(t)]_{q_j},\right.$$

$$\left.\max_{s\in[t-h,t]} w_j(s), [\max_{s\in[t-h,t]} w(s)]_{p_j}, [\max_{s\in[t-h,t]} v(s)]_{q_j}\right), \quad for \quad t \in [0, T],$$

$$v_j(t) \le \varphi_j(t), \quad w_j(t) \ge \varphi_j(t), \quad t \in [-h, 0], \quad j = 1, 2, \ldots, n. \tag{6.6}$$

Remark 6.1.1. *We will note that the pair of lower and upper quasisolutions is a generalization of the lower and upper solutions in the scalar case* $(n = 1, p_1 = q_1 = 0)$.

Definition 6.1.3. *The pair of functions* $v, w \in C([-h, T], \mathbb{R}^n) \cup C^1([0, T], \mathbb{R}^n)$, $v = (v_1, v_2, \ldots, v_n)$, $w = (w_1, w_2, \ldots, w_n)$ *is called a pair of quasisolutions of the initial value problem for the system of differential equations with "maxima" (6.1), (6.2) if (6.5), (6.6) are satisfied only for equalities.*

Definition 6.1.4. *The pair of functions* $v, w \in C([-h, T], \mathbb{R}^n) \cup C^1([0, T], \mathbb{R}^n)$, $v = (v_1, v_2, \ldots, v_n)$, $w = (w_1, w_2, \ldots, w_n)$ *is called a pair of minimal and maximal quasisolutions of the initial value problem for the system of differential equations with "maxima" (6.1), (6.2) if it is a pair of quasisolutions of the same problem,* $v(t) \le w(t)$ *and for any other pair* (μ, ν) *of quasisolutions of (6.1), (6.2), the inequalities* $v(t) \le \mu(t) \le w(t)$, $v(t) \le \nu(t) \le w(t)$ *hold for* $t \in [-h, T]$.

Remark 6.1.2. *We will note that the pair of the minimal and maximal quasisolutions is generalization of the minimal and maximal solutions in the scalar case ($n = 1, p_1 = q_1 = 0$).*

Remark 6.1.3. *We will note that if the pair of functions $v, w \in C([-h, T], \mathbb{R}^n) \cup C^1([0, T], \mathbb{R}^n)$ is a pair of minimal and maximal quasisolutions, then the inequality $v(t) \leq w(t)$ holds. Also, for any pair of quasisolutions this inequality could not be true.*

Remark 6.1.4. *We will note that for all natural numbers $j: 1 \leq j \leq n$ the equalities $p_j = n - 1$ and $q_j = 0$ hold and the pair $v, w \in C([-h, T], \mathbb{R}^n) \cup C^1([0, T], \mathbb{R}^n)$ is a pair of quasisolutions of (6.1), (6.2). Then the functions v and w are also solutions of the same problem. If the initial value problem (6.1), (6.2) has a unique solution $u(t)$, then the pair of minimal and maximal quasisolutions is (u, u).*

For all pairs of functions $v, w \in C([-h, T], \mathbb{R}^n)$ such that $v(t) \leq w(t)$ for $t \in [-h, T]$, we define the sets

$$S(v, w) = \{u \in C([-h, T], \mathbb{R}^n): \ v(t) \leq u(t) \leq w(t), t \in [-h, T]\}. \tag{6.7}$$

Lemma 6.1.1. *[Comparison result] Let the scalar function $m \in C([-h, T], \mathbb{R}) \cup C^1([0, T], \mathbb{R})$ satisfies the inequalities*

$$m'(t) \leq -Mm(t) - N \min_{s \in [t-h,t]} m(s) \quad \text{for } t \in [0, T], \tag{6.8}$$

$$m(t) = m(0), \quad t \in [-h, 0], \tag{6.9}$$

$$m(0) \leq 0, \tag{6.10}$$

where M, N are positive constants and

$$(M + N)T < 1. \tag{6.11}$$

Then the inequality $m(t) \leq 0$ holds for $t \in [-h, T]$.

Proof. Assume the contrary, i.e., there exists a point $\xi \in (0, T]$ such that $m(\xi) > 0$. Consider the following two cases:

Case 1. Let $m(0) < 0$. According to the assumptions there exists a point $\eta \in (0, T]$ such that $m(t) \leq 0$ for $t \in [-h, \eta]$, $m(\eta) = 0$, $m(t) > 0$ for $t \in (\eta, \eta + \epsilon)$, where $\epsilon > 0$ is a small enough constant. Denote

$$\inf\{m(t): \ t \in [-h, \eta]\} = -\lambda < 0.$$

Let $\varsigma \in [0, \eta)$ be such that $m(\varsigma) = -\lambda$. According to the mean value theorem there exists a point $\xi_0 \in (\varsigma, \eta)$ such that $m(\eta) - m(\varsigma) = m'(\xi_0)(\eta - \varsigma)$. Therefore from inequality (6.8) and $-\lambda \leq \min_{s \in [\xi_0 - h, \xi_0]} m(s)$ we get

$$\lambda = m(\eta) - m(\varsigma) = m'(\xi_0)(\eta - \varsigma) \leq (M + N)\lambda T. \qquad (6.12)$$

Inequality (6.12) contradicts inequality (6.11).

Case 2. Let $m(0) = 0$. Then from equality (6.9) follows that $m(t) \equiv 0$ for $t \in [-h, 0]$.

Consider the following two cases:

Case 2.1. There exist points $\xi_1, \xi_2 \in [0, T]$, $\xi_1 < \xi_2$ such that $m(t) = 0$ for $t \in [-h, \xi_1]$, and $m(t) > 0$ for $t \in (\xi_1, \xi_2)$. Without loss of generality we could assume $\xi_2 - \xi_1 < h$. Therefore, for $t \in [\xi_1, \xi_2]$ the equality $\min_{s \in [t-h, t]} m(s) = 0$ holds. From inequality (6.8) it follows that $m'(t) \leq -Mm(t) \leq 0$ for $t \in (\xi_1, \xi_2]$. Therefore the function $m(t)$ is continuous nonincreasing function on $[\xi_1, \xi_2]$, i.e., $m(t) \leq m(\xi_1) = 0$ for $t \in [\xi_1, \xi_2]$. The last inequality contradicts the assumption.

Case 2.2 There exists a point $\eta \in (0, T]$ such that $m(t) \leq 0$ for $t \in [-h, \eta]$, $m(\eta) = 0$, $m(t) > 0$ for $t \in (\eta, \eta + \epsilon)$, where $\epsilon > 0$ is a small enough constant. As in the proof of case 1, we obtain a contradiction.

The obtained contradictions prove the claim of Lemma 6.1.1.

□

6.1.1   Multidimensional Case

We will give an algorithm for constructing a sequence of successive approximations and we will prove the application of monotone-iterative technique to the initial value problem for a system of nonlinear differential equations with "maxima."

Theorem 6.1.1. *Let the following conditions be fulfilled:*

1. *The function $\varphi \in C([-h, 0], \mathbb{R}^n)$, $\varphi = (\varphi_1, \varphi_2, \dots, \varphi_n)$.*

2. *The pair of functions $\alpha, \beta \in C([-h, T], \mathbb{R}^n) \cup C^1([0, T], \mathbb{R}^n)$, where $\alpha = (\alpha_1, \alpha_2, \dots, \alpha_n)$, $\beta = (\beta_1, \beta_2, \dots, \beta_n)$, is a pair of lower and upper quasisolutions of the initial value problem (6.1), (6.2), such that $\alpha(t) \leq \beta(t)$ for $t \in [-h, T]$, and $\alpha(0) - \varphi(0) \leq \alpha(t) - \varphi(t)$, $\beta(0) - \varphi(0) \geq \beta(t) - \varphi(t)$ for $t \in [-h, 0]$.*

3. *The function $f : [0,T] \times \mathbb{R}^n \times \mathbb{R}^n \to \mathbb{R}^n$, $f = (f_1, f_2, \ldots, f_n)$, where*
   $f_j(t,x,y) = f_j(t, x_j, [x]_{p_j}, [x]_{q_j}, y_j, [y]_{p_j}, [y]_{q_j})$, is nondecreasing in
   $[x]_{p_j}$ and $[y]_{p_j}$, nonincreasing in $[x]_{q_j}$ and $[y]_{q_j}$, and for $x,y,u,v \in$
   $\mathbb{R}^n$, $y \le x$, $v \le u$ the inequality

$$f_j(t, x_j, [x]_{p_j}, [x]_{q_j}, u_j, [u]_{p_j}, [u]_{q_j}) - f_j(t, y_j, [y]_{p_j}, [y]_{q_j}, v_j, [v]_{p_j}, [v]_{q_j})$$
$$\ge -M_j(x_j - y_j) - N_j(u_j - v_j), \quad t \in [0,T], \quad j = 1, 2, \ldots, n$$

   holds, where M_j, N_j, $j = 1, 2, \ldots, n$ are positive constants.

4. *The inequalities $(M_j + N_j)T < 1$, $j = 1, 2, \ldots, n$ hold.*

   Then there exist two sequences of functions $\{\alpha^{(k)}(t)\}_0^\infty$ and $\{\beta^{(k)}(t)\}_0^\infty$ such that:

   (a) *The sequences are increasing and decreasing correspondingly;*

   (b) *The pair of functions $\alpha^{(k)}(t)$, $\beta^{(k)}(t)$ is a pair of lower and upper quasisolutions of the initial value problem for the system of nonlinear differential equations with "maxima"(6.1), (6.2);*

   (c) *Both sequences uniformly converge on $[-h, T]$;*

   (d) *The limits $V(t) = \lim\limits_{k \to \infty} \alpha^{(k)}(t)$, $W(t) = \lim\limits_{k \to \infty} \beta^{(k}(t)$ are a pair of minimal and maximal solutions of the initial value problem for the system of nonlinear differential equations with "maxima"(6.1), (6.2).*

   (e) *If $u(t) \in S(\alpha, \beta)$ is a solution of the initial value problem for the system of nonlinear differential equations with "maxima" (6.1), (6.2), then $V(t) \le u(t) \le W(t)$.*

Proof. We fix two arbitrary functions $\eta, \mu \in S(\alpha, \beta)$ and for all natural numbers $j : 1 \le j \le n$ we consider the initial value problem for the scalar linear differential equation with "maxima"

$$u'(t) + M_j u(t) + N_j \max_{s \in [t-h,t]} u(s) = \psi_j(t, \eta, \mu), \quad \text{for} \quad t \in [0,T], \quad (6.13)$$

$$u(t) = \varphi_j(t), \quad t \in [-h, 0], \quad (6.14)$$

where $u \in \mathbb{R}$,

$$
\begin{aligned}
\psi_j&(t,\eta,\mu)\\
=\ & f_j\big(t,\eta_j(t),[\eta(t)]_{p_j},[\mu(t)]_{q_j},\max_{s\in[t-h,t]}\eta_j(s),\\
&[\max_{s\in[t-h,t]}\eta(s)]_{p_j},[\max_{s\in[t-h,t]}\mu(s)]_{q_j}\big)\\
&+M_j\eta_j(t)+N_j\max_{s\in[t-h,t]}\eta_j(s),
\end{aligned}
$$

According to the results in Section 3.2 the initial value problems (6.13)-(6.14) have a unique solution for the fixed pair of functions $\eta,\mu \in S(\alpha,\beta)$ (see formula (3.35)).

For any two functions $\eta,\mu \in S(\alpha,\beta)$ such that $\eta(t) \le \mu(t)$ for $t \in [-h,T]$ we define the operator $\Omega : S(\alpha,\beta) \times S(\alpha,\beta) \to S(\alpha,\beta)$ by $\Omega(\eta,\mu) = x(t)$, where $x(t) = (x_1(t),x_2(t),\ldots,x_n(t))$ and $x_j(t)$ is the unique solution of the initial value problem for the scalar differential equation with "maxima" (6.13) and (6.14) for the pair of functions η,μ.

The operator $\Omega(\eta,\mu)$ possesses the following set (**P6.1**) of properties:

P6.1.1 $\alpha \le \Omega(\alpha,\beta)$ and $\beta \ge \Omega(\beta,\alpha)$;

P6.1.2 For any functions $\eta,\mu \in S(\alpha,\beta)$ such that $\eta(t) \le \mu(t)$ for $t \in [-h,T]$ and the pair (η,μ) is a pair of lower and upper quasisolutions of the initial value problem (6.1), (6.2), the inequality $\Omega(\eta,\mu) \le \Omega(\mu,\eta)$ holds.

Indeed, we will prove property (P6.1.1). We denote $m(t) = \alpha(t) - \alpha^{(1)}(t)$, where $\alpha^{(1)}(t) = \Omega(\alpha,\beta)$.

Then from condition 2 and equation (6.13) for any $j = 1,2,\ldots,n$ applying the inequality

$$
\min_{s\in[t-h,t]}\Big(\alpha_j(s) - \alpha_j^{(1)}(s)\Big) \le \max_{s\in[t-h,t]}\alpha_j(s) - \max_{s\in[t-h,t]}\alpha_j^{(1)}(s)
$$

we get

$$
\begin{aligned}
m_j'(t) \le\ & M_j\Big(\alpha_j^{(1)}(t) - \alpha_j(t)\Big) + N_j\Big(\max_{s\in[t-h,t]}\alpha_j^{(1)}(s) - \max_{s\in[t-h,t]}\alpha_j(s)\Big)\\
\le\ & -M_jm_j(t) - N_j\Big(\max_{s\in[t-h,t]}\alpha_j(s) - \max_{s\in[t-h,t]}\alpha_j^{(1)}(s)\Big)\\
\le\ & -M_jm_j(t) - N_j\min_{s\in[t-h,t]}\Big(\alpha_j(s) - \alpha_j^{(1)}(s)\Big),\qquad t \in [0,T]
\end{aligned}
$$

$$(6.15)$$

Therefore the function $m(t)$ satisfies the inequalities (6.8) and (6.10). According to Lemma 6.1.1 the function $m(t)$ is nonpositive, i.e., $\alpha \leq \Omega(\alpha, \beta)$.

Analogously the validity of the inequality $\beta \geq \Omega(\beta, \alpha)$ can be proved.

We will prove property (P6.1.2). Let $\eta, \mu \in S(\alpha, \beta)$ form a pair of lower and upper quasisolutions of the initial value problem (6.1), (6.2) and $\eta(t) \leq \mu(t)$ for $t \in [-h, T]$. Consider functions $x^{(1)}(t)$ and $x^{(2)}(t)$ for $t \in [-h, T]$, where $x^{(1)} = \Omega(\eta, \mu)$, $x^{(2)} = \Omega(\mu, \eta)$, $g(t) = x^{(1)}(t) - x^{(2)}(t)$, $g = (g_1, g_2, \ldots, g_n)$.

Then from condition 3 and equation (6.13) we get for any j : $j = 1, 2, \ldots, n$ and $t \in [0, T]$

$$
\begin{aligned}
g_j'(t) \leq & - M_j g_j(t) - N_j \left(\max_{s\in[t-h,t]} x_j^{(1)}(s) - \max_{s\in[t-h,t]} x_j^{(2)}(s) \right) \\
& + M_j \left(\eta_j(t) - \mu_j(t) \right) + N_j \left(\max_{s\in[t-h,t]} \eta_j(s) - \max_{s\in[t-h,t]} \mu_j(s) \right) \\
& + f_j(t, \eta_j(t), [\eta(t)]_{p_j}, [\mu(t)]_{q_j}, \max_{s\in[t-h,t]} \eta_j(s), [\max_{s\in[t-h,t]} \eta(s)]_{p_j}, \\
& \hspace{6cm} [\max_{s\in[t-h,t]} \mu(s)]_{q_j}) \\
& - f_j(t, \mu_j(t), [\mu(t)]_{p_j}, [\eta(t)]_{q_j}, \max_{s\in[t-h,t]} \mu_j(s), [\max_{s\in[t-h,t]} \mu(s)]_{p_j}, \\
& \hspace{6cm} [\max_{s\in[t-h,t]} \eta(s)]_{q_j}) \\
\leq & - M_j g_j(t) - N_j \left(\max_{s\in[t-h,t]} x_j^{(1)}(s) - \max_{s\in[t-h,t]} x_j^{(2)}(s) \right) \\
\leq & - M_j g_j(t) - N_j \min_{s\in[t-h,t]} g_j(s). \hspace{3cm} (6.16)
\end{aligned}
$$

Inequality (6.16) proves the validity of conditions of Lemma 6.1.1. According to Lemma 6.1.1 the functions $g_j(t)$, $j = 1, 2, \ldots, n$ are nonpositive, i.e., $\Omega(\eta, \mu) \leq \Omega(\mu, \eta)$.

We define the sequences of functions $\{\alpha^{(k)}(t)\}_0^\infty$ and $\{\beta^{(k)}(t)\}_0^\infty$ by the equalities

$$
\begin{aligned}
\alpha^{(0)} &\equiv \alpha, & \beta^{(0)} &\equiv \beta, \\
\alpha^{(k+1)} &= \Omega(\alpha^{(k)}, \beta^{(k)}), & \beta^{(k+1)} &= \Omega(\beta^{(k)}, \alpha^{(k)}).
\end{aligned}
$$

According to property P6.1.1 of the operator $\Omega(\eta, \mu)$ it follows that functions $\alpha^{(k)}(t)$ and $\beta^{(k)}(t)$ form a pair of lower and uper solutions.

According to the property P6.1.2 of the operator $\Omega(\eta, \mu)$ it follows that for $t \in [-h, T]$ the following inequalities

$$\alpha^{(0)}(t) \leq \alpha^{(1)}(t) \leq \cdots \leq \alpha^{(k)}(t) \leq \beta^{(k)}(t) \leq \cdots \leq \beta^{(1)}(t) \leq \beta^{(0)}(t)$$

(6.17)

hold.

Both sequences of functions $\{\alpha^{(k)}(t)\}_0^\infty$ and $\{\beta^{(k)}(t)\}_0^\infty$ are convergent on $[-h, T]$. Let $V_j(t) = \lim_{k \to \infty} v_j^{(k)}(t)$, $W_j(t) = \lim_{k \to \infty} w_j^{(k)}(t)$, $j = 1, 2, \ldots, n$. We will prove that the pair of functions $V(t)$ and $W(t)$, where $V = (V_1, V_2, \ldots, V_n)$ and $W = (W_1, W_2, \ldots, W_n)$, are a pair of minimal and maximal quasisolutions of the initial value problem (6.1), (6.2). From the definition of functions $\alpha^{(k)}(t)$, $\alpha^{(k)} = (\alpha_1^{(k)}, \alpha_2^{(k)}, \ldots, \alpha_n^{(k)})$ and $\beta^{(k)}(t)$, $\beta^{(k)} = (\beta_1^{(k)}, \beta_2^{(k)}, \ldots, \beta_n^{(k)})$ follows that these functions satisfy the initial value problem $(j = 1, 2, \ldots, n)$

$$(\alpha_j^{(k)}(t))' + M_j \alpha_j^{(k)}(t) + N_j \max_{s \in [t-h,t]} \alpha_j^{(k)}(s)) = \psi_j(t, \alpha^{(k-1)}, \beta^{(k-1)}),$$

(6.18)

$$(\beta_j^{(k)}(t))' + M_j \beta_j^{(k)}(t) + N_j \max_{s \in [t-h,t]} \beta_j^{(k)}(s) = \psi_j(t, \beta^{(k-1)}, \alpha^{(k-1)}),$$

(6.19)

for $t \in [0, T]$,

$$\alpha_j^{(k)}(t) = \alpha_j^{(k)}(0), \quad \beta_j^{(k)}(t) = \beta_j^{(k)}(0), \quad t \in [-h, 0],$$

(6.20)

From equations (6.18) and (6.20) it follows that the pair of functions $V(t)$ and $W(t)$ is a pair of quasisolutions of the initial value problem (6.1), (6.2). Let $u, z \in S(\alpha, \beta)$ be a pair of quasisolutions of the initial value problem (6.1), (6.2). From inequalities (6.17) it follows that there exists a natural number k such that $\alpha^{(k)}(t) \leq u(t) \leq \beta^{(k)}(t)$ and $\alpha^{(k)}(t) \leq z(t) \leq \beta^{(k)}(t)$ for $t \in [-h, T]$. We introduce the notation $g(t) = \alpha^{(k+1)}(t) - u(t)$, $g = (g_1, g_2, \ldots, g_n)$. According to Lemma 5.1.1 the inequalities $g_j(t) \leq 0$, $j = 1, 2, \ldots$ hold for $t \in [-h, T]$, i.e., $\alpha^{(k+1)}(t) \leq u(t)$.

Analogously the validity of inequalities $\beta^{(k+1)}(t) \geq u(t)$ and $\alpha^{(k+1)}(t) \leq z(t) \leq \beta^{(k+1)}(t)$ for $t \in [-h, T]$ can be proved.

Let $u(t) \in S(\alpha, \beta)$ be a solution of the initial value problem (6.1), (6.2). Consider the pair of functions (u, u) that is a pair of quasisolutions of the initial value problem (6.1), (6.2). According to the proof given above the inequality $V(t) \leq u(t) \leq W(t)$ holds for $t \in [-h, T]$.

$\square$

6.1.2   Scalar Case

In the scalar case $n = 1$ the problem (6.1) and (6.2) reduces to an initial value problem for a scalar differential equation with "maxima." In this case, we use lower and upper solutions:

Definition 6.1.5. *The function $v \in C([-h,T],\mathbb{R}) \cup C^1([0,T],\mathbb{R})$ is called a lower solution of the initial value problem for the differential equation with "maxima" (6.1), (6.2) $(n=1)$ if*

$$v \leq f(t, v(t), \max_{s \in [t-h,t]} v(s)) \tag{6.21}$$

$$v(t) \leq \varphi(t), \quad t \in [-h, 0]. \tag{6.22}$$

Analogously the upper solution of the initial value problem for the differential equation with "maxima" (6.1), (6.2) $(n = 1)$ is defined.

Then the following result is a partial case of the theorem proved above:

Theorem 6.1.2. *Let the following conditions be fulfilled:*

1. *The function $\varphi \in C([-h, 0], \mathbb{R})$.*

2. *The functions $\alpha, \beta \in C([-h, T], \mathbb{R}) \cup C^1([0, T], \mathbb{R})$ are lower and upper solutions of the initial value problems (6.1) and (6.2) for $n = 1$, correspondingly, and $\alpha(t) \leq \beta(t)$ for $t \in [-h, T]$, and $\alpha(0) - \varphi(0) \leq \alpha(t) - \varphi(t)$, $\beta(0) - \varphi(0) \geq \beta(t) - \varphi(t)$ for $t \in [-h, 0]$.*

3. *The function $f \in C([0, T] \times \mathbb{R} \times \mathbb{R}, \mathbb{R})$, and for $x, y, u, v \in \mathbb{R}$, $y \leq x$, $v \leq u$ the inequality*

$$f(t, x, u) - f(t, y, v) \geq -M(x - y) - N(u - v), \quad t \in [0, T],$$

   holds, where M, N are positive constants.

4. *The inequality $(M + N)T < 1$ holds.*

Then there exist two sequences of functions $\{\alpha^{(k)}(t)\}_0^\infty$ and $\{\beta^{(k)}(t)\}_0^\infty$ such that:

   (a) *the functions $\alpha^{(k)}(t), \beta^{(k)} \in C([-h, T], \mathbb{R})$ are solutions of the initial value problems for the following scalar equations*

$$\left(\alpha^{(k)}(t)\right)' + M\alpha^{(k)}(t) + N \max_{s \in [t-h,t]} \alpha^{(k)}(s) = \psi(t, \alpha^{(k-1)}),$$

$$for \quad t \in [0, T], \quad (6.23)$$

and

$$\left(\beta^{(k)}(t)\right)' + M\beta^{(k)}(t) + N \max_{s\in[t-h,t]} \beta^{(k)}(s) = \psi(t, \beta^{(k-1)}),$$

$$for \quad t \in [0, T], \quad (6.24)$$

with initial conditions

$$\alpha^{(k)}(t) = \varphi(t), \quad \alpha^{(k)}(t) = \varphi(t), \quad t \in [-h, 0], \quad (6.25)$$

where

$$\psi(t, \eta) = f(t, \eta(t), \max_{s\in[t-h,t]} \eta(s)) + M\eta(t) + N \max_{s\in[t-h,t]} \eta(s),$$

(b) *Both sequences are increasing and decreasing correspondingly;*

(c) *The functions $\alpha^{(k)}(t)$ are lower solutions and the functions $\beta^{(k)}(t)$ are upper solutions of the initial value problem for the nonlinear differential equation with "maxima"(6.1) and (6.2) (n=1);*

(d) *Both sequences uniformly converge on $[-h, T]$;*

(e) *The limits $V(t) = \lim_{k\to\infty} \alpha^{(k)}(t)$, $W(t) = \lim_{k\to\infty} \beta^{(k}(t)$ are minimal and maximal solutions, correspondingly, of the initial value problem for the nonlinear differential equation with "maxima"(6.1) and (6.2) (n=1).*

(f) *If $u(t) \in S(\alpha, \beta)$ is a solution of the initial value problem for the nonlinear differential equation with "maxima" (6.1) and (6.2), then $V(t) \le u(t) \le W(t)$.*

Remark 6.1.5. *As a particular case of the above results we obtain the monotone-iterative techniques for the initial value problem for nonlinear differential equations (scalar case as well as case of systems) considered by many authors (see the monograph [Ladde et al. 1985] and references cited therein).*

Example 6.1.1. *Consider the following scalar differential equation with "maxima"*

$$x' = \frac{1}{8}e^{-t}x(t) - \frac{1}{4}\max_{s\in[t-0.5,t]} x(s), \quad for \;\; t \in [0, 2], \quad (6.26)$$

with initial condition

$$x(t) = 0, \quad t \in [-0.5, 0]. \quad (6.27)$$

It is easy to check that the problems (6.26) and (6.27) has zero solution.

From the inequality $\frac{1}{8}e^{-t} - \frac{1}{4} \leq \frac{1}{8} - \frac{1}{4} < 0$ on $[0,2]$ it follows that the funtion $\alpha^{(0)}(t) \equiv -2$ is a lower solution of the initial value problems (6.26) and (6.27) and the function $\beta^{(0)}(t) \equiv 2$ is an upper solution of (6.26), (6.27), i.e., inequalities $(\alpha^{(0)}(t))' \leq \frac{1}{8}e^{-t}\alpha^{(0)}(t) - \frac{1}{4}\max_{s\in[t-0.5,t]}\alpha^{(0)}(s)$ and $(\beta^{(0)}(t))' \geq \frac{1}{8}e^{-t}\beta^{(0)}(t) - \frac{1}{4}\max_{s\in[t-0.5,t]}\beta^{(0)}(s)$ hold.

In this case $f(t,u,v) = \frac{1}{8}e^{-t}u - \frac{1}{4}v$ and $f(t,x,u) - f(t,y,v) = \frac{1}{8}e^{-t}(x-y) - \frac{1}{4}(u-v) \geq -M(x-y) - N(u-v)$ for $x \geq y$, $u \geq v$, where $M = \frac{1}{8}$ and $N = \frac{1}{4}$. Then $(M+N)T = (\frac{1}{8}+\frac{1}{4})2 = \frac{3}{4} < 1$.

Then the successive approximations to the zero solution of the initial value problems (6.26) and (6.27) are solutions of the linear differential equations with "maxima" (6.23) and (6.24), that are reduced in this case to the following equations

$$\left(\alpha^{(k)}(t)\right)' = -\frac{1}{8}\alpha^{(k)}(t) - \frac{1}{4}\max_{s\in[t-0.5,t]}\alpha^{(k)}(s) + \frac{1}{8}(e^{-t}+1)\alpha^{(k-1)}(t),$$

$$\text{for}\quad t \in [0,2], \quad (6.28)$$

and

$$\left(\beta^{(k)}(t)\right)' = -\frac{1}{8}\beta^{(k)}(t) - \frac{1}{4}\max_{s\in[t-0.5,t]}\beta^{(k)}(s) + \frac{1}{8}(e^{-t}+1)\beta^{(k-1)}(t),$$

$$\text{for}\quad t \in [0,2], \quad (6.29)$$

with initial conditions

$$\alpha^{(k)}(t) = 0, \quad \beta^{(k)}(t) = 0, \quad t \in [-0.5,0]. \qquad (6.30)$$

The solution of the initial value problems (6.28) and (6.30) is given by the formula

$$\alpha^{(k)}(t) = \left\{ \int_0^t \frac{1}{8}(e^{-t}+1)\alpha^{(k-1)}(s)ds \right.$$

$$\left. - 0.25\int_0^t \max_{\xi\in[s-0.5,s)}\alpha^{(k)}(\xi)ds \right\}\left(e^{0.125t} - 1\right) \quad \text{for}\quad t \in [0,2]$$

$$\alpha^{(k)}(t) = 0 \quad \text{for}\quad t \in [-0.5,0].$$

$$(6.31)$$

and the solution of the initial value problems (6.29) and (6.30) is given
by the formula

$$\beta^{(k)}(t) = \left\{ \int_0^t \frac{1}{8}(e^{-t} + 1)\beta^{(k-1)}(s)ds \right.$$

$$\left. - 0.25 \int_0^t \max_{\xi \in [s-0.5,s)} \beta^{(k)}(\xi)ds \right\} \left(e^{0.125t} - 1 \right) \quad \text{for} \quad t \in [0, 2]$$

$$\beta^{(k)}(t) = 0 \quad \text{for} \quad t \in [-0.5, 0].$$

$$(6.32)$$

It is easy to calculate that

$$\alpha^{(0)}(0.25) = -2, \qquad\qquad \beta^{(0)}(0.25) = 2,$$
$$\alpha^{(1)}(0.25) = -0.0037, \qquad \beta^{(1)}(0.25) = 0.003739,$$
$$\alpha^{(2)}(0.25) = -0.000036, \qquad \beta^{(2)}(0.25) = 0.0000367,$$
$$\alpha^{(3)}(0.25) = -0.00000036, \qquad \beta^{(3)}(0.25) = 0.000000363.$$

The above sequences are monotonic, increasing and decreasing correspondingly, and approach zero, which is the exact solution.

6.2 Monotone-Iterative Technique for Periodic Boundary Value Problems

Note that the differential equations with "maxima" generate functionals not having the property of linearity even if the equations are linear.
These equations can be integrated in a closed form only in exceptional
cases. In relation to this, it is necessary to elaborate approximate methods for their solution.

In the present section, a couple of minimal and maximal quasisolutions of a boundary value problem for a system of differential equations
with "maxima" is constructed by means of the monotone-iterative techniques of Lakshmikantham ([Ladde et al. 1985]).

Consider the boundary value problem

$$x' = f\left(t, x(t), \max_{s \in [t-h,t]} x(s)\right) \quad \text{for} \quad t \in [0, T], \qquad (6.33)$$

$$x(0) = x(T), \quad x(t) = x(0) \quad \text{for} \quad t \in [-h, 0], \qquad (6.34)$$

where $x \in \mathbb{R}^n$, $f : [0, T] \times \mathbb{R}^n \times \mathbb{R}^n \to \mathbb{R}^n$, $h = const > 0$.

For any $x, y, z \in \mathbb{R}^n$ and any two nonnegative integers p_j and q_j such that $p_j + q_j = n - 1$ we will use notation $(z_j, [x]_{p_j}, [y]_{q_j})$ introduced in Section 6.1 and we rewrite down the boundary value problem (6.33), (6.34) in the form

$$x'_j = f_j\left(t, x_j, [x]_{p_j}, [x]_{q_j}, \max_{s\in[t-h,t]} x_j(s), \left[\max_{s\in[t-h,t]} x(s)\right]_{p_j}, \right.$$

$$\left. \left[\max_{s\in[t-h,t]} x_j(s)\right]_{q_j}\right),$$

$$\text{for} \quad t \in [0, T],$$

$$x_j(0) = x_j(T), x_j(t) = x_j(0) \quad \text{for} \quad t \in [-h, 0], \quad j = \overline{1, n}.$$

Definition 6.2.1. *The functions* $v, w \in C([-h, T], \mathbb{R}^n) \cup C^1([0, T], \mathbb{R}^n)$ *are said to be a couple of lower and upper quasisolutions of the boundary value problem (6.33), (6.34) if:*

$$v'_j(t) \leq f_j\left(t, v_j, [v]_{p_j}, [w]_{q_j}, \max_{s\in[t-h,t]} v_j(s), \left[\max_{s\in[t-h,t]} v(s)\right]_{p_j}, \right.$$

$$\left. \left[\max_{s\in[t-h,t]} w_j(s)\right]_{q_j}\right), \quad \text{for} \quad t \in [0, T], \qquad (6.35)$$

$$w'_j(t) \geq f_j\left(t, w_j, [w]_{p_j}, [v]_{q_j}, \max_{s\in[t-h,t]} w_j(s), \left[\max_{s\in[t-h,t]} w(s)\right]_{p_j}, \right.$$

$$\left. \left[\max_{s\in[t-h,t]} v_j(s)\right]_{q_j}\right), \quad \text{for} \quad t \in [0, T],$$

$$v_j(0) \leq v_j(T), \quad w_j(0) \geq w_j(T), \qquad (6.36)$$

$$v_j(t) = v_j(0), w_j(t) = w_j(0) \quad \text{for} \quad t \in [-h, 0],$$

where $v = (v_1, v_2, \ldots, v_n)$, *and* $w = (w_1, w_2, \ldots, w_n)$.

Definition 6.2.2. *In the case when (6.33), (6.34) is a boundary value problem for a scalar differential equation, i.e., $n = 1$, $p_1 = q_1 = 0$, the couple of lower and upper quasisolutions of the same problem are said to be lower and upper solutions of (6.33), (6.34).*

Definition 6.2.3. *The functions* $v, w \in C([-h, T], \mathbb{R}^n) \cup C^1([0, T], \mathbb{R}^n)$ *are said to be a couple of quasisolutions of the boundary value problem (6.33), (6.34) if (6.35) and (6.36) are satisfied only as equalities.*

Definition 6.2.4. *The functions* v, $w \in C([-h, T], \mathbb{R}^n) \cup C^1([0, T], \mathbb{R}^n)$ *are said to be a couple of minimal and maximal quasisolutions of the boundary value problem (6.33), (6.34) if (v, w) is a couple of quasisolutions of the same problem and for any couple of quasisolutions (u, z) of (6.33), (6.34) the inequalities $v(t) \leq u(t) \leq w(t)$, $v(t) \leq z(t) \leq w(t)$ hold for $t \in [0, T]$.*

Remark 6.2.1. *For the couple (v, w) of minimal and maximal quasisolutions of problem (6.33), (6.34) the inequality $v(t) \leq w(t)$ holds for $t \in [0, T]$ while for an arbitrary couple of quasisolutions (u, z) an analogous inequality relating the functions $u(t)$ and $z(t)$ may not be valid.*

We shall note that for ordinary differential equations the notions of a couple of lower and upper quasisolutions, a couple of quasisolutions, and a couple of minimal and maximal quasisolutions were introduced by Lakshmikantham et al. ([Ladde et al. 1985]).

Consider the set $S(v, w)$ defined by (6.7).

In the further considerations, we shall use the following lemma.

Lemma 6.2.1. *Let F be a Banach space and $B = C([a, b], F)$. Let $S : B \to F$ be an operator for which*

$$\left\| S_\varphi - S_\psi \right\|_F \leq \alpha \|\varphi - \psi\|_B \qquad 0 \leq \alpha < 1.$$

Then for any point $\xi \in [a, b]$ there exists an element $\varphi \in B$ such that $S_\varphi = \varphi(\xi)$.

Lemma 6.2.2. *Let the function $\sigma \in C([0, T], \mathbb{R})$. Then the boundary value problem*

$$x' + Mx + N \max_{s \in [t-h, t]} x(s) = \sigma(t) \quad for \quad t \in [0, T], \tag{6.37}$$

$$x(0) = x(T), \tag{6.38}$$

$$x(t) = x(0) \quad for \quad t \in [-h, 0], \tag{6.39}$$

has a solution, where $x \in \mathbb{R}$ and the positive constants M and N are such that

$$N < M. \tag{6.40}$$

Proof. Equation (6.37) can be written in the form

$$x' + Mx = \sigma(t) + N \max_{s \in [t-h, t]} x(s).$$

By the variation of parameters formula we obtain

$$x(0) = \frac{1}{e^{MT} - 1} \int_0^T \left[\sigma(s) + N \max_{\xi \in [s-h,s]} x(\xi) \right] e^{Ms} ds.$$

Define the map $S : \Omega \to \mathbb{R}$ by the equality

$$S_u = \frac{1}{e^{MT} - 1} \int_0^T \left[\sigma(s) + N \max_{\xi \in [s-h,s]} u(\xi) \right] e^{Ms} ds,$$

where $\Omega = \{ u \in C([-h, T], \mathbb{R}^n) \cup C^1([0, T], \mathbb{R}^n) \; : \; u(t) = u(0) \text{ for } t \in [-h, 0] \}$ with norm

$$|u|_0 = \max_{t \in [-h,T]} |u(t)|.$$

Then the following inequality holds:

$$\left| S_u - S_v \right| \leq \frac{N}{e^{MT} - 1} \int_0^T |u - v|_0 e^{Ms} ds = \frac{N}{M} |u - v|_0. \qquad (6.41)$$

By Lemma 6.2.1 there exists a function $u \in \Omega$ for which $S_u = u(0)$, i.e., the function $u(t)$ is a solution of the boundary value problem (6.37), (6.38), (6.39).

This completes the proof of Lemma 6.2.2.

□

Lemma 6.2.3. *[Comparison result] Let* $m \in C([-h, T], \mathbb{R}^n) \cup C^1([0, T], \mathbb{R}^n)$ *satisfy the inequalities*

$$m' \leq -Mm(t) - N \min_{s \in [t-h,t]} m(s) \quad for \quad t \in [0, T], \qquad (6.42)$$

$$m(t) = m(0) \quad for \quad t \in [-h, 0], \qquad (6.43)$$

$$m(0) \leq m(T), \qquad (6.44)$$

where M *and* N *are positive constants such that*

$$(M + N)T < 1. \qquad (6.45)$$

Then the inequality $m(t) \leq 0$ *holds for* $t \in [-h, T]$.

Proof. We will consider the following two cases with respect to the possible values of $m(T)$:

Case 1. Let $m(0) \leq 0$. Then according to Lemma 6.1.1 the inequality $m(t) \leq 0$ holds for $t \in [-h, T]$.

Case 2. Let $m(0) > 0$. Then $m(T) > 0$. Consider the following two cases:

Case 2.1. Let $m(t) \geq 0$ for $t \in [0, T]$. Then from equality (6.43) it follows that $m(t) \geq 0$ for $t \in [-h, T]$. In view of inequality (6.42) the inequality $m'(t) \leq 0$ holds for $t \in [0, T]$ which shows that the function $m(t)$ is monotone nonincreasing in the interval $[0, T]$. Hence $m(t) \leq m(0)$ for $t \in [0, T]$. Inequality (6.44) proves that $m(t) \equiv c$ for $t \in [-h, T]$, where $c = $ const. From inequality (6.42) we obtain the inequality $0 \leq -(M + N)c$. Therefore, $c \leq 0$, or $m(T) = c \leq 0$. The obtained contradiction proves the impossibility of this case.

Case 2.2. Let a point $\eta \in (0, T)$ exist such that $m(\eta) < 0$. Introduce the notation

$$\min_{t \in [0,T]} m(t) = -\lambda, \qquad \lambda = \text{const} > 0.$$

From the continuity of the function $m(t)$ it follows that there exists a point $\zeta \in (0, T)$ such that $m(\zeta) = -\lambda$. According to (6.43) the equality $\min_{t \in [-h,T]} m(t) = -\lambda$ holds. Moreover, there exists a point $\nu \in (\zeta, T)$ such that

$$m'(\nu) = \frac{m(T) - m(\zeta)}{T - \zeta} \geq \frac{\lambda}{T}. \tag{6.46}$$

From inequality (6.42), $m(\nu) \geq -\lambda$, and $\min_{s \in [\nu-h,\nu]} m(s) \geq -\lambda$ it follows that

$$m'(\nu) \leq -Mm(\nu) - N \min_{s \in [\nu-h,\nu]} m(s) \leq -M(-\lambda) - N(-\lambda). \tag{6.47}$$

From (6.46) and (6.47) we can obtain the inequality

$$(M + N)T \geq 1. \tag{6.48}$$

Inequality (6.48) contradicts inequality (6.45).

The obtained contradiction shows us the impossibility of this case. □

Theorem 6.2.1. *Let the following conditions be fulfilled:*

1. *The functions v, $w \in C([-h, T], \mathbb{R}^n) \cup C^1([0, T], \mathbb{R}^n)$ are a couple of lower and upper quasisolutions of the boundary value problem (6.33), (6.34) and satisfy the conditions $v(t) \leq w(t)$ for $t \in [0, T]$ and $v(t) \equiv v(0)$, $w(t) \equiv w(0)$ for $t \in [-h, 0]$.*

2. *The function $f \in C([0,T] \times \mathbb{R}^n \times \mathbb{R}^n, \mathbb{R}^n)$, $f = (f_1, f_2, \ldots, f_n)$, $f_j(t, x, y) = f_j(t, x_j, [x]_{p_j}, [x]_{q_j}, y_j, [y]_{p_j}, [y]_{q_j})$ is monotone nondecreasing with respect to $[x]_{p_j}$ and $[y]_{p_j}$, and monotone non-increasing with respect to $[x]_{q_j}$ and $[y]_{q_j}$, and for $x, y \in S(v,w)$, $y(t) \le x(t)$ satisfies the inequalities*

$$f_j\left(t, x_j, [x]_{p_j}, [x]_{q_j}, \max_{s\in[t-h,t]} x_j(s), \left[\max_{s\in[t-h,t]} x(s)\right]_{p_j}, \left[\max_{s\in[t-h,t]} x(s)\right]_{q_j}\right)$$

$$- f_j\left(t, y_j, [x]_{p_j}, [x]_{q_j}, \max_{s\in[t-h,t]} y_j(s), \left[\max_{s\in[t-h,t]} x(s)\right]_{p_j}, \left[\max_{s\in[t-h,t]} x(s)\right]_{q_j}\right)$$

$$\ge - M_j\big(x_j(t) - y_j(t)\big) - N_j\left(\max_{s\in[t-h,t]} x_j(s) - \max_{s\in[t-h,t]} y_j(s)\right),$$

$$j = \overline{1,n},$$
$$(6.49)$$

where M_j and N_j ($j = \overline{1,n}$) are positive constants such that

$$N_j < M_j \le 1/(2T).$$

Then there exist two sequences of functions $\{v^{(k)}(t)\}_0^\infty$ and $\{w^{(k)}(t)\}_0^\infty$ such that:

(a) *The sequences are increasing and decreasing, correspondingly;*

(b) *The pair of functions $v^{(k)}(t)$, $w^{(k)}(t)$ is a pair of lower and upper quasisolutions of the boundary value problem for the system of nonlinear differential equations with "maxima" (6.33), (6.34);*

(c) *Both sequences uniformly converge on $[-h, T]$;*

(d) *The limits $V(t) = \lim\limits_{k\to\infty} v^{(k)}(t)$, $W(t) = \lim\limits_{k\to\infty} w^{(k)}(t)$ are a pair of minimal and maximal solutions of the boundary value problem for the system of nonlinear differential equations with "maxima" (6.33), (6.34).*

(e) *If $u(t) \in S(v,w)$ is a solution of the boundary value problem for the system of nonlinear differential equations with "maxima" (6.33), (6.34), then $V(t) \le u(t) \le W(t)$.*

Proof. Fix two arbitrary functions η, $\mu \in S(v, w)$ and consider the scalar boundary value problems

$$x_j' + M_j x_j + N_j \max_{s \in [t-h,t]} x_j(s) = \sigma_j(t, \eta, \mu) \quad \text{for} \quad t \in [0, T], \quad (6.50)$$

$$x_j(t) = x_j(0) \quad \text{for} \quad t \in [-h, 0], \tag{6.51}$$

$$x_j(0) = x_j(T), \quad j = \overline{1, n}, \tag{6.52}$$

where

$$\sigma_j(t, \eta, \mu)$$
$$= f_j \left(t, \eta_j, [\eta]_{p_j}, [\mu]_{q_j}, \max_{s \in [t-h,t]} \eta_j(s), \left[\max_{s \in [t-h,t]} \eta_j(s) \right]_{p_j}, \left[\max_{s \in [t-h,t]} \mu_j(s) \right]_{q_j} \right)$$
$$+ M_j \eta_j(t) + N_j \max_{s \in [t-h,t]} \eta_j(s), \quad j = \overline{1, n}.$$

By Lemma 6.2.2 for any fixed j boundary value problem (6.50)–(6.52) has a unique solution. Suppose that for a fixed couple of functions η, $\mu \in S(v, w)$ there exist two distinct solutions $x(t)$ and $y(t)$ of the boundary value problem (6.50)–(6.52). Define a function $m(t) = x(t) - y(t)$ for $t \in [-h, T]$, $m(t) = (m_1(t), m_2(t), \ldots, m_n(t))$. The functions $m_j(t)$ $(j = \overline{1, n})$ satisfy the inequalities

$$m_j'(t) = x_j'(t) - y_j'(t) = -M_j m_j(t) - N_j \left(\max_{s \in [t-h,t]} x_j(s) - \max_{s \in [t-h,t]} y_j(s) \right)$$
$$\leq -M_j m_j(t) - N_j \min_{s \in [t-h,t]} m_j(s) \quad \text{for} \quad t \in [0, T],$$

$$m_j(t) = m_j(0) \quad \text{for} \quad t \in [-h, 0],$$
$$m_j(0) = m_j(T), \quad j = \overline{1, n}.$$

By Lemma 6.2.3 for $t \in [-h, T]$ the inequalities $m_j(t) \leq 0$ or $x_j(t) \leq y_j(t)$ ($j = \overline{1, n}$) are valid. Analogously, if we consider the function $m(t) = y(t) - x(t)$, we obtain that $y_j(t) \leq x_j(t)$, $j = \overline{1, n}$. Consequently, $x_j(t) = y_j(t)$, $j = \overline{1, n}$, i.e., for any fixed couple of functions η, $\mu \in S(v, w)$ problem (6.50)–(6.52) has a unique solution.

Define the map $\mathcal{A} : S(v, w) \times S(v, w) \to S(v, w)$ by the equality $\mathcal{A}(\eta, \mu) = x$, where $x = (x_1, x_2, \ldots, x_n)$ and $x_j(t)$ is the unique solution of the boundary value problem (6.50)–(6.52) for the couple of functions η, $\mu \in S(v, w)$.

We shall show that $v \leq \mathcal{A}(v, w)$. Introduce the notation $v^{(1)} = \mathcal{A}(v, w)$, $\mathbf{g} = v - v^{(1)}$, $\mathbf{g} = (g_1, g_2, \ldots, g_n)$.

Then the following inequalities hold

$$(g_j(t))' = (v_j(t))' - (v_j^{(1)}(t))'$$

$$\leq f_j\Bigg(t, \; v_j, \; [v]_{p_j}, \; [w]_{q_j}, \max_{s \in [t-h,t]} v_j(s), \; \Big[\max_{s \in [t-h,t]} v_j(s)\Big]_{p_j},$$

$$\Big[\max_{s \in [t-h,t]} w_j(s)\Big]_{q_j}\Bigg)$$

$$+M_j v_j^{(1)} + N_j \max_{s \in [t-h,t]} v_j^{(1)}(s) - \sigma_j(t, \; v, \; w)$$

$$\leq -M_j\Big(v_j(t) - v_j^{(1)}(t)\Big) - N_j\Big(\max_{s \in [t-h,t]} v_j(s) - \max_{s \in [t-h,t]} v_j^{(1)}(s)\Big)$$

$$\leq -M_j g_j(t) - N_j \min_{s \in [t-h,t]} g_j(s) \quad \text{for} \quad t \in [0,T],$$

$$g_j(t) = g_j(0) \quad \text{for} \quad t \in [-h,0],$$
$$g_j(0) \leq g_j(T), \quad j = \overline{1,n}.$$

By Lemma 6.2.3 the functions $g_j(t)$, $j = \overline{1,n}$, are nonpositive, i.e., $v(t) \leq v^{(1)}(t)$ for $t \in [-h,T]$. In an analogous way, it is proved that the inequality $w \geq \mathcal{A}(w,v)$ holds.

Let η, $\mu \in S(v,w)$ be such that $\eta(t) \leq \mu(t)$ for $t \in [-h,T]$. From the definition of the map $\mathcal{A}$ and Lemma 6.2.3 it follows that the inequality $\mathcal{A}(\eta,\mu) \leq \mathcal{A}(\mu,\eta)$ is valid.

Define sequences of functions $\{v^{(k)}(t)\}_0^\infty$ and $\{w^{(k)}(t)\}_0^\infty$ by

$$v^{(0)}(t) = v(t), \qquad w^{(0)}(t) = w(t),$$
$$v^{(k+1)} = \mathcal{A}(v^{(k)}, w^{(k)}), \quad w^{(k+1)} = \mathcal{A}(w^{(k)}, v^{(k)}).$$

The functions $v^{(k)}(t)$ and $w^{(k)}(t)$ for $t \in [-h,T]$ satisfy the inequalities

$$v^{(0)}(t) \leq v^{(1)}(t) \leq \cdots \leq v^{(k)}(t) \leq \cdots \leq w^{(k)}(t) \leq \cdots \leq w^{(1)}(t) \leq w^{(0)}(t).$$
$$(6.53)$$

The sequences of functions $\{v^{(k)}(t)\}_0^\infty$ and $\{w^{(k)}(t)\}_0^\infty$ are uniformly convergent on the interval $[-h,T]$. Then there exist functions $\bar{v}(t)$ and $\bar{w}(t)$ such that $\bar{v}(t) = \lim_{k \to \infty} v^{(k)}(t)$ and $\bar{w}(t) = \lim_{k \to \infty} w^{(k)}(t)$ for $t \in [-h,T]$. From the definition of the functions $v^{(k)}(t)$ and $w^{(k)}(t)$ it follows that these functions satisfy the boundary value problems

$$(v_j^{(k)})' + M_j v_j^{(k)} + N_j \max_{s \in [t-h,t]} v_j^{(k)}$$
$$= \sigma_j \left(t, \ v^{(k-1)}, \ w^{(k-1)} \right) \quad \text{for} \quad t \in [0,T],$$
$$(w_j^{(k)})' + M_j w_j^{(k)} + N_j \max_{s \in [t-h,t]} w_j^{(k)}$$
$$= \sigma_j \left(t, \ w^{(k-1)}, \ v^{(k-1)} \right) \quad \text{for} \quad t \in [0,T],$$

(6.54)

$$v_j^{(k)}(t) = v_j^{(k)}(0) \qquad w_j^{(k)}(t) = w_j^{(k)}(0) \qquad \text{for} \quad t \in [-h,0],$$

(6.55)

$$v_j^{(k)}(0) = v_j^{(k)}(T), \qquad w_j^{(k)}(0) = w_j^{(k)}(T).$$

(6.56)

We pass to the limit in equalities (6.54)–(6.56) and obtain that the couple of functions $(\bar{v}, \bar{w})$ is a couple of quasisolutions of the boundary value problem (6.33), (6.34). Let $u, \ z \in S(v,w)$ be a couple of quasisolutions of problem (6.33), (6.34). From inequalities (6.53) and Definition 6.2.4 it follows that there exists an integer $k \geq 0$ such that $v^{(k)}(t) \leq z(t) \leq w^{(k)}(t)$ and $v^{(k)}(t) \leq u(t) \leq w^{(k)}(t)$ for $t \in [-h,T]$. Introduce the notation $g(t) = v^{(k+1)}(t) - z(t)$, $g = (g_1, \ g_2, \dots, \ g_n)$. According to the definition of the functions $v^{(k)}(t)$, $w^{(k)}(t)$ and condition 2 of Theorem 6.2.1, the function $g(t)$ satisfies the inequalities

$$(g_j(t))' \leq (v_j^{(k+1)}(t))' - (z_j(t))'$$
$$\leq - M_j v_j^{(k+1)}(t) - N_j \max_{s \in [t-h,t]} v_j^{(k+1)}(s) + \sigma_j \left(t, \ v^{(k)}, \ w^{(k)} \right)$$
$$- f_j \left(t, \ z_j(t), \ [z]_{p_j}, \ [u]_{q_j}, \ \max_{s \in [t-h,t]} z_j(s), \right.$$
$$\left. \left[\max_{s \in [t-h,t]} z_j(s) \right]_{p_j}, \left[\max_{s \in [t-h,t]} u_j(s) \right]_{q_j} \right)$$
$$= - M_j \left(v_j^{(k+1)}(t) - v_j^{(k)}(t) \right)$$
$$- N_j \left(\max_{s \in [t-h,t]} v_j^{(k+1)}(s) - \max_{s \in [t-h,t]} v_j^{(k)}(s) \right)$$
$$+ M_j \left(z_j^{(t)} - v_j^{(k)}(t) \right) + N_j \left(\max_{s \in [t-h,t]} z_j(s) - \max_{s \in [t-h,t]} v_j^{(k+1)}(s) \right)$$

$$\leq -M_j\left(v_j^{(k+1)}(t) - z_j(t)\right) - N_j \min_{s\in[t-h,t]}\left(V_j^{(k+1)}(s) - z_j(s)\right)$$

for $t \in [0, T]$.

By Lemma 6.2.3 the functions $g_j(t)$ for $j = \overline{1, n}$ are nonpositive, i.e., the inequalities $v^{(k+1)}(t) \leq z(t)$ hold for $t \in [-h, T]$.

In an analogous way it is proven that the inequalities $z(t) \leq w^{(k+1)}(t)$ and $v^{(k+1)}(t) \leq u(t) \leq w^{(k+1)}(t)$ hold for $t \in [-h, T]$ which shows that the couple of functions $(\bar{v}, \bar{w})$ is a couple of minimal and maximal quasisolutions of the boundary value problem (6.33), (6.34).

Let $u \in S(v, w)$ be a solution of the boundary value problems (6.33) and (6.34). Consider the couple of functions (u, u) which is a couple of quasisolutions of (6.33) and (6.34). In view of what was proven about the functions $\bar{v}$, $\bar{w}$ it follows that the inequalities $\bar{v}(t) \leq u(t) \leq \bar{w}(t)$ hold for $t \in [-h, T]$.

This completes the proof of Theorem 6.2.1.

□

In the case when (6.33), (6.34) is a boundary value problem for a scalar differential equation with "maxima", i.e., $n = 1$ and $p_1 = q_1 = 0$, as a corollary of Theorem 6.2.1 the following result is obtained:

Theorem 6.2.2. *Let the following conditions be fulfilled:*

1. *The functions v, $w \in C([0, T], \mathbb{R}^n) \cup C^1([0, T], \mathbb{R})$ are lower and upper quasi-solutions of the boundary value problem (6.33), (6.34) and satisfy the conditions $v(t) \leq w(t)$ for $t \in [-h, T]$ and $v(t) \equiv v(0)$, $w(t) \equiv w(0)$ for $t \in [-h, 0]$.*

2. *The function $f \in C([0, T] \times \mathbb{R} \times \mathbb{R}, \mathbb{R})$ and for x, $y \in S(v, w)$, $y(t) \leq x(t)$ satisfies the inequality*

$$f\left(t, \ x(t), \ \max_{s\in[t-h,t]} x(s)\right) - f\left(t, \ y(t), \ \max_{s\in[t-h,t]} y(s)\right)$$
$$> -M\big(x(t) - y(t)\big) - N\left(\max_{s\in[t-h,t]} x(s) - \max_{s\in[t-h,t]} y(s)\right),$$

where M and N are positive constants such that $N < M \leq 1/(2T)$.

Then there exist two sequences of functions $\{v^{(k)}(t)\}_0^\infty$ and $\{w^{(k)}(t)\}_0^\infty$ such that:

(a) The sequences are increasing and decreasing, correspondingly;

(b) The pair of functions $v^{(k)}(t)$, $w^{(k)}(t)$ is a pair of lower and upper quasisolutions of the boundary value problem for the system of nonlinear differential equations with "maxima"(6.33), (6.34);

(c) Both sequences uniformly converge on $[-h, T]$;

(d) The limits $V(t) = \lim\limits_{k \to \infty} v^{(k)}(t)$, $W(t) = \lim\limits_{k \to \infty} w^{(k)}(t)$ are a pair of minimal and maximal solutions of the boundary value problem for the system of nonlinear differential equations with "maxima"(6.33), (6.34).

(e) If $u(t) \in S(v, w)$ is a solution of the boundary value problem for the system of nonlinear differential equations with "maxima" (6.33), (6.34), then $V(t) \le u(t) \le W(t)$.

6.3 Monotone-Iterative Technique for Second Order Differential Equations with "Maxima"

Consider the following periodic boundary value problem for the scalar second order differential equation with "maxima"

$$-x''(t) = f\left(t, \ x(t), \ \max_{s \in [t-h,t]} x(s)\right) \quad \text{for} \quad t \in [0, T], \tag{6.57}$$

$$x(0) = x(T), \quad x'(0) = x'(T), \tag{6.58}$$

$$x(t) = x(0) \quad \text{for} \quad t \in [-h, 0], \tag{6.59}$$

where $x \in \mathbb{R}$, $f : [0, T] \times \mathbb{R} \times \mathbb{R} \to \mathbb{R}$, h and T are positive constants with $T > h$.

Introduce the following notation $E = C([-h, T], \mathbb{R}) \cup C^2([0, T], \mathbb{R})$.
Let $x \in E$. Denote $||x||_0 = \max_{t \in [-h,T]} |x(t)|$, $||x'||_0 = \max_{t \in [-h,T]} |x'(t)|$ and $||x||_1 = \max\left(||x||_0, ||x'||_0\right)$.

Definition 6.3.1. *The function $\alpha \in E$ is a lower solution of the boundary value problem (6.57), (6.58), (6.59) if the inequalities*

$$-\alpha''(t) \le f\Big(t,\ \alpha(t),\ \max_{s\in[t-h,t]} \alpha(s)\Big) \quad for \quad t \in [0,T], \tag{6.60}$$

$$\alpha(0) = \alpha(T), \quad \alpha'(0) \ge \alpha'(T), \tag{6.61}$$

$$\alpha(t) = \alpha(0) \quad for \quad t \in [-h,0] \tag{6.62}$$

hold.

Definition 6.3.2. *The function $\beta \in E$ is an upper solution of the boundary value problem (6.57), (6.58), (6.59) if inequalities (6.60), (6.61) and (6.62) hold in the opposite direction.*

Let the functions $\alpha,\ \beta \in C([-h,T],\mathbb{R})$ be such that $\alpha(t) \le \beta(t)$ for $t \in [-h,T]$. Define the sets

$$S(\alpha,\beta) = \Big\{x \in C([-h,T],\mathbb{R}) : \alpha(t) \le x(t) \le \beta(t) \ \text{for } t \in [-h,T]\Big\},$$

$$\Omega(\alpha,\beta) = \Big\{(t,x,y) : t \in [0,T],$$

$$\alpha(t) \le x \le \beta(t), \quad \max_{s\in[t-h,t]} \alpha(s) \le y \le \max_{s\in[t-h,t]} \beta(s)\Big\}.$$

Lemma 6.3.1. *[Comparison result] The function $u \in E$ satisfies the inequalities*

$$-u''(t) \le -Mu(t) - N \inf_{s\in[t-h,t]} u(s) \quad for \quad t \in [0,T], \tag{6.63}$$

$$u(0) = u(T), \quad u'(0) \ge u'(T), \tag{6.64}$$

$$u(t) = u(0) \quad for \quad t \in [-h,0], \tag{6.65}$$

where the constants M and N are positive and

$$2(M+N)T \le 1. \tag{6.66}$$

Then $u(t) \le 0$ for $t \in [-h,T]$.

Proof. Suppose the claim is not true. Consider the following two cases:
    Case 1. Let $u(t) \ge 0$ for $t \in [0,T]$ and there exists a $t^* \in [0,T]$ such that $u(t^*) > 0$;
    Inequality (6.63) implies that $u''(t) \ge 0$ on $[0,T]$. Therefore $u'(t)$ is nondecreasing on $[0,T]$ and $u'(0) \ge u'(T)$. Therefore $u'(t)$ must be a

constant: $u'(t) \equiv c$ ($t \in [0, T]$) and therefore $0 = -u''(t^*) \leq -Mu(t^*) < 0$, a contradiction.

Case 2. Let there exist t^*, $t_* \in [0, T]$ such that $u(t^*) > 0$ and $u(t_*) < 0$.

Denote $\min_{t \in [0,T]} u(t) = -\lambda$. Then $\lambda > 0$ and there exists a $t_* \in (0, T)$ such that $u(t_*) = -\lambda$. By (6.63) we have

$$-u''(t) \leq M\lambda + N\lambda = (M + N)\lambda. \tag{6.67}$$

We now show that there exists a $\bar{t} \in [0, T]$ such that

$$u'(\,\bar{t}\,) \leq 0. \tag{6.68}$$

In fact, if $u'(t) > 0$ for all $t \in [0, T]$, then $u(t)$ is strictly increasing on $[0, T]$, which contradicts $u(0) = u(T)$. Hence (6.68) holds.

From the mean value theorem we obtain

$$u'(\,\bar{t}\,) - u'(0) = u''(\xi_0)t_1, \qquad 0 < \xi_0 < \bar{t}. \tag{6.69}$$

By (6.67) we get

$$u'(0) - u'(\,\bar{t}\,) \leq (M + N)\lambda T. \tag{6.70}$$

From (6.68) and (6.70) it follows that

$$u'(0) \leq u'(\,\bar{t}\,) + \lambda(M + N)T \leq \lambda(M + N)T. \tag{6.71}$$

We now show that

$$u'(t) \leq 2\lambda(M + N)T \quad \text{for } t \in [0, T]. \tag{6.72}$$

In fact, let $t \in [0, T]$. Then according to the mean value theorem, we have

$$u'(T) - u'(t) = u''(\xi_1)(T - t), \quad 0 < \xi_1 < T$$

and so, similar to (6.70) and (6.71), we get

$$u'(t) \leq u'(T) + \lambda(M + N)T. \tag{6.73}$$

It is clear that (6.72) follows from (6.71) and (6.73) and the fact that $u'(0) \geq u'(T)$.

Now let $t^* \in [0, T]$. First assume that $t_* \le t^*$. Apply Mean Value Theorem and obtain

$$u(t^*) - u(t_*) = u'(\eta)(t^* - t_*), \quad t_* < \eta < t^*. \tag{6.74}$$

So, using (6.72)

$$u(t^*) - u(t_*) \le 2\lambda T^2 (M + N) \tag{6.75}$$

together with $u(t_*) = -\lambda$, we get

$$0 < u(t^*) \le -\lambda + 2\lambda(M + N)T,$$

which contradicts (6.66).

If $t_* > t^*$, the argument is similar.

Hence, we obtain that $u(t) \le 0$ for $t \in [0, T]$.

According to (6.65), $u(t) \le 0$ holds for $t \in [-h, T]$.

$\square$

Lemma 6.3.2. *Let $\sigma \in C([-h, T], \mathbb{R})$, $\varphi \in C[-h, 0]$, $M > 0$, $N > 0$ be constants and $u \in E$ be a solution of the linear integral equation*

$$u(t) = \int_0^T G(t, r)\sigma_1(r)dr + z(t) + \varphi(0), \quad t \in [0, T], \tag{6.76}$$
$$u(t) = \varphi(t), \quad t \in [-h, 0],$$

where

$$G(t, r) = \frac{1}{2\sqrt{M}\left(e^{T\sqrt{M}} - 1\right)} \begin{cases} e^{\sqrt{M}(r-t)} + e^{\sqrt{M}(T-r+t)}, & t \le r, \\ e^{\sqrt{M}(t-r)} + e^{\sqrt{M}(T-t+r)}, & t > r, \end{cases} \tag{6.77}$$

$$\sigma_1(t) = \sigma(t) - N \max_{s \in [t-h, t]} u(s). \tag{6.78}$$

Then $u(t) \in E$ is a solution of the following linear boundary value problem

$$-u''(t) = -Mu(t) - N \max_{s \in [t-h, t]} u(s) + \sigma(t) \quad for \quad t \in [0, T],$$

$$u(0) = u(T), \quad u'(0) = u'(T), \quad u(t) = \varphi(t) \quad for \quad t \in [-h, 0]. \tag{6.79}$$

Proof. Let $u \in E$ be a solution of (6.76). From $t \in [0, T]$ direct differentiation of (6.76) gives

$$u'(t) = \int_0^T G_t'(t, r)\sigma_1(r)dr \qquad (6.80)$$

and

$$u''(t) = M \int_0^T G(t, r)\sigma_1(r)dr - \sigma_1(t), \qquad (6.81)$$

where

$$G_t'(t, r) = \frac{1}{2\left(e^{T\sqrt{M}} - 1\right)} \begin{cases} -e^{\sqrt{M}(r-t)} + e^{\sqrt{M}(T-r+t)}, & t < r, \\ e^{\sqrt{M}(t-r)} - e^{\sqrt{M}(T-t+r)}, & t > r. \end{cases} \qquad (6.82)$$

From the above equalities, we get

$$-u''(t) = -Mu(t) + \sigma_1(t) = -Mu(t) - N \max_{s \in [t-h, t]} u(s) + \sigma(t).$$

Moreover, from (6.76) and (6.80), it is easy to find

$$u(T) = \frac{1}{2\sqrt{M}\left(e^{T\sqrt{M}} - 1\right)} \int_0^T \left(e^{\sqrt{M}(T-r)} + e^{\sqrt{M}r}\right)\sigma_1(t)dr + \varphi(0) = u(0),$$

$$u'(T) = \frac{1}{2\left(e^{T\sqrt{M}} - 1\right)} \int_0^T \left(e^{\sqrt{M}(T-r)} - e^{\sqrt{M}r}\right)\sigma_1(t)dr = u'(0).$$

Hence $u(t) \in E$ is a solution of the boundary value problem (6.79). $\square$

Lemma 6.3.3. *Let $u \in E$, $M > 0$, $N > 0$ be constants and the following equations hold*

$$\beta_1 = TG_0N < 1, \qquad (6.83)$$

$$\beta_2 = \frac{1}{2}TN, \qquad (6.84)$$

where G_0 will be defined below.
Then equation (6.76) has a unique solution $u(t) \in E$.

Proof. Define an operator F in E by

$$(Fu)(t) = \int_0^T G(t, r)\sigma_1(r)dr + \varphi(0), \quad t \in [0, T],$$
$$(Fu)(t) = \varphi(t), \quad t \in [-h, 0], \qquad (6.85)$$

where $G(t, r)$ and $\sigma_1(t)$ are defined by (6.77) and (6.78). We have

$$(Fu)'(t) = \int_0^T G'_t(t, r)\sigma_1(r)dr \quad \text{for } t \in [0, T]. \qquad (6.86)$$

From (6.77) and (6.82), we have

$$\max_{t, r \in [0, T]} G(t, r) = \frac{1 + e^{T\sqrt{M}}}{2\sqrt{M}\left(e^{T\sqrt{M}} - 1\right)} = G_0, \qquad (6.87)$$

$$\max_{t, r \in [0, T]} G'_t(t, r) = \frac{1}{2}, \quad t \neq r. \qquad (6.88)$$

For any u, $v \in E$, (6.85) and (6.87) imply that

$$\left\| (Fu)(t) - (Fv)(t) \right\|_0 \leq TG_0 N \|u - v\|_0$$
$$\leq TG_0 N \|u - v\|_1 \leq \beta_1 \|u - v\|_1.$$

Similarly, from (6.86) and (6.88), we have

$$\left\| (Fu)'(t) - (Fv)'(t) \right\|_0 \leq \beta_2 \|u - v\|_1.$$

Hence

$$\left\| (Fu)(t) - (Fv)(t) \right\|_1 \leq \beta \|u - v\|_1 \quad \text{for any } u, \ v \in E,$$

where $\beta = \max\{\beta_1, \beta_2\} < 1$. The Banach fixed point theorem implies that F has a unique fixed point in E and the lemma is proved.

$\square$

We now give a procedure for constructing two sequences of functions that are respectively monotone-increasing and monotone-decreasing which converge to the extremal solutions of the boundary value problem (6.57), (6.58), (6.59).

Theorem 6.3.1. *Let the following conditions be fulfilled:*

1. *The functions α, $\beta \in E$ are lower and upper solutions, respectively, of the boundary value problem (6.57), (6.58), (6.59) such that $\alpha(t) \leq \beta(t)$ for $t \in [-h, T]$;*

2. *The function $f \in C([0, T] \times \mathbb{R} \times \mathbb{R}, \mathbb{R})$ and for $(t, \ x_1, \ y_1)$, $(t, \ x_2, \ y_2) \in \Omega(\alpha, \beta)$, $x_1 \geq x_2$, $y_1 \geq y_2$, the inequality $f(t, \ x_1, \ y_1) - f(t, \ x_2, \ y_2) \geq -M(x_1 - x_2) - N(y_1 - y_2)$ holds;*

3. Inequalities (6.66), (6.83) and (6.84) hold.

Then there exist two sequences of functions $\{\alpha_k(t)\}_0^\infty$ and $\{\beta_k(t)\}_0^\infty$ such that:

(a) The sequences are increasing and decreasing, correspondingly;

(b) The pair of functions $\alpha_k(t)$, $\beta_k(t)$ is a pair of lower and upper quasisolutions of the boundary value problem for the system of nonlinear differential equations with "maxima" (6.57), (6.58), (6.59);

(c) Both sequences uniformly converge on $[-h, T]$;

(d) The limits $V(t) = \lim_{k \to \infty} \alpha_k(t)$, $W(t) = \lim_{k \to \infty} \beta_k(t)$ are a pair of minimal and maximal solutions of the boundary value problem for the system of nonlinear differential equations with "maxima" (6.57), (6.58), (6.59).

(e) If $u(t) \in S(\alpha, \beta)$ is a solution of the boundary value problem for the system of nonlinear differential equations with "maxima" (6.57), (6.58), (6.59), then $V(t) \le u(t) \le W(t)$.

Proof. Fix a function $\eta \in S(\alpha, \beta)$ and consider the following boundary value problem for the linear impulsive differential equation with "maxima"

$$-u''(t) = -Mu(t) - N \max_{s \in [t-h,t]} u(s) + \sigma(t, \eta) \quad \text{for} \quad t \in [0, T], \quad (6.89)$$

$$u(t) = u(0) \quad \text{for} \quad t \in [-h, 0], \quad (6.90)$$

$$u(0) = u(T), \quad (6.91)$$

where $\sigma(t, \eta) = f\left(t, \eta(t), \max_{s \in [t-h,t]} \eta(s)\right) + M\eta(t) + N \max_{s \in [t-h,t]} \eta(s)$.

By Lemma 6.3.2 and Lemma 6.3.3, the boundary value problem (6.89), (6.90), (6.91) has a solution $x \in E$ which is the unique solution of equation (6.76) in $C([-h, T], \mathbb{R}) \cap C^1([0, T], \mathbb{R})$. We define a map W by the equality $W(\eta) = x$. Let η, $\mu \in S(\alpha, \beta)$ such that $\eta(t) \le \mu(t)$ and $x = W(\eta)$, $y = W(\mu)$. Consider the function $v(t) = x(t) - y(t)$.

The function $v(t)$ satisfies the inequalities

$$
\begin{aligned}
- \; v''(t) &= -Mv(t) - N\Big\{ \max_{s\in[t-h,t]} x(s) - \max_{s\in[t-h,t]} y(s) \Big\} \\
&\quad + f\Big(t,\, \eta(t),\, \max_{s\in[t-h,t]} \eta(s) \Big) \\
&\quad - f\Big(t,\, \mu(t),\, \max_{s\in[t-h,t]} \mu(s) \Big) + M\big(\eta(t) - \mu(t)\big) \\
&\quad + N\Big\{ \max_{s\in[t-h,t]} \eta(s) - \max_{s\in[t-h,t]} \mu(s) \Big\} \\
&\le \; -Mv(t) - N \inf_{s\in[t-h,t]} v(s) \qquad \text{for } t \in [0,T],
\end{aligned}
$$

$$
v(0) = v(T), \qquad v'(0) = v'(T), \qquad v(t) = v(0) \text{ for } t \in [-h,0].
$$

According to Lemma 6.3.1, the function $v(t)$ is nonpositive, i.e., $x(t) \le y(t)$ and the map W is nondecreasing.

Now starting with $\alpha_0 = \alpha$, $\beta_0 = \beta$, we can recursively define two sequences $\{\alpha_k\}$ and $\{\beta_k\}$ by the equalities

$$
\alpha_m = W(\alpha_{m-1}), \qquad \beta_m = W(\beta_{m-1}), \qquad m \ge 1.
$$

From the relationships above, we see that the inequalities

$$
\alpha_0(t) \le \alpha_1(t) \le \cdots \le \alpha_m(t) \le \beta_m(t) \le \cdots \le \beta_0(t), \quad t \in [-h,T]
$$

hold.

By the standard arguments, we can see that the sequences $\{\alpha_k\}$ and $\{\beta_k\}$ are uniformly bounded and completely continuous. Therefore the sequences are uniformly convergent in $[-h,T]$.

Let $\alpha_*(t) = \lim_{m\to\infty} \alpha_m(t)$ and $\beta^*(t) = \lim_{m\to\infty} \beta_m(t)$, we see that the functions $\alpha_*(t)$ and $\beta^*(t)$ are the solutions of the boundary value problem (6.57), (6.58), (6.59), $\alpha_*(t) \le \beta^*(t)$ and $\alpha_*(t), \beta^*(t) \in S(\alpha,\beta)$.

Let $x \in S(\alpha,\beta)$ solve the boundary value problem (6.57), (6.58), (6.59). Now consider the function $w(t) = \alpha_*(t) - x(t)$. The function $w(t)$ satisfies the conditions of Lemma 6.3.1 and therefore we have $w(t) \le 0$ for $t \in [-h,T]$. Using Lemma 6.3.1 again, we conclude that $\alpha_*(t) \le x(t)$. These inequalities imply that the solutions $\alpha_*(t)$ and $\beta^*(t)$ are the minimal and maximal solutions of the boundary value problem (6.57), (6.58), (6.59), respectively, in $S(\alpha,\beta)$. The proof is complete.

$\square$

6.4   Method of Quasilinearization for Initial Value Problems

The method of quasilinearization is a practically useful method for obtaining approximate solutions of nonlinear problems. The origin of this method lies in the theory of dynamic programming [Bellman and Kalaba 1965]. The quasilinearization method is a Taylor series numerical method in which the truncation is chosen so that the convergence of the iterates is quadratic. Many authors have applied this method to finding approximate solutions of various types of first and second order ordinary differential equations (see the monograph [Lakshmikantham and Vatsala 1998] and references cited therein).

In this section an application of the method of quasilinearization to the initial value problem for a first order scalar differential equation with "maxima" is presented.

Consider the following initial value problem for the nonlinear differential equation with "maxima" (IVP)

$$x' = f(t, x(t), \max_{s \in [t-h,t]} x(s)) \quad \text{for} \quad t \geq 0, \tag{6.92}$$

$$x(t) = \varphi(t) \quad \text{for} \quad t \in [-h, 0], \tag{6.93}$$

where $x \in \mathbb{R}$, $f : [0, T] \times \mathbb{R} \times \mathbb{R} \to \mathbb{R}$, $\varphi(t) : [-h, 0] \to \mathbb{R}$, $h > 0$, $T > 0$ are fixed constants.

Definition 6.4.1. *The function* $\alpha \in C([-h, T], \mathbb{R}) \cup C^1([0, T], \mathbb{R})$ *is called a* lower solution *of the IVP (6.92), (6.93), if the following inequalities are satisfied:*

$$\alpha'(t) \leq f(t, \alpha(t), \max_{s \in [t-h,t]} \alpha(s)) \quad \text{for} \quad t \geq 0,$$
$$\alpha(t) \leq \varphi(t) \quad \text{for} \quad t \in [-h, 0]. \tag{6.94}$$

Definition 6.4.2. *The function* $\alpha \in C([-h, T], \mathbb{R}) \cup C^1([0, T], \mathbb{R})$ *is called an* upper solution *of the IVP (6.92), (6.93), if the following inequalities are satisfied:*

$$\alpha'(t) \geq f(t, \alpha(t), \max_{s \in [t-h,t]} \alpha(s)) \quad \text{for} \quad t \geq 0,$$
$$\alpha(t) \geq \varphi(t) \quad \text{for} \quad t \in [-h, 0]. \tag{6.95}$$

Let the functions $\alpha, \beta \in C([-h, T], \mathbb{R})$ be such that $\alpha(t) \leq \beta(t)$. Consider the sets:

$$
\begin{aligned}
S(\alpha, \beta) &= \{u \in C([-h, T], \mathbb{R}) : \ \alpha(t) \leq u(t) \leq \beta(t) \ \text{ for } \ t \in [-h, T]\}, \\
\Omega(\alpha, \beta) &= \{(t, x) \in [-h, T] \times \mathbb{R} : \ \alpha(t) \leq x \leq \beta(t)\}.
\end{aligned}
$$

We will prove a comparison result for the lower and upper solutions of IVP (6.92), (6.93).

Theorem 6.4.1. *Let the following conditions be fulfilled:*

1. *The functions $\alpha, \beta \in C([-h, T], \mathbb{R}) \cup C^1([0, T], \mathbb{R})$ are lower and upper solutions of the IVP (6.92), (6.93), correspondingly.*

2. *The function $f \in C([0, T] \times \mathbb{R}^2, \mathbb{R}$ is nondecreasing in its last argument and for any $x \geq y$, $u \geq v$ and $t \in [0, T]$ the inequality*

$$
f(t, x, u) - f(t, y, v) \leq M(x - y) + N(u - v)
$$

holds, where $M, N > 0$ are constants.

Then $\alpha(t) \leq \beta(t)$ for $t \in [0, T]$.

Proof. Consider the following three cases:

Case 1. Let both inequalities (6.94) be strict. Assume the claim of Theorem 6.4.1 is not true. We will prove $\alpha(t) < \beta(t)$ on $[0, T]$. If not, there exists a point t_0 such that $\alpha(t) < \beta(t)$ on $[-h, t_0)$, $\alpha(t_0) = \beta(t_0)$ and $\alpha'(t_0 - 0) \geq \beta(t_0 - 0)$. Then from condition 2 follows that $f(t, \alpha(t_0), \max_{s \in [t_0-h,t_0]} \alpha(s)) \leq f(t, \alpha(t_0), \max_{s \in [t_0-h,t_0]} \beta(s))$ and we obtain the following contradiction:

$$
f(t, \alpha(t_0), \max_{s \in [t_0-h,t_0]} \alpha(s)) > \alpha'(t_0) \geq \beta'(t_0) \geq f(t, \beta(t_0), \max_{s \in [t_0-h,t_0]} \beta(s))
$$
$$
= f(t, \alpha(t_0), \max_{s \in [t_0-h,t_0]} \beta(s)).
$$

Case 2. Let both inequalities (6.95) be strict. As in Case 1 we obtain a contradiction.

Case 3. Let there exist a point such that at least one of the inequalities (6.94) and (6.95) is not strict. Choose a small enough number $\epsilon > 0$ and define a function $w(t) = \beta(t) + \epsilon e^{2(M+N)t}$. Then $w(t) > \varphi(t)$ on

$[-h, 0]$, $\max_{s\in[t-h,t]} w(s) \leq \max_{s\in[t-h,t]} \beta(s) + \epsilon e^{2(M+N)t}$ and

$$
\begin{aligned}
w'(t) &\geq f(t, \beta, \max_{s\in[t-h,t]} \beta(s)) + 2(M+N)\epsilon e^{2(M+N)t} \\
&= f(t, w(t), \max_{s\in[t-h,t]} w(s)) + 2(M+N)\epsilon e^{2(M+N)t} \\
&\quad - \left(f(t, w(t), \max_{s\in[t-h,t]} w(s)) - f(t, \beta(t), \max_{s\in[t-h,t]} \beta(s)) \right) \\
&\geq f(t, w(t), \max_{s\in[t-h,t]} w(s)) + 2(M+N)\epsilon e^{2(M+N)t} \\
&\quad - \left(M+N \right)\epsilon e^{2(M+N)t} \\
&> f(t, w(t), \max_{s\in[t-h,t]} w(s)), \quad t \in [0, T].
\end{aligned}
$$

According to Case 2, the inequality $\alpha(t) < w(t)$ holds for $t \in [0, T]$. Taking a limit as ϵ approaches 0 we prove the claim of Theorem 6.4.1.

□

In our further investigations we will use some results for differential inequalities with "maxima." In the case of a nonnegative coefficient before the maximum function the following result is true:

Lemma 6.4.1. *[Comparison result] Let the following conditions be fulfilled:*

1. *The functions $g_1 \in C([0, T], \mathbb{R})$, $g_2 \in C([0, T], \mathbb{R}_+)$ and $g_2 \not\equiv 0$ for $t \in [0, T]$.*

2. *The inequality*
$$(M+N)T < 1, \tag{6.96}$$
holds, where $M = max\{|g_1(t)| : t \in [0, T]\} > 0$, $N = max\{g_2(t) : t \in [0, T]\} > 0$.

3. *The function $u \in C([-h, T], \mathbb{R}) \cup C^1([0, T], \mathbb{R})$ satisfies the inequalities*

$$u' \leq g_1(t)u(t) + g_2(t) \max_{s\in[t-h,t]} u(s) \quad for \quad t \geq 0 \tag{6.97}$$

$$u(t) \leq 0 \quad for \quad t \in [-h, 0]. \tag{6.98}$$

Then the function $u(t)$ is nonpositive on the interval $[0, T]$.

Proof. Assume the claim of Lemma 6.4.1 is not true.

Denote $\lambda = \max_{t \in [0,T]} u(t)$, $\quad \lambda > 0$.

Then there exist points $\xi, \eta \in (0,T]$ such that $\xi > \eta$, $u(\eta) = 0$, $u(\xi) = \lambda > 0$, $0 < u(t) \leq u(\xi)$ for $t \in (\eta, \xi]$.

Therefore there exists a point $\tau \in (\eta, \xi)$ such that

$$\lambda = u(\xi) - u(\eta) = u'(\tau)(\xi - \eta) \tag{6.99}$$

According to (6.99) $u'(\tau) > 0$ and $\lambda T \leq u'(\tau)T$. From (6.97) and the inequality $\max_{s \in [\tau - h, \tau]} u(s) \leq \lambda$ it follows that

$$\lambda \leq u'(\tau)T \leq g_1(\tau)u(\tau)T + g_2(\tau)T \max_{s \in [\tau - h, \tau]} u(s) \leq Mu(\tau)T + g_2(\tau)\lambda T$$

$$\leq (M + N)\lambda T. \tag{6.100}$$

The inequality (6.100) contradicts the inequality (6.96).

$\square$

We will apply the method of quasilinearization for approximate finding of a solution of the initial value problem for a nonlinear differential equation with "maxima." We will prove the convergence of the sequence of successive approximations is quadratic.

Theorem 6.4.2. *Let the following conditions be fulfilled:*

1. *The functions $\alpha_0(t), \beta_0(t) \in C([-h,T], \mathbb{R}) \cup C^1([0,T], \mathbb{R})$ are lower and upper solutions of the IVP (6.92), (6.93), and $\alpha_0(t) \leq \beta_0(t)$ for $t \in [0,T]$.*

2. *The function $\varphi \in C([-h,0], \mathbb{R})$.*

3. *There exist functions $F, g \in C^{0,2,2}(\Omega(\alpha_0, \beta_0), \mathbb{R})$ such that*

$$F(t,x,y) = f(t,x,y) + g(t,x,y)$$

and

$$F_{xx}(t,x,y) \geq 0, \quad F_{xy}(t,x,y) \geq 0, \quad F_{yy}(t,x,y) \geq 0,$$
$$g_{xx}(t,x,y) \geq 0, \quad g_{xy}(t,x,y) \geq 0, \quad g_{yy}(t,x,y) \geq 0,$$

$$F_y(t, \alpha_0(t), \max_{s \in [t-h,t]} \alpha_0(s)) \geq g_y(t, \beta_0(t), \max_{s \in [t-h,t]} \beta_0(s)),$$

$$(M + N)T \leq 1,$$

where

$$
\begin{aligned}
M &= max\{|F_x(t, \alpha_0(t), \max_{s\in[t-h,t]} (\alpha_0(s))) - g_x(t, \beta_0(t), \\
&\quad \max_{s\in[t-h,t]} (\beta_0(s)))|, \quad t \in [0, T]\}, \\
N &= max\{F_y(t, \beta_0(t), \max_{s\in[t-h,t]} \beta_0(s)) - g_y(t, \alpha_0(t), \\
&\quad \max_{s\in[t-h,t]} \alpha_0(s)), \quad t \in [0, T]\}.
\end{aligned}
$$

Then there exist two sequences of functions $\{\alpha_n(t)\}_0^\infty$ and $\{\beta_n(t)\}_0^\infty$ such that:

a. The sequences are increasing and decreasing, correspondingly;

b. Both sequences uniformly converge in the interval $[0, T]$ and the limits are equal to the unique solution $x(t)$ of the IVP (6.92), (6.93) in $S(\alpha_0, \beta_0)$;

c. The convergence is quadratic, i.e., $\|x - \alpha_{n+1}\| \leq \lambda_1\|x - \alpha_n\|^2 + \mu_1\|x - \beta_n\|^2$ and $\|x - \beta_{n+1}\| \leq \lambda_2\|x - \alpha_n\|^2 + \mu_2\|x - \beta_n\|^2$, where $\|u\| = \max_{s\in[-h,T]} |u(s)|$ for any function $u \in C([-h, T], \mathbb{R})$.

Proof. From the condition 3 of Theorem 6.4.2 it follows that for (t, x_1, y_1), $(t, x_2, y_2) \in \Omega(\alpha_0, \beta_0)$ and $x_1 \geq x_2, y_1 \geq y_2$ the inequalities

$$
\begin{aligned}
f(t, x_1, y_1) \geq f(t, x_2, y_2) + F_x(t, x_2, y_2)(x_1 - x_2) + F_y(t, x_2, y_2)(y_1 - y_2) \\
+ g(t, x_2, y_2) - g(t, x_1, y_1),
\end{aligned}
$$
$$(6.101)$$

$$
\begin{aligned}
f(t, x_2, y_2) \leq f(t, x_1, y_1) + F_x(t, x_2, y_2)(x_2 - x_1) + F_y(t, x_2, y_2)(y_2 - y_1) \\
+ g(t, x_1, y_1) - g(t, x_2, y_2).
\end{aligned}
$$
$$(6.102)$$

hold.

Let $L_0 = \min_{s\in[-h,0]} (\varphi(s) - \alpha_0(s)) \geq 0$. Choose a number $k_0 \in [0, 1)$ such that

$$k_0 \leq L_0.$$
$$(6.103)$$

We consider the linear differential equation with "maxima"

$$x'(t) = f(t, \alpha_0(t), \max_{s \in [t-h,t]} (\alpha_0(s))) + Q_0(t)(x(t) - \alpha_0(t))$$

$$+ q_0(t)(\max_{s \in [t-h,t]} (x(s)) - \max_{s \in [t-h,t]} (\alpha_0(s))) \text{ for } t \in [0, T],$$

$$(6.104)$$

with initial condition

$$x(t) = \varphi(t) - k_0 L_0 \text{ for } t \in [-h, 0], \tag{6.105}$$

where

$$Q_0(t) = F_x(t, \alpha_0(t), \max_{s \in [t-h,t]} (\alpha_0(s))) - g_x(t, \beta_0(t), \max_{s \in [t-h,t]} (\beta_0(s))),$$

$$q_0(t) = F_y(t, \alpha_0(t), \max_{s \in [t-h,t]} (\alpha_0(s))) - g_y(t, \beta_0(t), \max_{s \in [t-h,t]} (\beta_0(s))) \geq 0.$$

According to the results in Section 3.2 the IVP (6.104),(6.105) has a unique solution $\alpha_1(t)$ on $[-h, 0]$.

We will prove that $\alpha_1(t) \in S(\alpha_0, \beta_0)$.

Consider the function $u(t) = \alpha_0(t) - \alpha_1(t)$, $t \in [-h, T]$. From the initial condition (6.105) and condition 1 of Theorem 6.4.2 the inequality $\alpha_1(t) = \varphi(t) - k_0 L_0 \geq \alpha_0(t)$ holds, i.e., $u(t) \leq 0$ on $[-h, 0]$.

Let $t \in [0, T]$. Then the inequalities

$$\max_{s \in [t-h,t]} \alpha_0(s) - \max_{s \in [t-h,t]} \alpha_1(s) = \alpha_0(\xi) - \max_{s \in [t-h,t]} \alpha_1(s)$$

$$\leq \alpha_0(\xi) - \alpha_1(\xi) \leq \max_{s \in [t-h,t]} (\alpha_0(s) - \alpha_1(s)) = \max_{s \in [t-h,t]} u(s)$$

hold and since $q_0(t) \geq 0$ we get

$$u'(t) \leq Q_0(t)u(t) + q_0(t) \max_{s \in [t-h,t]} u(s), \quad t \in [0, T].$$

According to Lemma 6.4.1 the inequality $\alpha_0(t) \leq \alpha_1(t)$ holds on $[0, T]$.

We will prove that $\alpha_1(t) \leq \beta_0(t)$. Consider the function $u(t) = \alpha_1(t) - \beta_0(t)$, $t \in [-h, T]$.

For $t \in [-h, 0]$ according to condition 1 of Theorem 6.4.2 we obtain $\alpha_1(t) = \varphi(t) - k_0 L_0 \leq \varphi(t) \leq \beta_0(t)$, i.e., $u(t) \leq 0$.

According to the definition of the functions α_1 and β_0, the inequality (6.102) and the inequality

$$\max_{s \in [t-h,t]} \alpha_1(s) - \max_{s \in [t-h,t]} \beta_0(s) \leq \max_{s \in [t-h,t]} (\alpha_1(s) - \beta_0(s))$$

we get

$$u'(t) \leq f(t, \alpha_0(t), \max_{s \in [t-h,t]} \alpha_0(s)) + Q_0(t)(\alpha_1(t) - \alpha_0(t))$$

$$+ q_0(t)(\max_{s \in [t-h,t]} \alpha_1(s) - \max_{s \in [t-h,t]} \alpha_0(s)) - f(t, \beta_0(t), \max_{s \in [t-h,t]} \beta_0(s))$$

$$\leq Q_0(t)u(t) + q_0(t) \max_{s \in [t-h,t]} u(s).$$

$$(6.106)$$

According to Lemma 6.4.1 we obtain that $u(t) \leq 0$, i.e., the inequality $\alpha_1(t) \leq \beta_0(t)$ holds on $[0, T]$.

Let $C_0 = \min_{s \in [-h,0]} (\beta_0(s) - \varphi(s)) \geq 0$. Choose a number $p_0 \in [0, 1)$ such that

$$p_0 \leq C_0. \tag{6.107}$$

We consider the linear differential equation with "maxima"

$$x'(t) = f(t, \beta_0(t), \max_{s \in [t-h,t]} \beta_0(s)) + Q_0(t)(x(t) - \beta_0(t))$$

$$+ q_0(t)(\max_{s \in [t-h,t]} x(s) - \max_{s \in [t-h,t]} \beta_0(s)) \quad \text{for} \quad t \in [0, T] \quad (6.108)$$

with initial condition

$$x(t) = \varphi(t) + p_0 C_0 \quad \text{for} \quad t \in [-h, 0]. \tag{6.109}$$

There exists a unique solution $\beta_1(t)$ of the IVP (6.108), (6.109). The inclusion $\beta_1(t) \in S(\alpha_0, \beta_0)$ is valid.

We will prove that $\alpha_1(t) \leq \beta_1(t)$ for $t \in [-h, T]$.

Define the function $u(t) = \alpha_1(t) - \beta_1(t)$ for $t \in [-h, T]$. From the initial conditions (6.105) and (6.109) and condition 1 of Theorem 6.4.2 the inequality $\alpha_1(t) = \varphi(t) - k_0 L_0 \leq \varphi(t) \leq \beta_1(t)$ holds, i.e., $u(t) \leq 0$ on $[-h, 0]$.

From the choice of the functions $\alpha_1(t)$ and $\beta_1(t)$ and the inequality (6.102) we obtain that the function $u(t)$ satisfies the inequalities

$$u' = f(t, \alpha_0(t), \max_{s \in [t-h,t]} \alpha_0(s)) - f(t, \beta_0(t), \max_{s \in [t-h,t]} \beta_0(s)) + Q_0(t)u(t)$$

$$+ Q_0(t)(\beta_0(t) - \alpha_0(t)) + q_0(t)(\max_{s \in [t-h,t]} \alpha_1(s) - \max_{s \in [t-h,t]} \beta_1(s))$$

$$+ q_0(t)(\max_{s \in [t-h,t]} \beta_0(s) - \max_{s \in [t-h,t]} \alpha_0(s))$$

$$\leq Q_0(t)u(t) + q_0(t) \max_{s \in [t-h,t]} u(s) \quad \text{for} \quad t \in [0, T].$$

$$(6.110)$$

According to the Lemma 6.4.1 and inequality (6.110) it follows that the inequality $u(t) \leq 0, \quad t \in [0, T]$ holds, i.e., $\alpha_1(t) \leq \beta_1(t)$.

Similarly, step by step, we can construct two sequences of functions $\{\alpha_n(t)\}_0^\infty$ and $\{\beta_n(t)\}_0^\infty$. Assume the functions $\alpha_n(t)$ and $\beta_n(t)$ are constructed and let $L_n = \min_{s \in [-h,0]} (\varphi(s) - \alpha_n(s))$, $C_n = \min_{s \in [-h,0]} (\beta_n(s) - \varphi(s))$. Choose numbers $k_n, p_n \in [0, 1)$ such that

$$k_n \leq L_n \quad \text{and} \quad p_n \leq C_n. \tag{6.111}$$

Then the function $\alpha_{n+1}(t)$ is the unique solution of the initial value problem for the linear differential equation with "maxima"

$$x' = f(t, \alpha_n(t), \max_{s \in [t-h,t]} \alpha_n(s)) + Q_n(t)(x - \alpha_n(t))$$
$$+ q_n(t)(\max_{s \in [t-h,t]} x(s) - \max_{s \in [t-h,t]} \alpha_n(s)) \quad \text{for} \quad t \in [0, T], \tag{6.112}$$

$$x(t) = \varphi(t) - k_n L_n \quad \text{for} \quad t \in [-h, 0], \tag{6.113}$$

and the function $\beta_{n+1}(t)$ is the unique solution of the initial value problem

$$x'(t) = f(t, \beta_n(t), \max_{s \in [t-h,t]} \beta_n(s)) + Q_n(t)(x - \beta_n(t))$$
$$+ q_n(t)(\max_{s \in [t-h,t]} x(s) - \max_{s \in [t-h,t]} \beta_n(s)) \quad \text{for} \quad t \in [0, T], \tag{6.114}$$

$$x(t) = \varphi(t) + p_n C_n \quad \text{for} \quad t \in [-h, 0], \tag{6.115}$$

where

$$Q_n(t) = F_x(t, \alpha_n(t), \max_{s \in [t-h,t]} \alpha_n(s)) - g_x(t, \beta_n(t), \max_{s \in [t-h,t]} \beta_n(s)),$$

$$q_n(t) = F_y(t, \alpha_n(t), \max_{s \in [t-h,t]} \alpha_n(s)) - g_y(t, \beta_n(t),$$

$$\max_{s \in [t-h,t]} \beta_n(s)) \geq q_0(t) \geq 0.$$

Analogously as in the case $n = 0$ it can be proved that the functions $\alpha_n(t)$ and $\beta_n(t)$ are lower and upper solutions of the IVP (6.92),(6.93), inclusion $\alpha_n, \beta_n \in S(\alpha_{n-1}, \beta_{n-1})$ is valid and the inequalities

$$\alpha_0(t) \leq \alpha_1(t) \leq \cdots \leq \alpha_n(t) \leq \beta_n(t) \leq \cdots \leq \beta_0(t), t \in [-h, T] \tag{6.116}$$

hold.

Therefore both sequences $\{\alpha_n(t)\}_0^\infty$ and $\{\beta_n(t)\}_0^\infty$ are uniformly convergent on $[-h, T]$.

Denote

$$\lim_{n \to \infty} \alpha_n(t) = x(t), \quad \lim_{n \to \infty} \beta_n(t) = \tilde{x}(t).$$

From the uniform convergence and the definition of the functions $\alpha_n(t)$ and $\beta_n(t)$ it follows the validity of the inequalities

$$\alpha_0(t) \le x(t) \le \tilde{x}(t) \le \beta_0(t). \tag{6.117}$$

From equations (6.112), (6.114) as n approaches infinity we obtain that the functions $u(t)$ and $v(t)$ are solutions of the IVP (6.92), (6.93). According to the conditions of Theorem 6.4.2 the IVP (6.92), (6.93) has a unique solution on $S(\alpha, \beta)$, i.e., $x(t) = \tilde{x}(t)$.

We will prove the convergence is quadratic.

Define the functions $a_{n+1}(t) = x(t) - \alpha_{n+1}(t)$ and $b_{n+1}(t) = \beta_{n+1}(t) - x(t)$, $t \in [-h, T]$. Both functions are nonnegative.

From the choice of the constants k_n it follows the validity of the inequalities

$$x(t) - \alpha_{n+1}(t) = k_n L_n \le (L_n)^2 = \left(\min_{s \in [-h,0]} (\varphi(s) - \alpha_n(s)) \right)^2$$

$$= \left(\min_{s \in [-h,0]} (x(s) - \alpha_n(s)) \right)^2 \le \|x - \alpha_n\|_0^2$$

for any $t \in [-h, 0]$, i.e.,

$$\|x - \alpha_{n+1}\|_0 \le \|x - \alpha_n\|_0^2. \tag{6.118}$$

For $t \in [0, T]$ we obtain the inequalities

$$a'_{n+1} \le Q_n(t) a_{n+1}(t) + q_n(t) \max_{s \in [t-h,t]} a_{n+1}(s)$$

$$+ [F_x(t, x(t), \max_{s \in [t-h,t]} x(s)) - g_x(t, \alpha_n(t), \max_{s \in [t-h,t]} \alpha_n(s)))$$

$$- Q_n(t)] a_n(t)$$

$$+ [F_y(t, x(t), \max_{s \in [t-h,t]} x(s))) - g_y(t, \alpha_n(t), \max_{s \in [t-h,t]} \alpha_n(s))$$

$$- q_n(t)] a_n(t_k)$$

$$= Q_n(t)a_{n+1}(t) + q_n(t)\max_{s\in[t-h,t]} a_{n+1}(s) + F_{xx}(t,\xi_1,\xi_2)a_n^2(t)$$

$$+ F_{xy}(t,\xi_1,\xi_2)a_n(t)(\max_{s\in[t-h,t]} x(s) - \max_{s\in[t-h,t]}\alpha_n(s))$$

$$+ g_{xx}(t,\eta_1,\eta_2)a_n(t)(\beta_n(t) - \alpha_n(t))$$

$$+ g_{xy}(t,\eta_1,\eta_2)a_n(t)(\max_{s\in[t-h,t]}\beta_n(s) - \max_{s\in[t-h,t]}\alpha_n(s))$$

$$+ F_{xy}(t,\xi_3,\xi_4)(\max_{s\in[t-h,t]} x(s)) - \max_{s\in[t-h,t]}\alpha_n(s))a_n(t)$$

$$+ F_{yy}(t,\xi_3,\xi_4)(\max_{s\in[t-h,t]} x(s)) - \max_{s\in[t-h,t]}\alpha_n(s))^2$$

$$+ g_{yy}(t,\eta_3,\eta_4)(\max_{s\in[t-h,t]} x(s)$$

$$- \max_{s\in[t-h,t]}\alpha_n(s))(\max_{s\in[t-h,t]}\beta_n(s)$$

$$- \max_{s\in[t-h,t]}\alpha_n(s))$$

$$+ g_{xy}(t,\eta_3,\eta_4)(\max_{s\in[t-h,t]} x(s))$$

$$- \max_{s\in[t-h,t]}\alpha_n(s))(\beta_n(t) - \alpha_n(t)), \qquad (6.119)$$

where $u(t) \leq \xi_i \leq \alpha_n(t)$, $\max_{s\in[t-h,t]} x(s) \leq \xi_l \leq \max_{s\in[t-h,t]}\alpha_n(s)$, $\alpha_n(t) \leq \eta_i \leq \beta_n(t)$, $\max_{s\in[t-h,t]}\alpha_n(s) \leq \eta_l \leq \max_{s\in[t-h,t]}\beta_n(s)$, $i = 1,3$, $l = 2,4$.

It is easy to verify that the inequalities

$$a_n(t)(\beta_n(t) - \alpha_n(t)) = a_n(b_n + a_n) \leq \frac{1}{2}b_n^2(t) + \frac{3}{2}a_n^2(t),$$

$$a_n(t)(\max_{s\in[t-h,t]}\beta_n(s) - \max_{s\in[t-h,t]}\alpha_n(s)) \leq \frac{1}{2}\|b_n\|^2 + \frac{3}{2}\|a_n\|^2,$$

$$\max_{s\in[t-h,t]} a(s)(\beta_n(t) - \alpha_n(t)) \leq \frac{1}{2}\|b_n\|^2 + \frac{3}{2}\|a_n\|^2,$$

$$\max_{s\in[t-h,t]} a_n(s)(\max_{s\in[t-h,t]}\beta_n(s) - \max_{s\in[t-h,t]}\alpha_n(s)) \leq \frac{1}{2}\|b_n\|^2 + \frac{3}{2}\|a_n\|^2$$

$$(6.120)$$

where $\|a\| = max\{|a(t)| : t \in [0,T]\}$.

From inequalities (6.119) and (6.120) and $q_n(t) \leq 0$ it follows that for $t \in [0,T]$ the inequalities

$$a'_{n+1}(t) \leq Q_n(t)a_{n+1}(t) + q_n(t)\max_{s\in[t-h,t]} a_{n+1}(s) + \sigma_n(t) \qquad (6.121)$$

hold, where

$$\sigma_n(t) = [F_{xx}(t,\xi_1,\xi_2) + \frac{3}{2}g_{xx}(t,\eta_1,\eta_2) + \frac{3}{2}g_{xy}(t,\eta_1,\eta_2)$$
$$+ F_{xy}(t,\xi_1,\xi_2) + F_{xy}(t,\xi_3,\xi_4) + \frac{3}{2}g_{xy}(t,\eta_3,\eta_4)$$
$$+ \frac{3}{2}g_{yy}(t,\eta_3,\eta_4) + F_{yy}(t,\xi_3,\xi_4)]\|a_n\|^2 \tag{6.122}$$
$$+ \frac{1}{2}[g_{xx}(t,\eta_1,\eta_2) + g_{xy}(t,\eta_1,\eta_2) + g_{xy}(t,\eta_3,\eta_4)$$
$$+ g_{yy}(t,\eta_3,\eta_4)]\|b_n\|^2.$$

From inequalities (6.121) and (6.118) we obtain

$$a_{n+1}(t) \leq a_{n+1}(0) + \int_0^t Q_n(s)a_{n+1}(s)ds + \int_0^t q_n(s)\max_{\xi\in[s-h,s]} a_{n+1}(\xi)ds$$
$$+ \int_0^t \sigma_n(s)ds$$
$$\leq B_n(t) + \int_0^t Q_n(s)a_{n+1}(s)ds$$
$$+ \int_0^t q_n(s)\max_{\xi\in[s-h,s]} a_{n+1}(\xi)ds \quad \text{for} \quad t \in [0,T],$$
$$a_{n+1}(t) \leq \|u - \alpha_n\|_0^2 \quad \text{for} \quad t \in [-h,0],$$

$$\tag{6.123}$$

where $B_n(t) = \|u - \alpha_n\|_0^2 + \int_0^t \sigma_n(s)ds$.

According to Corollary 2.1.2 from inequality (6.123) we get for $t \in [0,T)$ the inequality

$$a_{n+1}(t) \leq B_n(t)\exp\left(\int_0^t \left[Q_n(s) + q_n(s)\right]ds\right). \tag{6.124}$$

From the properties of the functions $F(t,x,y)$ and $g(t,x,y)$, the definition
(6.122) of the $\sigma(t)$ and inequality (6.124) it follows that there exist constants $\lambda_1 > 0$ and $\lambda_2 > 0$ such that

$$\|a_{n+1}\| \leq \lambda_1\|a_n\|^2 + \lambda_2\|b_n\|^2. \tag{6.125}$$

Analogously it can be proven that there exist constants $\mu_1 > 0$ and $\mu_2 > 0$ such that

$$\|b_{n+1}\| \leq \mu_1\|b_n\|^2 + \mu_2\|a_n\|^2. \tag{6.126}$$

The inequalities (6.125) and (6.126) prove that the convergence is quadratic.

□

Remark 6.4.1. *Sometimes it is difficult to find both lower and upper solutions of the given problem. Then we utilize the solution $x(t)$ as missing lower or upper solution and obtain only one monotone sequence approaching the solution $x(t)$.*

Now we will illustrate the application of the method of quasilinearization to a nonlinear scalar differential equation with "maxima."

Example 6.4.1. *Consider the following scalar nonlinear differential equation with "maxima"*

$$x' = \frac{1}{1-x(t)} + \frac{1}{1-\max_{s\in[t-0.5,t]} x(s)} - 2, \quad for \quad t \in [0,0.32], \quad (6.127)$$

with initial condition

$$x(t) = 0, \quad t \in [h,0], \tag{6.128}$$

where $x \in \mathbb{R}$, $h \in (0,0.32)$ is a fixed constant.

It is easy to check that the initial value problem (6.127), (6.128) has zero solution.

In this case $F(t,x,y) \equiv f(t,x,y) \equiv \frac{1}{1-x} + \frac{1}{1-y} - 2$ and $g(t,x,y) \equiv 0$. It is easy to check that $f'_x(t,x,y) = \frac{1}{(1-x)^2}$ and $f'_y(t,x,y) = \frac{1}{(1-y)^2}$, the function $\beta_0(t) \equiv \frac{1}{4}$ is an upper solution of (6.127), (6.128), $M = 1, N = \frac{16}{9}$ and the conditions of Theorem 6.4.2 are satisfied for $\alpha_0(t) \equiv 0$. We will construct a decreasing sequence of functions that is quadratically convergent to 0.

Choose $p_0 = \frac{1}{4}$ and consider the initial value problem

$$x' = \frac{1}{1-\frac{1}{4}} + \frac{1}{1-\frac{1}{4}} - 2 + \left(\frac{1}{(1-\frac{1}{4})^2}\right)(x-\frac{1}{4})$$
$$+ \left(\frac{1}{(1-\frac{1}{4})^2}\right)\left(\max_{s\in[t-h,t]} x(s) - \frac{1}{4}\right)$$
$$= -\frac{2}{9} + \frac{16}{9}x + \frac{16}{9}\max_{s\in[t-h,t]} x(s), \tag{6.129}$$
$$x(t) = \frac{1}{16}, \quad t \in [-h,0].$$

The function $\beta_1(t) \equiv \frac{1}{16}$ is a solution of the initial value problem (6.129).

Choose $p_1 = \frac{1}{16}$ and consider the initial value problem

$$x' = \frac{1}{1 - \frac{1}{16}} + \frac{1}{1 - \frac{1}{16}} - 2 + \left(\frac{1}{(1 - \frac{1}{16})^2}\right)(x - \frac{1}{16})$$

$$+ \left(\frac{1}{(1 - \frac{1}{16})^2}\right)\left(\max_{s\in[t-h,t]} x(s) - \frac{1}{16}\right)$$

$$= -\frac{2}{225} + \frac{256}{225}x + \frac{256}{225}\max_{s\in[t-h,t]} x(s), \qquad (6.130)$$

$$x(t) = \frac{1}{16^2}, \quad t \in [-h, 0].$$

The function $\beta_2(t) \equiv \frac{1}{256}$ is a solution of the initial value problem (6.130).

The sequence of successive approximations $\{\frac{1}{2^{2k}}\}_{k=1}^{\infty}$ is decreasing, it approaches zero, and the convergence is quadratical. The zero limit is the exact solution of the initial value problem (6.127), (6.128).

6.5 Method of Quasilinearization for Periodic Boundary Value Problems

Consider the following boundary value problem (PBVP) for the nonlinear differential equation with "maxima" (6.92) with boundary conditions

$$x(0) = x(T), \qquad x(t) = x(0) \quad \text{for } t \in [-h, 0], \qquad (6.131)$$

where $x \in \mathbb{R}$, $f : [0, T] \times \mathbb{R} \times \mathbb{R} \to \mathbb{R}$, $h > 0$ is a fixed constant.

Definition 6.5.1. *The function $\alpha(t) \in C([-h, T], \mathbb{R}) \cup C^1([0, T], \mathbb{R})$ is called a lower solution of the PBVP (6.92), (6.131), if the following inequalities are satisfied:*

$$\alpha'(t) \leq f(t, \alpha(t), \max_{s\in[t-h,t]} \alpha(s)) \quad \text{for } t \geq 0,$$

$$\alpha(0) \leq \alpha(T) \qquad \alpha(t) = \alpha(0) \quad \text{for } t \in [-h, 0].$$

Definition 6.5.2. *The function $\alpha(t) \in C([-h, T], \mathbb{R}) \cup C^1([0, T], \mathbb{R})$ is called a lower solution of the PBVP (6.92), (6.131), if the following* *inequalities are satisfied:*

$$\alpha'(t) \geq f(t, \alpha(t), \max_{s \in [t-h,t]} \alpha(s)) \quad for \quad t \geq 0,$$

$$\alpha(0) \geq \alpha(T) \qquad \alpha(t) = \alpha(0) \quad for \quad t \in [-h, 0].$$

Let the functions $\alpha, \beta \in C([-h, T], \mathbb{R})$ be such that $\alpha(t) \leq \beta(t)$. Consider the sets:

$$S(\alpha, \beta) = \{u \in C([-h, T], \mathbb{R}) : \quad \alpha(t) \leq u(t) \leq \beta(t) \quad for \quad t \in [-h, T]\},$$
$$\Omega(\alpha, \beta) = \{(t, x) \in [-h, T] \times \mathbb{R} : \quad \alpha(t) \leq x \leq \beta(t)\}.$$

Lemma 6.5.1. *[Comparison result] Let the following conditions be fulfilled:*

1. *The functions $g_1 \in C([0, T], (-\infty, 0]), g_2 \in C([0, T], \mathbb{R}_+)$ and $g_1(t) \leq -g_2(t)$ for $t \in [0, T]$.*

2. *The inequality*

$$(M + N)T < 1, \tag{6.132}$$

   holds, where $M = max\{|g_1(t)| : t \in [0, T]\} > 0$, $N = max\{g_2(t) : t \in [0, T]\} > 0$.

3. *The function $u \in C([-h, 0], \mathbb{R}) \cup C^1([0, T], \mathbb{R})$ satisfies the inequalities*

$$u' \leq g_1(t)u(t) + g_2(t) \max_{s \in [t-h,t]} u(s) \quad for \quad t \geq 0,$$

$$u(t) \leq u(0) \quad for \quad t \in [-h, 0], \quad u(0) \leq u(T). \tag{6.133}$$

Then the function $u(t)$ is nonpositive on the interval $[0, T]$.

Proof. We consider the following two cases:

   Case 1. Let $u(0) \leq 0$. Then according to Lemma 6.4.1 it follows the validity of the inequality $u(t) \leq 0$ for $t \in [0, T]$.

   Case 2. Let $u(0) > 0$. Then $u(T) > 0$. Let $u(\mu) = max_{t \in [0,T]} u(t) = A > 0$, where $\mu \in [0, T]$.

   Case 2.1. Let $\mu > 0$.

   Case 2.1.1. Let there exist a point $t_1 \in (0, \mu)$ such that $u(t_1) < 0$. Then there exists a point $\eta \in (0, \mu)$ such that $u(\eta) = 0$.

   Therefore, there exists a point $\zeta \in (\eta, \mu)$ such that

$$A = u(\mu) - u(\eta) = u'(\zeta)(\mu - \eta). \tag{6.134}$$

From equality (6.134) it follows that $u'(\zeta) > 0$ and $u'(\zeta)(\mu - \eta) \leq u'(\zeta)T$, i.e.,

$$A \leq u'(\zeta)T. \tag{6.135}$$

From inequalities (6.133) and (6.135) we obtain

$$A \leq u'(\zeta)T \leq g_1(\zeta)u(\zeta)T + g_2(\zeta) \max_{s \in [\zeta-h,\zeta]} u(s)T \leq (M + N)AT. \tag{6.136}$$

Inequality (6.136) contradicts (6.132).

Case 2.1.2. Let $u(t) > 0$ for $t \in [0, \mu]$. Then there exists a point $\xi \in [0, \mu)$ such that $u(\xi) = min_{t \in [0,\mu]} u(t) = B > 0$.

Therefore, there exists a point $\zeta \in (\xi, \mu)$ such that

$$A - B = u(\mu) - u(\eta) = u'(\zeta)(\mu - \eta), \tag{6.137}$$

where $\zeta \in (\eta, \mu)$.

From inequality (6.137) it follows that $u'(\zeta) > 0$ and $u'(\zeta)(\mu - \eta) \leq u'(\zeta)T$, i.e.,

$$A - B \leq u'(\zeta)T. \tag{6.138}$$

From inequalities (6.133) and (6.138) we obtain

$$A - B \leq u'(\zeta)T \leq \left(g_1(\zeta)u(\zeta) + g_2(\zeta) \max_{s \in [\zeta-h,\zeta]} u(s) \right)T \leq g_1(\zeta)BT + g_2(\zeta)AT$$

$$\leq \left(- g_2(\zeta)B + g_2(\zeta)A \right)T \leq N(A - B)T < (N + M)(A - B)T. \tag{6.139}$$

Inequality (6.139) contradicts (6.132).

Case 2.2. Let $\mu = 0$, i.e., $m(0) = m(T) = max_{t \in [0,T]} u(t) = A > 0$ or $\mu = T$. As in the proof of Case 2.1., we obtain a contradiction. $\qquad\square$

We will apply the method of quasilinearization for approximate finding of a solution of the periodic boundary value problem for a non-linear differential equation with "maxima." We will prove that the convergence of the successive approximations is quadratic.

Theorem 6.5.1. *Let the following conditions be fulfilled:*

1. *The functions $\alpha_0(t), \beta_0(t) \in C([-h, 0], \mathbb{R}) \cup C^1([0, T], \mathbb{R})$ are lower and upper solutions of the PBVP (6.92), (6.131) and $\alpha_0(t) \leq \beta_0(t)$ for $t \in [0, T]$.*

2. *There exist functions* $F, g \in C^{0,2,2}(\Omega(\alpha_0, \beta_0), \mathbb{R})$ *such that*

$$F(t, x, y) = f(t, x, y) + g(t, x, y),$$

   and

$$F_{xx}(t, x, y) \geq 0, \qquad F_{xy}(t, x, y) \geq 0, \qquad F_{yy}(t, x, y) \geq 0,$$
$$g_{xx}(t, x, y) \geq 0, \qquad g_{xy}(t, x, y) \geq 0, \qquad g_{yy}(t, x, y) \geq 0,$$

$$F_x(t, \beta_0(t), \max_{s \in [t-h,t]} (\beta_0(s))) \leq g_x(t, \alpha_0(t), \max_{s \in [t-h,t]} (\alpha_0(s))),$$

$$F_y(t, \alpha_0(t), \max_{s \in [t-h,t]} (\alpha_0(s))) \geq g_y(t, \beta_0(t), \max_{s \in [t-h,t]} (\beta_0(s))),$$

$$F_x(t, \beta_0(t), \max_{s \in [t-h,t]} (\beta_0(s))) + F_y(t, \beta_0(t), \max_{s \in [t-h,t]} (\beta_0(s)))$$

$$\geq g_x(t, \alpha_0(t), \max_{s \in [t-h,t]} (\alpha_0(s))) + g_y(t, \alpha_0(t), \max_{s \in [t-h,t]} (\alpha_0(s)))$$

$$\tag{6.140}$$

$$(M + N)T \leq 1,$$

   where

$$
\begin{aligned}
M \; &= \; \max\{|F_x(t, \beta_0(t), \lambda_k(\beta_0(t_k))) - g_x(t, \alpha_0(t), \lambda_k(\alpha_0(t_k)))|, \\
&\qquad t \in [0, T]\}, \\
N \; &= \; \max\{g_y(t, \alpha_0(t), \lambda_k(\alpha_0(t_k))) - F_y(t, \beta_0(t), \lambda_k(\beta_0(t_k))), \\
&\qquad t \in [0, T]\}.
\end{aligned}
$$

Then there exist two sequences of functions $\{\alpha_n(t)\}_0^\infty$ *and* $\{\beta_n(t)\}_0^\infty$ *such that:*

(a) *The sequences are increasing and decreasing correspondingly;*

(b) *Both sequences uniformly converge in the interval* $[0, T]$ *and the limits are equal to the unique solution of the PBVP (6.92), (6.131) in* $S(\alpha_0, \beta_0)$;

(c) *The convergence is quadratic.*

Proof. From condition 2 of Theorem 6.5.1 it follows that for (t, x_1, y_1), $(t, x_2, y_2) \in \Omega(\alpha_0, \beta_0)$ and $x_1 \geq x_2, y_1 \geq y_2$ the inequalities (6.101) and (6.102) hold.

   We consider the linear differential equation with "maxima" (6.104) with a boundary condition (6.131).

According to the results in Section 3.2 the PBVP (6.104),(6.131) has an unique solution $\alpha_1(t)$. We will prove that $\alpha_1(t) \in S(\alpha_0, \beta_0)$.

Indeed, the function $u(t) = \alpha_0(t) - \alpha_1(t), \quad t \in [-h, T]$ satisfies the inequalities

$$u'(t) \leq Q_0(t)u(t) + q_0(t)max_{s\in[t-h,t]}u(s), \quad t \in [0,T], \qquad (6.141)$$

$$u(0) \leq u(T), \quad u(t) = u(0) \quad \text{for} \quad t \in [-h,0]. \qquad (6.142)$$

According to Lemma 6.5.1 the inequality $\alpha_0(t) \leq \alpha_1(t)$ holds.

We will prove that $\alpha_1(t) \leq \beta_0(t)$. We define the function $u(t) = \alpha_1(t) - \beta_0(t), t \in [-h, T]$. According to the definition of the functions α_1 and β_0, the inequalities (6.140),

$$\max_{s\in[t-h,t]} \alpha_1(s) - \max_{s\in[t-h,t]} \beta_0(s) \leq \min_{s\in[t-h,t]} (\alpha_1(s) - \beta_0(s)),$$

and the equation (6.104) we obtain the inequality (6.106) and (6.142).

From (6.106) and (6.142) according to Lemma 6.5.1 we obtain that $u(t) \leq 0$, i.e., the inequality $\alpha_1(t) \leq \beta_0(t), t \in [0, T]$ holds.

We consider the linear differential equation (6.108) with a boundary condition (6.131).

The PBVP (6.108) and (6.131) has an unique solution $\beta_1(t)$: $\beta_1(t) \in S(\alpha_0, \beta_0)$.

We will prove that $\alpha_1(t) \leq \beta_1(t)$ for $t \in [-h, T]$.

Define the function $u(t) = \alpha_1(t) - \beta_1(t)$ for $t \in [-h, T]$. From the choice of the functions $\alpha_1(t)$ and $\beta_1(t)$ and the inequalities (6.140) we obtain that the function $u(t)$ satisfies the inequalities (6.141) and (6.142). According to the Lemma 6.5.1 it follows that the inequality $u(t) \leq 0, \quad t \in [0, T]$ holds, i.e., $\alpha_1(t) \leq \beta_1(t)$.

Similarly, we can construct two sequences of functions $\{\alpha_n(t)\}_0^\infty$ and $\{\beta_n(t)\}_0^\infty$, where $\alpha_n, \beta_n \in S(\alpha_{n-1}, \beta_{n-1})$. The function $\alpha_{n+1}(t)$ is the unique solution of the periodic boundary value problem for the linear differential equation with "maxima" (6.112), (6.131), and the function $\beta_{n+1}(t)$ is the unique solution of the periodic boundary value problem (6.114), (6.131).

Analogously as in the case $n = 0$ it can be proved that the functions $\alpha_n(t)$ and $\beta_n(t)$ are lower and upper solutions of the PBVP (6.92), (6.131) and the inequalities (6.116) hold.

Therefore, the sequences $\{\alpha_n(t)\}_0^\infty$ and $\{\beta_n(t)\}_0^\infty$ are uniformly bounded and equicontinuous in the interval $[0, T]$ and uniformly convergent.

Denote $\lim_{n\to\infty} \alpha_n(t) = u(t)$, $\lim_{n\to\infty} \beta_n(t) = v(t)$.

From the uniform convergence and the definition of the functions $\alpha_n(t)$ and $\beta_n(t)$ it follows the validity of the inequalities $\alpha_0(t) \le u(t) \le v(t) \le \beta_0(t)$. From the equalities (6.112) and (6.114) we obtain that the functions $u(t)$ and $v(t)$ are solutions of the PBVP (6.92) and (6.131). According to the conditions of Theorem 6.5.1 the PBVP (6.92) and (6.131) has a unique solution on $S(\alpha, \beta)$, i.e., $u(t) = v(t)$.

We will prove the convergence is quadratic.

Similarly to the proof of Theorem 6.4.2 we define functions $a_{n+1}(t) = u(t) - \alpha_{n+1}(t)$ and $b_{n+1}(t) = \beta_{n+1}(t) - u(t)$, $t \in [0, T]$. Both functions are nonnegative, $a_{n+1}(t) = a_{n+1}(0) = a_{n+1}(T)$, $t \in [-h, 0]$ and for $t \in [0, T]$ the inequalities (6.119) and (6.121) are satisfied.

From the inequality (6.121) and the periodic boundary condition we obtain the following estimate for the function $a_{n+1}(t)$:

$$a_{n+1}(t) \le$$

$$\left\{ a_{n+1}(0) + \int_0^t \left(\sigma_n(s) + q_n(s) \max_{\xi \in [s-h,s]} a_{n+1}(\xi) \right) \right.$$

$$\times \exp\left(-\int_0^s Q_n(\tau)d\tau \right) ds \Big\} \qquad (6.143)$$

$$\times \exp\left(\int_0^t Q_n(s)ds \right),$$

where

$$a_{n+1}(0) \le \left[1 - \exp\left(\int_0^T Q_n(s)ds \right) \right]^{-1}$$

$$\times \int_0^T \left(\sigma_n(s) + q_n(s) \max_{\xi \in [s-h,s]} a_{n+1}(\xi) \right) \exp\left(\int_s^T Q_n(\tau)d\tau \right) ds.$$

$$(6.144)$$

From (6.143) and the definition of the function $a_{n+1}(t)$ we obtain

$$a_{n+1}(t) \le f_1(t) + \int_0^t f_2(t, s) \max_{\xi \in [s-h,s]} a_{n+1}(\xi)ds, \qquad \text{for } t \in [0, T],$$

$$a_{n+1}(t) = a_{n+1}(0) \quad \text{for } t \in [-h, 0],$$

$$(6.145)$$

where

$$f_1(t) = a_{n+1}(0)exp\left(\int_0^t Q_n(s)ds \right),$$

$$f_2(t,s) = \sigma_n(t)exp\left(\int_s^t Q_n(s)ds\right),$$

and the function $\sigma(t)$ is defined by (6.122).

According to Theorem 2.1.2 from (6.145) we obtain the following bound

$$a_{n+1}(t) \le a_{n+1}(0)exp\left(\int_0^t Q_n(s)ds\right)exp\left(\int_0^t f_2(t,s)ds\right)$$

$$= a_{n+1}(0)exp\left(\int_0^t Q_n(s)ds\right)exp\left(\int_0^t \sigma_n(s)exp\left(\int_\xi^s Q_n(\xi)d\xi\right)ds\right).$$

$$(6.146)$$

From condition 2 of Theorem 6.5.1 and the inequalities (6.144) and (6.146) it follows that there exist constants $\lambda_1 > 0$ and $\lambda_2 > 0$ such that inequalities (6.125) and (6.126). These inequalities prove the quadratic convergence.

□

Chapter 7

Averaging Method

In this chapter, the averaging method for several types of differential equations with "maxima" will be presented. First order differential equations with "maxima" and neutral differential equations with "maxima" are considered. Different schemes of averaging are applied to initial value problems and boundary value problems. The application of the averaging method for differential equations with "maxima" gives the possibilities to considerably simplify them. The presence of maximum of the function in the right side of the equation has significant influence on the application of the considered method.

Note that in this chapter, if $x \in \mathbb{R}^n$, $x = (x_1, x_2, \ldots, x_n)$ then we use the following norm $\|x\| = \sqrt{\sum_{i=1}^{n} x_i^2}$.

7.1 Averaging Method for Initial Value Problems

This section is devoted to the justification of averaging method for an initial value problem associated to a vector differential equation with "maxima." In applications usually the maxima arises when the control law corresponds to the maximal deviation of the regulated quantity. If the control law also takes into account the maximal velocity of deviation of this quantity, then the process is governed by a differential equation, containing the maximum of the unknown functions as well as the maximum of its derivative.

Consider the system of differential equations with "maxima"

$$x'(t) = \epsilon X\Big(t, \ x(t), \ \max_{s \in [t-h,t]} x(s), \ \max_{s \in [t-h,t]} x'(s)\Big), \quad t \geq 0, \qquad (7.1)$$

with initial condition

$$x(t) = \varphi(t), \quad x'(t) = \varphi'(t), \qquad t \in [-h, 0], \tag{7.2}$$

where $x \in \mathbb{R}^n$, $x = (x_1, x_2, \ldots, x_n)$, $h > 0$ is a constant, $\varphi(t)$ is the initial function, $\epsilon > 0$ is a small parameter and the following notation

$$\max_{s \in [t-h,t]} x'(s) = \Big(\max_{s \in [t-h,t]} x_1'(s), \max_{s \in [t-h,t]} x_2'(s), \ldots, \max_{s \in [t-h,t]} x_n'(s) \Big)$$

is used.

Suppose that there exists the limit

$$\lim_{T \to \infty} \frac{1}{T} \int_0^T X(t, x, x, 0) dt = \bar{X}(x) \tag{7.3}$$

uniformly with respect to x.

Then the averaged system is the following system of ordinary differential equations

$$\xi'(t) = \epsilon \bar{X}(\xi(t)), \quad \text{with initial condition} \quad \xi(0) = \varphi(0), \tag{7.4}$$

where $\xi \in \mathbb{R}^n$.

The following theorem gives conditions for proximity between the solution $x(t)$ of the initial value problem (7.1), (7.2) and the solution $\xi(t)$ of (7.4).

Theorem 7.1.1. *Let the following conditions be fulfilled:*

1. *The functions $X(t, x, y, z) \in C(W, \mathbb{R}^n)$, $\varphi(t) \in C^1([-h, 0], \Omega)$, and the derivative $\varphi'(t) \in \Upsilon$ for $t \in [-h, 0]$, where the domains $W = [0, \infty) \times \Omega \times \Omega \times \Upsilon$, and $\Omega \subseteq \mathbb{R}^n$, $\Upsilon \subseteq \mathbb{R}^n$ are open domains.*

2. *The following inequalities*

$$\|X(t, x, y, z)\| \le M \quad \text{for} \quad (t, x, y, z) \in W,$$
$$\|X(t, x, y, z)\| - \|X(t, x_1, y_1, z_1)\|$$
$$\le \lambda (\|x - x_1\| + \|y - y_1\| + \|z - z_1\|),$$
$$(t, x, y, z), (t, x_1, y_1, z_1) \in W$$

hold, where M and λ are positive constants.

3. *The limit (7.3) exists uniformly for $x \in \Omega$ and the function $\bar{X}(x) \in C(\Omega, \mathbb{R}^n)$.*

4. *For each $\epsilon \in (0, \mathcal{E}]$ the initial value problem (7.1), (7.2) has a unique solution $x(t) \in \Omega$ defined on $\mathbb{R}_+$, where $\mathcal{E} = const > 0$.*

5. *For each $\epsilon \in (0, \mathcal{E}]$ the initial value problem (7.4) has a unique solution $\xi(t)$, such that $U_\xi(t) \subseteq \Omega$ for $t \in \mathbb{R}_+$, where $U_\xi(t)$ is a ρ-neighborhood of $\xi(t)$, $(\rho = const > 0)$.*

Then for each $\eta > 0$ and $L > 0$ there exists a number $\epsilon_0 \in (0, \mathcal{E}]$, $(\epsilon_0 = \epsilon_0(\eta, L))$ such that for $0 < \epsilon \leq \epsilon_0$ the inequality

$$\left\| x(t) - \xi(t) \right\| < \eta, \quad 0 \leq t \leq Le^{-1}$$

holds true.

Proof. The solution of (7.1), (7.2) satisfies the following integral equation

$$x(t) = \varphi(0) + \epsilon \int_0^t X\big(s, x(s), \max_{\theta \in [s-h,s]} x(\theta), \max_{\theta \in [s-h,s]} x'(\theta)\big) ds, \quad t \geq 0,$$

$$\tag{7.5}$$

$$x(t) = \varphi(t), \quad x'(t) = \varphi'(t), \quad -h \leq t \leq 0, \tag{7.6}$$

and the solution of (7.4) satisfies

$$\xi(t) = \varphi(0) + \epsilon \int_0^t \bar{X}\big(\xi(s)\big) ds. \tag{7.7}$$

Subtract (7.7) from (7.5) and obtain for $t \geq 0$

$\left\| x(t) - \xi(t) \right\|$

$$\leq \epsilon \left\| \int_0^t \big[X\big(s, x(s), \max_{\theta \in [s-h,s]} x(\theta), \max_{\theta \in [s-h,s]} x'(\theta)\big) - \bar{X}\big(\xi(s)\big) \big] ds \right\|$$

$$\leq \epsilon \int_0^t \left\| X\big(s, x(s), \max_{\theta \in [s-h,s]} x(\theta), \max_{\theta \in [s-h,s]} x'(\theta)\big) - X\big(s, \xi(s), \xi(s), 0\big) \right\| ds$$

$$+ \epsilon \left\| \int_0^t \big[X\big(s, \xi(s), \xi(s), 0\big) - \bar{X}\big(\xi(s)\big) \big] ds \right\| \equiv I_1 + I_2. \tag{7.8}$$

Now, let $L > 0$ be a fixed number, and $\epsilon > 0$ be a small parameter, whose value will be found later.

Then for $0 \le t \le L\epsilon^{-1}$ according to the condition 2 of Theorem 7.1.1 we get

$$I_1 = \epsilon \int_0^t \left\| X\left(s, x(s), \max_{\theta \in [s-h,s]} x(\theta), \max_{\theta \in [s-h,s]} x'(\theta)\right) - X\left(s, \xi(s), \xi(s), 0\right) \right\| d\theta$$

$$\le \epsilon \lambda \int_0^t \left[\|x(s) - \xi(s)\| + \left\| \max_{\theta \in [s-h,s]} x(\theta) - \xi(s) \right\| + \left\| \max_{\theta \in [s-h,s]} x'(\theta) \right\| \right] ds$$

$$< 2\epsilon \lambda \int_0^t \|x(s) - \xi(s)\| ds + \epsilon \lambda \int_0^t \left\| \max_{\theta \in [s-h,s]} x'(\theta) \right\| ds$$

$$+ \epsilon \lambda \int_0^t \left\| \max_{\theta \in [s-h,s]} x(\theta) - x(s) \right\| ds \tag{7.9}$$

Without loss of generality we will asumme that $t > h$.

From conditions 1 and 2 of Theorem 7.1.1 we obtain the inequality

$$\int_0^t \left\| \max_{\theta \in [s-h,s]} x'(\theta) \right\| ds$$

$$\le \int_0^h \left\| \max_{\theta \in [s-h,s]} x'(\theta) \right\| ds + \int_h^t \left\| \max_{\theta \in [s-h,s]} x'(\theta) \right\| ds \tag{7.10}$$

$$\le Bh + \int_h^t \left[\sum_1^n \left(x'(\gamma_i(s)) \right)^2 \right]^{\frac{1}{2}} ds \le Bh + \epsilon(t-h) M \sqrt{n}$$

$$\le Bh + LM\sqrt{n},$$

where $\gamma_i(s) \in [s-h, s]$, $i = 1, 2, \ldots, n$ are such that $\max_{\theta \in [s-h,s]} x_i'(\theta) = x_i'(\gamma_i(s))$ and $B = \max_{t \in [-h, 0]} \|\varphi'(t)\|$.

From (7.5) it follows the inequality $\|x(t)\| \le A + \epsilon Mt$. Then for $t \in [0, L\epsilon^{-1}]$ we get

$$\int_0^t \left\| \max_{\theta \in [s-h,s]} x(\theta) - x(s) \right\| ds$$

$$\le \int_0^h \left\| \max_{\theta \in [s-h,s]} x(\theta) \right\| ds + \int_0^h \|x(s)\| ds + \int_h^t \left\| \max_{\theta \in [s-h,s]} x(\theta) - x(s) \right\| ds$$

$$\le 2Ah + \epsilon M \frac{h^2}{2}$$

$$+ \int_h^t \left[\sum_{i=1}^n \left(\int_{\eta_i(s)}^s \epsilon X(\xi, x(\xi), \max_{\tau \in [\xi-h,\xi]} x(\tau), \max_{\tau \in [\xi-h,\xi]} x'(\tau)) d\xi \right)^2 \right]^{\frac{1}{2}} ds$$

$$\le 2Ah + \epsilon M \frac{h^2}{2} + (t-h)\epsilon M \sqrt{n} \le 2Ah + \epsilon M \frac{h^2}{2} + LM\sqrt{n}$$

$$\tag{7.11}$$

where $\eta_i(s) \in [s-h, s]$, $i = 1, 2, \ldots, n$ are such that $\max_{\theta \in [s-h,s]} x_i(\theta) = x_i(\eta_i(s))$ and $A = \max_{t \in [-h,\ 0]} ||\varphi(t)||$.

Fom inequalities (7.10) and (7.11) we obtain

$$I_1 \le 2\epsilon\lambda \int_0^t ||x(s)-\xi(s)|| ds + \epsilon\lambda \left(Bh + 2LM\sqrt{n} + 2Ah + \epsilon M \frac{h^2}{2} \right). \quad (7.12)$$

For $0 \le t \le L\epsilon^1$ we have

$$I_2 = \epsilon \left|\left| \int_0^t \left[X\big(s, \xi(s), \xi(s), 0\big) - \bar{X}\big(\xi(s)\big) \right] ds \right|\right|$$

$$\le \epsilon \left|\left| \int_0^{\frac{L}{\epsilon}} \left[X\big(s, \xi(s), \xi(s), 0\big) - \bar{X}\big(\xi(s)\big) \right] ds \right|\right| \le L\Phi\left(\frac{L}{\epsilon}, \xi \right), \quad (7.13)$$

where

$$\Phi(t, \xi) \quad = \quad \left|\left| \frac{1}{t} \int_0^t \left[X\big(s, \xi, \xi, 0\big) - \bar{X}\big(\xi\big) \right] ds \right|\right|$$

$$= \quad \left|\left| \frac{1}{t} \int_0^t X\big(s, \xi, \xi, 0\big) ds - \bar{X}\big(\xi\big) \right|\right| \quad (7.14)$$

According to condition 3 of Theorem 7.1.1 $\lim_{t \to \infty} \Phi(t, \xi) = 0$ for each $\xi \in \Omega$. Thus it follows from (7.8), (7.12) and (7.13) that

$$\left|\left| x(t) - \xi(t) \right|\right| \le \delta(\epsilon) + 2\epsilon\lambda \int_0^t \left|\left| x(\theta) - \xi(\theta) \right|\right| d\theta, \quad (7.15)$$

where

$$\delta(\epsilon) = \epsilon\lambda \left(Bh + 2LM\sqrt{n} + 2Ah + \epsilon M \frac{h^2}{2} \right) + L\Phi\left(\frac{L}{\epsilon}, \xi \right), \qquad \lim_{\epsilon \to 0} \delta(\epsilon) = 0. \quad (7.16)$$

Applying Gronwall inequality to (7.15) and we get

$$\left|\left| x(t) - \xi(t) \right|\right| \le \delta(\epsilon) \exp\{2\lambda L\} \quad \text{for} \quad 0 \le t \le L\epsilon^{-1}. \quad (7.17)$$

Since $\lim_{\epsilon \to 0} \delta(\epsilon) = 0$, for any $\eta > 0$ we could choose $\epsilon > 0$ such that $\left|\left| x(t) - \xi(t) \right|\right| \le \eta$ for $t \in [0, L\epsilon^{-1}]$.

$\square$

7.2 Averaging Method for Multipoint Boundary Value Problems

In this section we will apply an appropriate averaging scheme to differential equations with "maxima" with a boundary condition given at the fixed points $0 \leq t_0 < t_1 < \cdots < t_N < \infty$.

Consider the differential equations with "maxima" (7.1) with initial conditions (7.2) and a boundary condition

$$A_0 x(t_0) + \sum_{i=1}^{N} A_i(\epsilon) x(t_i) = \Gamma\big(x(t_0), \ldots, x(t_N), \epsilon\big), \qquad (7.18)$$

where $x \in \mathbb{R}^n$, $\epsilon \in (0, \mathcal{E}]$ is a small parameter, $\Gamma : \Omega \times \Omega \times \ldots \Omega \times (0, \mathcal{E}] \to \mathbb{R}^m$, A_0, $A_i(\epsilon)$, $(i = 1, 2, \ldots, N)$ are $m \times n$-dimentional matrices , $\mathcal{E} = const > 0$.

Assume the limit (7.3) exists. Then the averaged problem corresponding to (7.1), (7.2) and (7.18) is the following boundary value problem for the system of ordinary differential equations

$$\xi'(t) = \epsilon \bar{X}\big(\xi(t)\big), \qquad (7.19)$$

$$\xi(0) = \varphi(0), \qquad A_0 \xi(t_0) + \sum_{i=1}^{N} A_i \xi(t_i) = \Gamma\big(\xi(t_0), \ldots, \xi(t_N), \epsilon\big), \quad (7.20)$$

where the function $\bar{X}(t)$ is defined by (7.3).

For any matrix $A = \big(a_{jk}\big)_{m,n}$ we will use the following norms

$$||A|| = \left[\sum_{k=1}^{n} \sum_{j=1}^{m} a_{jk}^2 \right]^{\frac{1}{2}}.$$

The following theorem for proximity between the solution $x(t)$ of the boundary value problems (7.1), (7.2) and (7.18) and the solution $\xi(t)$ of (7.19) and (7.20) is valid.

Theorem 7.2.1. *Let the following conditions be fulfilled:*

1. *The function $X\big(t, \ x, \ y, \ z\big) \in C(W, \mathbb{R}^n)$, where $W = [0, \infty) \times \Omega \times \Omega \times \Upsilon$, $\Omega \subseteq \mathbb{R}^n$, $\Upsilon \subseteq \mathbb{R}^n$ are open domains.*

2. *The function $\varphi(t) \in C^1([-h, 0], \Omega)$, and $\varphi'(t) \in \Upsilon$ for $t \in [-h, 0]$.*

3. *The function $\Gamma : \Omega \times \Omega \times \ldots \Omega \times (0, \mathcal{E}] \to \mathbb{R}^m$, $\mathcal{E} = const > 0$.*

4. *The following inequalities*

$$\left\| X(t,\ x,\ y,\ z) \right\| \le M, \qquad ,(t,\ x,\ y,\ z) \in W$$

$$\left\| X(t,\ x,\ y,\ z) - X(t,\ x_1,\ y_1,\ z_1) \right\|$$

$$\le \lambda \Big(||x - x_1|| + ||y - y_1|| + ||z - z_1|| \Big),$$

$$(t,\ x,\ y,\ z), (t,\ x_1,\ y_1,\ z_1) \in W,$$

$$\left\| \Gamma(w_0, w_1, \ldots, w_N, \epsilon) \right.$$

$$\left. - \Gamma(w_0', w_1', \ldots, w_N', \epsilon) \right\| \le \mu_0 ||w_0 - w_0'||$$

$$+ \sum_{i=1}^{N} \mu_i(\epsilon) ||w_i - w_i'||$$

hold, where M, λ, μ_0 are positive constants, $\mu_i(\epsilon) > 0, (i = 1, 2, \ldots, N))$ are such that the function $b(\epsilon) = \max_{1 \le i \le N} \mu_i(\epsilon)$ is continuous in $(0, \mathcal{E}]$ and $\lim_{\epsilon \to 0} b(\epsilon) = 0$.

5. *The matrix A_0 is a constant matrix and $\det A_0 \ne 0$.*

6. *The matrices $A_i(\epsilon), (i = 1, 2, \ldots, N)$: $d(\epsilon) = \max_{1 \le i \le N} ||A_i(\epsilon)||$, $d(\epsilon)$ is a continuous function in $(0, \mathcal{E}]$ and $\lim_{\epsilon \to 0} d(\epsilon) = 0$.*

7. *The inequality*

$$\left\| \left(A_0 + \sum_{i=1}^{N} A_i(\epsilon) \right)^{-1} \right\| \left(\mu_0 + \sum_{i=1}^{N} \mu_i(\epsilon) \right) < 1$$

is fulfilled in $(0, \mathcal{E}]$.

8. *The limit (7.3) exists uniformly for $x \in \Omega$ and the function $\bar{X}(x) \in C(\Omega, \mathbb{R}^n)$.*

9. *For each $\epsilon \in (0, \mathcal{E}]$ the boundary value problems (7.1), (7.2) and (7.18) have a unique solution $x(t) \in \Omega$ defined on $\mathbb{R}_+$, where $\mathcal{E} = const > 0$.*

10. *For each $\epsilon \in (0, \mathcal{E}]$ problem (7.19), (7.20) has a unique solution $\xi(t)$, such that $U_\xi(t) \subseteq \Omega$ for $t \in \mathbb{R}_+$, where $U_\xi(t)$ is a ρ-neighborhood of $\xi(t)$, $(\rho = const > 0)$.*

Then for each $\eta > 0$ and $L > 0$ there exists a number $\epsilon_0 \in (0, \mathcal{E}]$, $(\epsilon_0 = \epsilon_0(\eta, L))$ such that for $0 < \epsilon \leq \epsilon_0$ the inequality

$$||x(t) - \xi(t)|| < \eta, \quad 0 \leq t \leq \frac{L}{\epsilon}$$

holds true.

Proof. The solution of the boundary value problem (7.1), (7.2), (7.18) satisfies the integral equation

$$x(t) = x_0 + \epsilon \int_{t_0}^{t} X\left(\theta, \ x(\theta), \ \max_{s \in [\theta - h, \theta]} x(s), \ \max_{s \in [\theta - h, \theta]} x'(s)\right) d\theta, \quad t > t_0,$$
$$(7.21)$$

$$x(t) = \varphi(t), \quad x'(t) = \varphi'(t), \quad -h \leq t \leq 0,$$

$$A_0(x_0 + \epsilon\beta_0) + \sum_{i=1}^{N} A_i(\epsilon)(x_0 + \epsilon\beta_i) = \Gamma(x_0 + \epsilon\beta_0, \ldots, x_0 + \epsilon\beta_N, \epsilon), \quad (7.22)$$

where $x_0 = x(t_0)$, $\beta_i = \int_{t_0}^{t_i} X\left(\theta, \ x(\theta), \max_{s \in [\theta - h, \theta]} x(s), \max_{s \in [\theta - h, \theta]} x'(s)\right) d\theta$, $(i = 0, 1, 2, \ldots, N)$.

The solution of the averaged boundary value problem (7.19), (7.20) satisfies the integral equation

$$\xi(t) = \xi_0 + \epsilon \int_{t_0}^{t} \bar{X}(\xi(\theta)) d\theta, \quad t \geq 0$$

$$(7.23)$$

$$\xi(0) = \varphi(0),$$

$$A_0(\xi_0 + \epsilon\bar{\beta}_0) + \sum_{i=1}^{N} A_i(\epsilon)(\xi_0 + \epsilon\bar{\beta}_i) = \Gamma(\xi_0 + \epsilon\bar{\beta}_0, \ldots, \xi_0 + \epsilon\bar{\beta}_{N_0}, \epsilon),$$

$$(7.24)$$

where $\xi_0 = \xi(t_0)$, $\bar{\beta}_i = \int_{t_0}^{t_i} \bar{X}(\xi(\theta)) d\theta$, $i = 0, 1, 2, \ldots, N$.

Subtract the equality (7.23) from (7.21) and obtain

$$||x(t) - \xi(t)|| \le ||x_0 - \xi_0||$$
$$+ \epsilon \left|\left| \int_{t_0}^t \left[X\big(\theta, \, x(\theta), \max_{s \in [\theta - h, \theta]} x(s), \max_{s \in [\theta - h, \theta]} x'(s)\big) - \bar{X}\big(\xi(\theta)\big) \right] d\theta \right|\right|$$

$$\le ||x_0 - \xi_0|| + \epsilon \int_{t_0}^t \left|\left| X\big(\theta, \, x(\theta), \max_{s \in [\theta - h, \theta]} x(s), \max_{s \in [\theta - h, \theta]} x'(s)\big) \right.\right.$$
$$- X\big(\theta, \, \xi(\theta), \, \xi(\theta), \, 0\big) \left.\right|\right| d\theta$$

$$+ \epsilon \left|\left| \int_{t_0}^t \left[X\big(\theta, \, \xi(\theta), \, \xi(\theta), \, 0\big) - \bar{X}\big(\xi(\theta)\big) \right] d\theta \right|\right|$$

$$\equiv ||x_0 - \xi_0|| + I_1 + I_2. \tag{7.25}$$

Now let $L > 0$ be a fixed number, and $\epsilon \in (0, \mathcal{E}]$ be a small number whose value will be set up later.

Similarly to the inequalities (7.12) and (7.13) we obtain for $t \in [0, L\epsilon^{-1}]$

$$I_1 = \epsilon \int_{t_0}^t \left|\left| X\big(\theta, \, x(\theta), \max_{s \in [\theta - h, \theta]} x(s), \max_{s \in [\theta - h, \theta]} x'(s)\big) \right.\right.$$
$$- X\big(\theta, \, \xi(\theta), \, \xi(\theta), \, 0\big) \left.\right|\right| d\theta$$

$$\le 2\epsilon\lambda \int_0^t ||x(s) - \xi(s)|| ds + \epsilon\lambda \big(Bh + 2LM\sqrt{n} + 2Ah + \epsilon M \frac{h^2}{2}\big). \tag{7.26}$$

and

$$I_2 = \epsilon \left|\left| \int_{t_0}^t \left[X\big(\theta, \, \xi(\theta), \, \xi(\theta), \, 0\big) - X\big(\xi(\theta)\big) \right] d\theta \right|\right|$$

$$\le L\Phi\big(\frac{L}{\epsilon}, \xi\big), \tag{7.27}$$

where $\Phi(t, \xi)$ is defined by (7.14).

According to condition 8 of Theorem 7.2.1 $\lim_{t \to \infty} \Phi(t, \xi) = 0$ for each $\xi \in \Omega$.

From (7.22) and (7.24) we get the validity of the inequality

$$
\left(A_0 + \sum_{i-1}^{N} A_i(\epsilon)\right)(x_0 + \xi_0) + \epsilon A_0\left(\beta_0 - \bar{\beta}_0\right) + \epsilon \sum_{i-1}^{N} A_i(\epsilon)\left(\beta_i - \bar{\beta}_i\right)
$$

$$
\leq \left(\mu_0 + \sum_{i-1}^{N} \mu_i(\epsilon)\right)(x_0 + \xi_0) + \epsilon \mu_0\left(\beta_0 - \bar{\beta}_0\right) + \epsilon \sum_{i-1}^{N} \mu_i(\epsilon)\left(\beta_i - \bar{\beta}_i\right)
$$

$$\tag{7.28}$$

or

$$
(x_0 + \xi_0)
$$

$$
\leq \left(A_0 + \sum_{i-1}^{N} A_i(\epsilon)\right)^{-1}\left(\mu_0 + \sum_{i-1}^{N} \mu_i(\epsilon)\right)(x_0 + \xi_0)
$$

$$
+ \epsilon\left(A_0 + \sum_{i-1}^{N} A_i(\epsilon)\right)^{-1}\left(\mu_0\left(\beta_0 - \bar{\beta}_0\right) + \sum_{i-1}^{N} \mu_i(\epsilon)\left(\beta_i - \bar{\beta}_i\right)\right) \tag{7.29}
$$

$$
- \epsilon\left(A_0 + \sum_{i-1}^{N} A_i(\epsilon)\right)^{-1}\left(A_0\left(\beta_0 - \bar{\beta}_0\right) + \sum_{i-1}^{N} A_i(\epsilon)\left(\beta_i - \bar{\beta}_i\right)\right)
$$

or

$$
\left|\left|x_0 - \xi_0\right|\right| \leq \epsilon\, G(\epsilon) \sum_{i=1}^{N} \left|\left|\beta_i - \bar{\beta}_i\right|\right| \tag{7.30}
$$

where

$$
G(\epsilon) = \left(b(\epsilon) + d(\epsilon)\right)\left|\left|\left(A_0 + \sum_{i=1}^{N} A_i(\epsilon)\right)^{-1}\right|\right| \times \tag{7.31}
$$

$$
\left(1 - \left|\left|\left(A_0 + \sum_{i=1}^{N} A_i(\epsilon)\right)^{-1}\right|\right|\left(\mu_0 + \sum_{i=1}^{N} \mu_i\right)\right)^{-1}.
$$

According to the conditions of Theorem 7.2.1, the definitions of $\beta_i, \bar{\beta}_i$, similar to the inequality (7.13) we get

$$
\epsilon\left|\left|\beta_i - \bar{\beta}_i\right|\right| = \epsilon \int_{t_0}^{t_i} \left|\left|X\left(\theta,\ x(\theta),\ \max_{s\in[\theta-h,\theta]} x(s),\ \max_{s\in[\theta-h,\theta]} x'(s)\right)d\theta - \bar{X}\left(\xi(\theta)\right)\right|\right|d\theta
$$

$$
\leq L\Phi\left(\frac{L}{\epsilon}, \xi\right).
$$

$$\tag{7.32}$$

The relations (7.25), (7.26), (7.27), (7.30), and (7.32) yield

$$||x(t) - \xi(t)|| \leq \sigma(\epsilon) + 2\epsilon\lambda \int_0^t ||x(\theta) - \xi(\theta)||d\theta, \qquad (7.33)$$

where $\sigma(\epsilon) = \epsilon\lambda\left(Bh + 2LM\sqrt{n} + 2Ah + \epsilon M\frac{h^2}{2}\right) + L(1 + G(\epsilon)N)\Phi\left(\frac{L}{\epsilon}, \xi\right)$.

Apply Gronwall-Bellman inequality to (7.33) and obtain for $t \in [0, L\epsilon^{-1}]$ the estimate

$$||x(t) - \xi(t)|| \leq \sigma(\epsilon)\exp\{2\epsilon\lambda t\}. \qquad (7.34)$$

Since $\lim_{\epsilon \to 0}\sigma(\epsilon) = 0$, for any $\eta > 0$ we could choose a number $\epsilon >$ such that $||x(t) - \xi(t)|| \leq \eta$ for $t \in [0, L\epsilon^{-1}]$.

$\square$

7.3 Partial Averaging Method

Now we will apply the partially averaging method to the initial value problem for the system of differential equations with "maxima" (7.1), (7.2), where $x \in \mathbb{R}^n$.

Let the function $\bar{X}(t, x)$ exist such that

$$\lim_{T \to \infty}\frac{1}{T}\int_0^T \left[X(t, x, x, 0) - \bar{X}(t, x)\right]dt = 0 \qquad (7.35)$$

uniformly with respect to x.

Then the averaged system of the system of differential equations with "maxima" is the following system of ordinary differential equations:

$$\dot{\xi} = \epsilon\bar{X}(t, \xi(t)) \quad \text{for} \quad t \geq 0 \qquad (7.36)$$

with initial condition

$$\xi(0) = \varphi(0), \qquad (7.37)$$

where $\xi \in \mathbb{R}^n$.

We shall prove the solutions of initial value problems (7.1) and (7.2) are close enough to the solutions of the initial value problem (7.36), (7.37).

Theorem 7.3.1. *Let the following conditions be fulfilled:*

1. *The function $X(t, x, y, z) \in C(W, \mathbb{R}^n)$, where $W = [0, \infty) \times \Omega \times \Omega \times \Upsilon$, $\Omega \subseteq \mathbb{R}^n$, $\Upsilon \subseteq \mathbb{R}^n$ are open domains.*

2. *The function $\varphi(t) \in C^1([-h, 0], \Omega)$, and $\varphi'(t) \in \Upsilon$ for $t \in [-h, 0]$.*

3. *The following inequalities*

$$\|X(t, \ x, \ y, \ z)\| + \|\bar{X}(t, x)\| \leq M$$
$$\text{for } (t, x, y, z) \in W, \ (t, x) \in \mathbb{R}_+ \times \Omega$$
$$\|X(t, x, y, z)\| - \|X(t, x_1, y_1, z_1)\|$$
$$\leq \lambda(\|x - x_1\| + \|y - y_1\| + \|z - z_1\|),$$
$$(t, x, y, z), (t, x_1, y_1, z_1) \in W$$
$$\left\|\bar{X}(t, x) - \bar{X}(t, x_1)\right\| \leq \mu\|x - x_1\|, \quad t \in \mathbb{R}_+, \ x, x_1 \in \Omega$$

hold, where M, μ and λ are positive constants.

4. *The limit (7.35) exists uniformly for $x \in \Omega$ and the function $\bar{X}(t, x) \in C(\mathbb{R}_+ \times \Omega, \mathbb{R}^n)$.*

5. *For each $\epsilon \in (0, \mathcal{E}]$ the initial value problem (7.1), (7.2) has a unique solution $x(t) \in \Omega$ defined on $\mathbb{R}_+$, where $\mathcal{E} = const > 0$.*

6. *For each $\epsilon \in (0, \mathcal{E}]$ the initial value problem (7.36),(7.37) has a unique solution $\xi(t)$, such that $U_\xi(t) \subseteq D \subset \Omega$ for $t \in \mathbb{R}_+$, where $U_\xi(t)$ is a ρ-neighborhood of $\xi(t)$, D is a compact, $(\rho = const > 0)$.*

Then for each $\eta > 0$ and $L > 0$ there exists a number $\epsilon_0 \in (0, \mathcal{E}]$, $(\epsilon_0 = \epsilon_0(\eta, L))$ such that for $0 < \epsilon \leq \epsilon_0$ the inequality

$$\|x(t) - \xi(t)\| < \eta, \quad 0 \leq t \leq L\epsilon^{-1}$$

holds.

Proof. From the condition 4 of Theorem 7.3.1 it follows that for the compact $D \subset \Omega$ there exists a continuous function $\alpha(T)$ such that $\lim_{T \to \infty} \alpha(T) = 0$ monotonically and for $x \in D$ the inequality

$$\left\|\int_0^T \left[X(t, x, x, 0) - \bar{X}(t, x)\right] d\tau\right\| \leq T\alpha(T) \qquad (7.38)$$

holds.

The solution $x(t)$ of (7.1) and (7.2) satisfies the integral equation

$$x(t) = \varphi(0) + \epsilon \int_0^t X\Big(\tau, x(\tau), \max_{s\in[\tau-h,\tau]} x(s), \max_{s\in[\tau-h,\tau]} x'(s)\Big)d\tau, \quad t \geq 0,$$
(7.39)

and the solution $\xi(t)$ of (7.36) and (7.37) satisfies

$$\xi(t) = \varphi(0) + \epsilon \int_0^t \bar{X}\big(\tau, \xi(\tau)\big)d\tau, \quad t \geq 0, \tag{7.40}$$

where $x(t) = \varphi(t)$ and $x'(t) = \varphi'(t)$ for $t \in [-h, 0]$.

Subtracting (7.40) from (7.39) for $t \geq 0$ we obtain

$$\big\|x(t) - \xi(t)\big\|$$

$$\leq \epsilon \int_0^t \Big\| X\Big(\tau, x(\tau), \max_{s\in[\tau-h,\tau]} x(s), \max_{s\in[\tau-h,\tau]} x'(s)\Big) - X\big(\tau, \xi(\tau), \xi(\tau), 0\big) \Big\| d\tau$$

$$+ \epsilon \left\| \int_0^t \Big[X\big(\tau, \xi(\tau), \xi(\tau), 0\big) - \bar{X}\big(\tau, \xi(\tau)\big) \Big] d\tau \right\|$$

$$\equiv I_1 + I_2. \quad (7.41)$$

Now let $L > 0$ be a fixed number and $\epsilon > 0$ be a number whose value will be obtained later.

Similar to inequalities (7.9) and (7.12) we obtain for $0 \leq t \leq L\epsilon^{-1}$ the bound

$$I_1 = \epsilon \int_0^t \Big\| X\Big(\tau, x(\tau), \max_{s\in[\tau-h,\tau]} x(s), \max_{s\in[\tau-h,\tau]} x'(s)\Big) - X\big(\tau, \xi(\tau), \xi(\tau), 0\big) \Big\| d\tau$$

$$\leq 2\epsilon\lambda \int_0^t \big\| x(s) - \xi(s) \big\| ds + \epsilon\lambda\Big(Bh + 2LM\sqrt{n} + 2Ah + \epsilon M\frac{h^2}{2}\Big)$$
(7.42)

From inequality (7.38) we get for $t \in [0, L\epsilon^{-1}]$

$$I_2 = \epsilon \left\| \int_0^t \Big[X\big(\tau, \xi(\tau), \xi(\tau), 0\big) - \bar{X}\big(\tau, \xi(\tau)\big) \Big] d\tau \right\| \leq L\alpha\Big(\frac{L}{\epsilon}\Big). \tag{7.43}$$

From inequalities (7.41), (7.42), and (7.43) it follows that for $t \in [0, \frac{L}{\epsilon}]$ the inequality

$$\big\| x(t) - \xi(t) \big\| \leq \sigma(\epsilon) + 2\epsilon\lambda \int_0^t \big\| x(\tau) - \xi(\tau) \big\| d\tau \tag{7.44}$$

holds, where

$$\sigma(\epsilon) = \epsilon\lambda\Big(Bh + 2LM\sqrt{n} + 2Ah + \epsilon M\frac{h^2}{2}\Big) + L\alpha(\frac{L}{\epsilon}).$$

Apply Gronwall-Bellman inequality to inequality (7.44) and obtain

$$||x(t) - \xi(t)|| \leq \sigma(\epsilon)\exp\{2\lambda L\}. \tag{7.45}$$

Inequality (7.45) and $\lim_{\epsilon\to 0}\sigma(\epsilon) = 0$ proves that for any $\eta > 0$ we could choose a number $\epsilon > 0$ such that $||x(t) - \xi(t)|| \leq \eta$.

□

Now we will apply the result of Theorem 7.3.1 to introduce a scheme for partial averaging for differential equations with "maxima."

Consider the initial value problem for the system of differential equations with "maxima" (7.1), (7.2), where $x \in \mathbb{R}^n$.

Introduce the following notations:

Let l, m be fixed natural numbers such that $l + m = n$. Then we assign to any vector $x \in \mathbb{R}^n$, $x = (x_1, x_2, \ldots, x_n)$ two vectors $z \in \mathbb{R}^l$ and $y \in \mathbb{R}^m$ such that

$$(z, y) = \underbrace{(x_1, x_2, \ldots, x_l,}_{z} \underbrace{x_{l+1}, \ldots, x_n)}_{y},$$

or $z_j = x_j$, $(j = 1, 2, \ldots, l)$ and $y_k = x_{k+l}$, $(k = 1, 2, \ldots m)$.

According to the above-introduced notation, we could rewrite the initial value problem (7.1), (7.2) in the following way:

$$
\begin{aligned}
z'(t) &= \epsilon Z\big(t, \ x(t), \ \max_{s\in[t-h,t]} x(s), \ \max_{s\in[t-h,t]} x'(s)\big), \quad t \geq 0,\\
y(t) &= \epsilon Y\big(t, \ x(t), \ \max_{s\in[t-h,t]} x(s), \ \max_{s\in[t-h,t]} x'(s)\big), \quad t \geq 0,\\
z(t) &= \psi(t), \quad z'(t) = \psi'(t), \qquad t \in [-h, 0],\\
z(t) &= \bar{\omega}(t), \quad y'(t) = \bar{\omega}'(t), \qquad t \in [-h, 0],
\end{aligned}
\tag{7.46}
$$

where

$$X(t, u, v, w) = \begin{pmatrix} Z(t, u, v, w) \\ Y(t, u, v, w) \end{pmatrix}, \quad \varphi(t) = \begin{pmatrix} \psi(t) \\ \varpi(t) \end{pmatrix},$$

$\psi \in \mathbb{R}^l$, $\varpi \in \mathbb{R}^m$, $u, v, w \in \mathbb{R}^n$.

Let the following limit exist

$$\lim_{T\to\infty} \frac{1}{T} \int_0^T Z(t,\ x,\ x,\ 0)dt = \bar{Z}(x). \qquad (7.47)$$

Then with the initial value problem (7.46) we associate the partially averaged system of ordinary differential equations

$$\begin{aligned}
z'(t) &= \epsilon\bar{Z}(x), \\
y'(t) &= \epsilon Y(t,\ x,\ x,\ 0) \quad \text{for} \quad t \ge 0, \qquad (7.48) \\
z(0) &= \psi(0), \qquad y(0) = \bar{\omega}(0),
\end{aligned}$$

where $x = \begin{pmatrix} z \\ y \end{pmatrix}$.

In order to distinguish the used norms in different spaces, we will use a subscript, i.e., $\|.\|_m$ will denote a norm in $\mathbb{R}^m$.

Theorem 7.3.2. *Let the following conditions be fulfilled:*

1. *The functions $Z(t,u,v,w) \in C(W,\mathbb{R}^l)$, $Y(t,u,v,w) \in C(W,\mathbb{R}^m)$, where $W = [0,\infty)\times\Omega\times\Omega\times\Upsilon$, $\Omega \subseteq \mathbb{R}^n$, $\Upsilon \subseteq \mathbb{R}^n$ are open domains.*

2. *The functions $\psi(t) \in C^1([-h,0],\mathbb{R}^l)$, $\bar{\omega}(t) \in C^1([-h,0],\mathbb{R}^m)$, and the inclusions $(\psi(t),\bar{\omega}(t)) \in \Omega$, $(\psi'(t),\bar{\omega}'(t)) \in \Upsilon$ hold for $t \in [-h,0]$.*

3. *The following inequalities*

$$\|Z(t,u,v,w)\|_l + \|Y(t,u,v,w)\|_m \le M \quad \text{for} \quad (t,u,v,w) \in W,$$
$$\|Z(t,u,v,w) - Z(t,u_1,v_1,w_1)\|_l \le \lambda(\|u-u_1\|_n + \|v-v_1\|_n$$
$$+\|w-w_1\|_n),$$
$$\|Y(t,u,v,w) - Y(t,u_1,v_1,w_1)\|_m \le \lambda(\|u-u_1\|_n + \|v-v_1\|_n$$
$$+\|w-w_1\|_n),$$
$$(t,u,v,w),(t,u_1,v_1,w_1) \in W$$

   hold, where M and λ are positive constants.

4. *The limit (7.47) exists uniformly for $x \in \Omega$ and the function $\bar{X} \in C(\mathbb{R}_+ \times \Omega, \mathbb{R}^n)$, where $\bar{X}(t,x) = \begin{pmatrix} \bar{Z}(x) \\ Y(t,x,x,0) \end{pmatrix}$.*

5. *For each $\epsilon \in (0,\mathcal{E}]$ the initial value problem (7.46) has a unique solution $x(t) = (z(t),y(t))$, $x(t) \in \Omega$, defined on $\mathbb{R}_+$, where $\mathcal{E} = const > 0$.*

6. For each $\epsilon \in (0, \mathcal{E}]$ the initial value problem (7.48) has a unique solution $\xi(t)$, such that $U_\xi(t) \subseteq D \subset \Omega$ for $t \in \mathbb{R}_+$, where $U_\xi(t)$ is a ρ-neighborhood of $\xi(t)$, ($\rho = const > 0$), D is a compact set.

Then for each $\eta > 0$ and $L > 0$ there exists a number $\epsilon_0 \in (0, \mathcal{E}]$, ($\epsilon_0 = \epsilon_0(\eta, L)$) such that for $0 < \epsilon \leq \epsilon_0$ the inequality

$$||x(t) - \xi(t)||_n < \eta, \quad 0 \leq t \leq L\epsilon^{-1}$$

holds.

Proof. From the limit (7.47) and the conditions of Theorem 7.3.2 it follows that in the domain Ω the function $\bar{Z}(x)$ is bounded, continuous and satisfies the Lipschitz condition. Hence conditions 1 and 2 of Theorem 7.3.1 are satisfied for the functions $X(t, u, v, w) \equiv \begin{pmatrix} Z(t, u, v, w) \\ Y(t, u, v, w) \end{pmatrix}$ and $\bar{X}(t, x) \equiv \begin{pmatrix} \bar{Z}(x) \\ Y(t, \ x, \ x, \ 0) \end{pmatrix}$. Further, the proof of Theorem 7.3.2 is similar to the one of Theorem 7.3.1 and we omit it.

$\square$

7.4 Partially Additive and Partially Multiplicative Averaging Method

We will apply a partially additive averaging scheme to the initial value problem for a system of differential equations with "maxima."

Consider the following system of differential equations with "maxima"

$$x'(t) = \epsilon \left[X_1\left(t, x(t), \max_{s \in [t-h,t]} x(s), \max_{s \in [t-h,t]} x'(s)\right) \right.$$

$$\left. + X_2\left(t, x(t), \max_{s \in [t-h,t]} x(s), \max_{s \in [t-h,t]} x'(s)\right) \right] \quad \text{for} \quad t \geq 0,$$

$$\tag{7.49}$$

with initial condition

$$x(t) = \varphi(t), \quad x'(t) = \varphi'(t) \quad \text{for} \quad t \in [-h, 0], \tag{7.50}$$

where $x \in \mathbb{R}^n$, $\varphi(t)$ is an initial function, $\epsilon > 0$ is a small parameter.

Let the following limit exist

$$\lim_{T \to \infty} \frac{1}{T} \int_0^T X_1(t, \ x, \ x, \ 0) dt = X_{10}(x). \tag{7.51}$$

Then with the initial value problem for differential equations with "maxima" (7.49), (7.50) we associate the averaged system of ordinary differential equations

$$\xi'(t) = \epsilon \Big[X_{10}(\xi(t)) + X_2 \big(t, \ \xi(t), \ \xi(t), \ 0\big) \Big], \tag{7.52}$$

with initial condition

$$\xi(0) = \varphi(0), \tag{7.53}$$

where $\xi \in \mathbb{R}^n$.

We shall prove a theorem of nearness of the solutions of the initial value problem (7.49),(7.50) and the averaged system (7.52) with initial condition (7.53).

Theorem 7.4.1. *Let the following conditions be fulfilled:*

1. *The functions $X_1(t, x, y, z)$, $X_2(t, x, y, z) \in C(W, \mathbb{R}^n)$, where $W = [0, \infty) \times \Omega \times \Omega \times \Upsilon$, $\Omega \subseteq \mathbb{R}^n$, $\Upsilon \subseteq \mathbb{R}^n$ are open domains.*

2. *The function $\varphi(t) \in C^1([-h, 0], \mathbb{R}^n)$, and $\varphi(t) \in \Omega$, $\varphi'(t) \in \Upsilon$ for $t \in [-h, 0]$.*

3. *The following inequalities*

$$\left\| X_1(t, x, y, z) \right\| + \left\| X_2(t, x, y, z) \right\| \le M \quad for \ (t, x, y, z) \in W,$$

$$\left\| X_1(t, x, y, z) - X_1(t, x_1, y_1, z_1) \right\|$$
$$+ \left\| X_2(t, x, y, z) - X_2(t, x_1, y_1, z_1) \right\|$$
$$\le \lambda \big(\|x - x_1\| + \|y - y_1\| + \|z - z_1\| \big),$$
$$(t, x, y, z), (t, x_1, y_1, z_1) \in W$$

   hold, where M and λ are positive constants.

4. *The limit (7.51) exists uniformly for $x \in \Omega$ and the function $X_{10} \in C(\Omega, \mathbb{R}^n)$.*

5. *For each $\epsilon \in (0, \mathcal{E}]$ the initial value problem (7.49),(7.50) has a unique solution $x(t) \in \Omega$, defined on $\mathbb{R}_+$, where $\mathcal{E} = const > 0$.*

6. *For each $\epsilon \in (0, \mathcal{E}]$ the initial value problem (7.52),(7.53) has a unique solution $\xi(t)$, such that $U_\xi(t) \subseteq D \subset \Omega$ for $t \in \mathbb{R}_+$, where $U_\xi(t)$is a ρ-neighborhood of $\xi(t)$, $(\rho = const > 0)$, D is a compact set.*

Then for each $\eta > 0$ and $L > 0$ there exists a number $\epsilon_0 \in (0, \mathcal{E}]$, $(\epsilon_0 = \epsilon_0(\eta, L))$ such that for $0 < \epsilon \le \epsilon_0$ the inequality

$$||x(t) - \xi(t)|| < \eta, \quad 0 \le t \le L\epsilon^{-1}$$

holds

Proof. The solution of the initial value problem (7.49), (7.50) satisfies

$$x(t) = x(0) + \epsilon \int_0^t \Big[X_1\big(\tau, x(\tau), \max_{s\in[\tau-h,\tau]} x(s), \max_{s\in[\tau-h,\tau]} x'(s)\big)$$
$$+ X_2\big(\tau, x(\tau), \max_{s\in[\tau-h,\tau]} x(s), \max_{s\in[\tau-h,\tau]} x'(s)\big) \Big] d\tau, \quad t \ge 0, \quad (7.54)$$

$$x(t) = \varphi(t) \text{ and } x'(t) = \varphi'(t)v \quad \text{for} \quad t \in [-h, 0], \quad (7.55)$$

and the solution of (7.52) and (7.53) satisfies

$$\xi(t) = \xi(0) + \epsilon \int_0^t \Big[X_{10}(\xi(\tau)) + X_2\big(\tau, \xi(\tau), \xi(\tau), 0\big) \Big] d\tau, \quad t \ge 0. \quad (7.56)$$

Subtract (7.56) from (7.54) and get for $t \ge 0$

$$||x(t) - \xi(t)||$$
$$\le \epsilon \int_0^t \Big|\Big| X_1\big(\tau, x(\tau), \max_{s\in[\tau-h,\tau]} x(s), \max_{s\in[\tau-h,\tau]} x'(s)\big) - X_1\big(\tau, \xi(\tau), \xi(\tau), 0\big) \Big|\Big| d\tau$$
$$+ \epsilon \int_0^t \Big|\Big| X_2\big(\tau, x(\tau), \max_{s\in[\tau-h,\tau]} x(s), \max_{s\in[\tau-h,\tau]} x'(s)\big) - X_2\big(\tau, \xi(\tau), \xi(\tau), 0\big) \Big|\Big| d\tau$$
$$+ \epsilon \Big|\Big| \int_0^t \Big[X_1\big(\tau, \xi(\tau), \xi(\tau), 0\big) - X_{10}(\xi(\tau)) \Big] d\tau \Big|\Big|.$$

$$(7.57)$$

Choose an arbitrary $L > 0$ and let $\epsilon > 0$ be a small enough number whose value will be defined later.

From condition 3 of Theorem 7.4.1 we obtain similarly to (7.9) and (7.12) the following inequalities

$$\int_0^t \left\| X_1\big(\tau, x(\tau),\ \max_{s\in[\tau-h,\tau]} x(s),\ \max_{s\in[\tau-h,\tau]} x'(s)\big) - X_1\big(\tau,\ \xi(\tau),\ \xi(\tau),\ 0\big)\right\| d\tau$$

$$\leq 2\lambda \int_0^t \|x(s) - \xi(s)\| ds + \lambda\big(Bh + 2LM\sqrt{n} + 2Ah + \epsilon M \tfrac{h^2}{2}\big),$$

$$\text{for } t \in [0, L\epsilon^{-1}].$$

$$(7.58)$$

and

$$\int_0^t \left\| X_2\big(\tau, x(\tau),\ \max_{s\in[\tau-h,\tau]} x(s),\ \max_{s\in[\tau-h,\tau]} x'(s)\big) - X_2\big(\tau,\ \xi(\tau),\ \xi(\tau),\ 0\big)\right\| d\tau$$

$$\leq 2\lambda \int_0^t \|x(s) - \xi(s)\| ds + \lambda\big(Bh + 2LM\sqrt{n} + 2Ah + \epsilon M \tfrac{h^2}{2}\big).$$

$$(7.59)$$

As in the proof of inequality (7.13) we get

$$\left\| \int_0^t \big[X_1\big(\tau,\ \xi(\tau),\ \xi(\tau),\ 0\big) - X_{10}\big(\xi(\tau)\big)\big] d\tau \right\| \leq \frac{L}{\epsilon} \Phi\big(\frac{L}{\epsilon}, \xi\big), \quad (7.60)$$

where $\Phi(t, \xi)$ is defined by (7.14) and $\lim_{t\to\infty} \Phi(t, \xi) = 0$.

Substitute inequalities (7.58), (7.59), and (7.60) in inequality (7.57) and obtain

$$\|x(t) - \xi(t)\| \leq 4\epsilon\lambda \int_0^t \|x(s) - \xi(s)\| ds + \sigma(\epsilon) \qquad \text{for } t \in [0, L\epsilon^{-1}],$$

$$(7.61)$$

where $\sigma(\epsilon) = 2\epsilon\lambda\big(Bh + 2LM\sqrt{n} + 2Ah + \epsilon M \tfrac{h^2}{2}\big) + L\Phi\big(\frac{L}{\epsilon}, \xi\big)$.

From inequality (7.61) according to Gronwall inequality it follows that

$$\|x(t) - \xi(t)\| < \sigma(\epsilon) exp(4L).$$

Since $\lim_{\epsilon\to 0} \sigma(\epsilon) = 0$ for any $\eta > 0$ we could find ϵ such that for $t \in [0, L\epsilon^{-1}]$ the estimate $\|x(t) - \xi(t)\| < \eta$ is valid.

$\square$

Now we will suggest a scheme for partially multiplicative averaging for an initial value problem for differential equations with "maxima."

Consider the following system of differential equations with "maxima"

$$x'(t) = \epsilon A\Big(t, x(t), \max_{s\in[t-h,t]} x(s), \max_{s\in[t-h,t]} x'(s)\Big)$$
$$\times X\Big(t, x(t), \max_{s\in[t-h,t]} x(s), \max_{s\in[t-h,t]} x'(s)\Big) \qquad (7.62)$$
$$\text{for} \quad t \geq 0,$$

with initial condition

$$x(t) = \varphi(t) \quad \text{and} \quad x'(t) = \varphi'(t), \qquad t \in [-h, 0], \qquad (7.63)$$

where $x \in \mathbb{R}^n$, $X : \mathbb{R} \times \mathbb{R}^n \times \mathbb{R}^n \times \mathbb{R}^n \to \mathbb{R}^m$, the matrix A is $n \times m$-dimensional with elements $a_{ij}\big(t, x(t), \bar{x}(t), \hat{x}(t)\big)$, $i = 1, 2, \dots, n$, $j = 1, 2, \dots, m$, $\varphi(t) : [-h, 0] \to \mathbb{R}^n$ is an initial function, $\epsilon > 0$ is a small parameter.

Let the following limit exist

$$\lim_{T\to\infty} \frac{1}{T} \int_0^T A(t, x, x, 0)dt = A_0(x). \qquad (7.64)$$

Then we assign to the system of differential equations with "maxima" the following averaged system of ordinary differential equations

$$\xi'(t) = \epsilon A_0\big(\xi(t)\big) X\big(t, \xi(t), \xi(t), 0\big) \qquad (7.65)$$

with initial condition

$$\xi(0) = \varphi(0), \qquad (7.66)$$

where $\xi \in \mathbb{R}^n$.

Theorem 7.4.2. *Let the following conditions be fulfilled:*

1. *The function $X(t, x, y, z) \in C(W, \mathbb{R}^m)$, where $W = [0, \infty) \times \Omega \times \Omega \times \Upsilon$, $\Omega \subseteq \mathbb{R}^n$, $\Upsilon \subseteq \mathbb{R}^n$ are open domains.*

2. *The $n \times m$ dimensional matrix $A(t, x, y, z)$ is continuous in W.*

3. *The function $\varphi(t) \in C^1([-h, 0], \mathbb{R}^n)$, and $\varphi(t) \in \Omega$, $\varphi'(t) \in \Upsilon$ for $t \in [-h, 0]$.*

4. *The following inequalities*

$$\left\|A(t,x,y,z)\right\| + \left\|X(t,x,y,z)\right\| \leq M \quad \text{for } (t,x,y,z) \in W,$$

$$\left\|A(t,x,y,z) - A(t,x_1,y_1,z_1)\right\| \leq \lambda(\|x - x_1\| + \|y - y_1\| + \|z - z_1\|),$$

$$\left\|X(t,x,y,z) - X(t,x_1,y_1,z_1)\right\| \leq \lambda(\|x - x_1\| + \|y - y_1\| + \|z - z_1\|),$$

$$(t,x,y,z), (t,x_1,y_1,z_1) \in W$$

hold, where M and λ are positive constants.

5. *The limit (7.64) exists uniformly for $x \in \Omega$ and the matrix $\bar{A}_0(x)$ is continuous in Ω.*

6. *The matrix $B(t,x) = A(t,x,x,0) - \bar{A}_0(x)$ is defined for $t \in \mathbb{R}_+$ and $x \in \Omega$ and its elements do not change their signs in $\mathbb{R}_+$.*

7. *For each $\epsilon \in (0,\mathcal{E}]$ the initial value problem (7.62), (7.63) has a unique solution $x(t) \in \Omega$, defined on $\mathbb{R}_+$, where $\mathcal{E} = const > 0$.*

8. *For each $\epsilon \in (0,\mathcal{E}]$ the initial value problem (7.65), (7.66) has a unique solution $\xi(t)$, such that $U_\xi(t) \subseteq D \subset \Omega$ for $t \in \mathbb{R}_+$, where $U_\xi(t)$ is a ρ-neighborhood of $\xi(t)$, $(\rho = const > 0)$, D is a compact set.*

Then for each $\eta > 0$ and $L > 0$ there exists a number $\epsilon_0 \in (0,\mathcal{E}]$, $(\epsilon_0 = \epsilon_0(\eta,L))$ such that for $0 < \epsilon \leq \epsilon_0$ the inequality

$$\left\|x(t) - \xi(t)\right\| < \eta, \quad 0 \leq t \leq L\epsilon^{-1}$$

holds.

The proof the Theorem 7.4.2 is similar to the proof of Theorem 7.3.1 and Theorem 7.4.1 and we omit it.

Chapter 8

Notes and Comments

The results in Chapter 2 are mainly obtained by S. Hristova, as part of the results in Section 2.1 are published in [Hristova and Stefanova 2010a], and part of the results in Section 2.3 are adopted by [Hristova and Stefanova 2010b]. A part of the results in Section 2.2 are proved by S. Hristova and J. Henderson.

The contents of Section 3.1 are by V. Angelov and D. Bainov ([Angelov and Bainov 1983]), the results of Section 3.2 are due to E. Stepanov ([Stepanov 1997]), and the results of Section 3.3 are adopted from D. Otrocol and I.A. Rus ([Otrocol and Ioan 2008a]).

Chapter 4 contains results of S. Hristova. Several of the results in this chapter are similar to the results for impulsive differential equations with "supremum" ([Hristova 2010b], [Hristova 2010c], [Hristova 2009c], and [Hristova 2009d]). A part of the results in Subsection 4.4.1 are published in [Hristova 2009b], while the results of Subsection 4.4.2 are proved by J. Henderson and S. Hristova ([Henderson and Hristova 2010]).

The results of Section 5.1 and Section 5.2 are adapted by G. Zhang and M. Migda [Zhang and Migda 2005] and B.G. Zhang and G. Zhang [Zhang and Zhang 2000]. The results of Section 5.3 are obtained by T. Donchev, S. Hristova and N. Markova ([Dontchev et al. 2010a]), while Section 5.4 and Section 5.5 contain the results of N. Markova and P. Simeonov.

The results of Section 6.1 are obtained by S. Hristova and are similar to the results for impulsive differential equations with "supremum" by D.D. Bainov and S.G. Hristova [Hristova and Bainov 1991]. The Section 6.2 contains results of D. D. Bainov and S. G. Hristova [Bainov and

Hristova 1995]. Section 6.3 is written by S. G. Hristova and it is similar to results for impulsive differential equations with "supremum" by Jian Ping Cai [Cai 2003]. The results in Section 6.4 and Section 6.5 are proved by S. Hristova.

The results in Section 7.1 and and Section 7.2 are adopted by D. Bainov and S. Milusheva ([Bainov and Milusheva 1983], [Milusheva and Bainov 1986b]). Section 7.3 and Section 7.4 are written by S. Hristova and the results are similar to [Bainov et al. 1996], [Bainov et al. 1994a], and [Milusheva and Bainov 1986a].

Bibliography

[Agarwal et al. 2000] Agarwal, R., Grace, S., O'Regan, S. 2000. *Oscillation Theory for Difference and Functional Differential Equations*. Kluwer: Dordrecht.

[Angelov and Bainov 1983] Angelov, V., and Bainov, D. 1983. On the functional-differential equations with "maximums". *Applicable Anal.* 16(3): 187–194.

[Angelov and Bainov 1981] Angelov, V., and Bainov, D. 1981. Existence and uniqueness of the global solutions of some integral-functional equations. *Sibirisk. Mat. J.* 22(2): 31–37 (In Russian).

[Appell and Zabrejko 1990] Appell, J., and Zabrejko, P. 1990. *Nonlinear Superposition Operators*. Volume 95 of *Cambridge Tracts in Math*. Cambridge: Cambridge University Press.

[Appleby and Wu 2008] Appleby, J.A.D., and Wu, H. 2008. Exponential growth and Gaussianlike fluctuations of solutions of stochastic differential equations with maximum functionals. *J. Phys.: Conf. Ser.* 138. 012002.

[Arolska and Bainov 1980] Arolska, M., and Bainov, D. 1980. Periodic solutions of differential equations with maxima. *Plovdiv. Univ. Nauchn. Trud.* 18(1): 143–153 (In Bulgarian).

[Atici et al. 2006] Atici, F., Cabada, A., and Ferreiro, J. 2006. First order difference equations with maxima and nonlinear functional boundary value conditions. *J. Difference Equat. Appl.* 12(6): 565–576.

[Atici et al. 2002]   Atici, F., Cabada, A., and Ferreiro, J. 2002. Existence and comparison results for first order periodic implicit difference equations with maxima. *J. Difference Equa. Appl.* 8(4): 357–369.

[Bainov et al. 1997]   Bainov, D., Petrov, V., and Proytcheva, V. 1997. Existence and asymptotic behavior of nonoscillatory solutions of second-order neutral differential equations with "maxima". *J. Comput. Appl. Math.* 83(2): 237–249.

[Bainov et al. 1996]   Bainov, D., Domshlak, Y., and Milusheva, S. 1996. Partial averaging for impulsive differential equations with supremum. *Georgian Math. J.* 3(1): 11–26.

[Bainov et al. 1995a] Bainov, D., Petrov, V., and Proicheva, V. 1995. Oscillation of neutral differential equations with "maxima". *Rev. Mat. Univ. Complut. Madrid* 8(1): 171–180.

[Bainov et al. 1995b] Bainov, D., Petrov, V., and Projcheva, V. 1995. Oscillation of neural differential equations with "maxima". *Revista Math.* 1: 171–180.

[Bainov et al. 1995c] Bainov, D., Petrov, V., and Proycheva, V. 1995. Oscillation and nonoscillation of first order neutral differential equations with maxima. *SUT J. Math.* 31(1): 17–28.

[Bainov et al. 1995d] Bainov, D., Petrov, V., and Proytcheva, V. 1995. Asymptotic behaviour of second order neutral differential equations with "maxima". *Tamkang J. Math.* 26(3): 267–275.

[Bainov et al. 1995e] Bainov, D., Petrov, V., and Proytcheva, V. 1995. Oscillatory and asymptotic behavior of second order neutral differential equations with "maxima". *Dynam. Systems Appl.* 4(1): 137–146.

[Bainov et al. 1994a]   Bainov, D., Domshlak, Y., and Milusheva, S. 1994. Method of partially-additive averaging for impulsive differential equations with supremum. *Dynam. Systems Appl.* 3 (3): 395–404.

[Bainov et al. 1994b]   Bainov, D., Milusheva, S., and Nieto, J. 1994. Partially multiplicative averaging for impulsive differential equations with supremum. *Proc. A. Razmadze Math. Inst.* 110: 7–18.

[Bainov et al. 1993]   Bainov, D., Milusheva, S., and Nieto, J. 1993. Averaging method for impulsive integro-differential equations with supremum. *Rend. Sem. Mat. Messina Ser.* II 2(16): 11–30.

[Bainov and Hristova 2010] Bainov, D., and Hristova S. 2010. Practical stability in terms of two measures for impulsive differential equations with "supremum". *Commun. Appl. Anal.* (accepted).

[Bainov and Hristova 1995] Bainov, D., and Hristova S. 1995. Monotone-iterative techniques of Lakshmikantham for a boundary value problem for systems of differential equations with maxima. *J. Math. Anal. Appl.* 190(2): 391–401.

[Bainov and Kazakova 1992] Bainov, D., and Kazakova N. 1992. A finite-difference method for solving the periodic problem for autonomous differential equations with maxima. *Math. J. Toyama Univ.* 15: 1–13.

[Bainov and Milusheva 1983] Bainov, D., and Milusheva S. 1983. Justification of the averaging method for functional-differential equations with maximums. *Hadronic J.* 6(4): 1034–1039.

[Bainov and Minchev 1999] Bainov, D., and Minchev, E. 1999. Forced oscillations of solutions of hyperbolic equations of neutral type with maxima. *Applicable Anal.* 70(3-4): 259–267.

[Bainov and Minchev 1998] Bainov, D., and Minchev, E. 1998. Forced oscillations of solutions of parabolic equations of neutral type with maxima. *Proc. of the 8th Colloq. Diff. Eq. Aug. 18-23*, 1997: 43–49. Bulgaria: Plovdiv.

[Bainov and Mishev 1991] Bainov, D., and Mishev, D. 1991. *Oscillation Theory for Neutral Differential Equations with Delay*. Adam Hilger, Bristol, Philadelphia and New York.

[Bainov and Sarafova 1981] Bainov, D., and Sarafova G. 1981. Application of the numerical-analytic method to the investigation of periodic systems of ordinary differential equations with maxima. *Rev. Roumaine Sci. Tech. Ser. Mec. Appl.* 26(3): 371–382 (In Russian).

[Bainov and Simeonov 1989] Bainov, D., and Simeonov P. 1989. *Integral inequalities and applications*. Kluwer, Boston.

[Bainov and Zahariev 1984] Bainov, D., and Zahariev A. 1984. Oscillating and asymptotic properties of a class of functional-differential equations with maxima. *Czechoslovak Math. J.* 34(2): 247–251.

[Bantsur and Trofimchuk 1998] Bantsur, N., and Trofimchuk O. 1998. On the existence of t-periodic solutions of essentially nonlinear scalar differential equations with maxima. *Nelineini Koliv.* (1): 1–5 (In Ukrainian).

[Bantsur and Trofimchuk 1998] Bantsur, N., and Trofimchuk, E. 1998. On the existence and stability of the periodic and almost periodic solutions of quasilinear systems with maxima. *Ukrainian Math. J.* 6: 747–754 (In Ukrainian).

[Bellman 1943] Bellman, R. 1943. The stability of solutions of linear differential equations. *Duke. Math. J.* **10**(4):643–647.

[Bellman and Kalaba 1965] Bellman, R., and Kalaba R. 1965. *Quasilinearization and Nonlinear Boundary Value Problems*. Elsevier: New York.

[Berenhaut et al. 2006] Berenhaut, K. S., Foley, J. D., and Stevic, S. 2006. Boundedness character of positive solutions of a max difference equation. *J. Difference Eq. Appl.* 12 (12): 1193-1199.

[Bihari 1956] Bihari, I. 1956. A generalization of a lemma of Bellman and its application to uniqueness problems of differential equations. *Acta Math. Hungarica* 7(1):81–94.

[Caballero et al. 2005] Caballero, J., Lopez, B., and Sadarangan, K. 2005. On monotonic solutions of an integral equation of Volterra type with supremum. *J. Math. Anal. Appl.* 305(1): 304–315.

[Cai 2003] Cai, J. 2003. Periodic boundary value problems for differential equations with "supremum". *Ann. Differential Equations* 19(2): 127–135.

[Castaing and Valadier 1977] Castaing, C., and M. Valadier. 1977. *Convex Analysis and Measurable Multifunctions*. Volume 8 of *Notes in Math*. Springer-Verlag: Berlin.

[Cinar et al. 2005] Cinar, C., Stevic, S., and Yalcinkaya, I. 2005. On positive solutions of a reciprocal difference equation with minimum. *J. Appl. Math. Comput.* 17 (1-2): 307-314.

[Chuanxi et al. 1990] Chuanxi, Q., Ladas, G., Zhang, B., Zhao, T. 1990. Sufficient conditions for oscillations and existence of positive solutions. *Appl. Anal.* 35: 187–194.

[Darwish 2008] Darwish, M. 2008. On monotonic solutions of a singular quadratic integral equations with supremum. *Dynamic Syst. Appl.* 17: 539–550.

[Dontchev et al. 2010a] Dontchev, T., Hristova, S., Markova, N. 2010. Asymptotic and oscillatory behavior of n-th order forced differential equations with "maxima". *PanAmer. Math. J.* 20(2): 37–51.

[Donchev et al. 2010b] Donchev, T., Kitanov, N., Kolev, D. 2010. Stability for the solutions of parabolic equations with "maxima". *PanAmer. Math. J.* 20(2): 1–19.

[El-Borai at al. 2006] El-Borai, M. M., Abbas, M. I. 2006. Fractional orders integral equations with supremum. *Int. J. of Pure and Applied Math.* 32: 365–374.

[El'sgol'ts and Norkin 1973] El'sgol'ts, L., and Norkin, S. 1973. *Introduction to the theory and application of differential equations with deviating arguments.* New York and London: *Academic Press.*

[Erbe et al. 1987] Erbe, L., Kong, Q., Zhang, B. 1987. *Oscillation Theory for Functional Differential Equations.* Marcel Dekker, Inc. New York.

[Fan 2004] Fan C. 2004. Impulsive difference equations with maxima. *J. Shanxi Educ. College* iss. 03.

[Fan et al. 2005] Fan C., Zhao A., Deng S. 2005. Oscillation of neutral difference equations with "maxima". *J. Shanxi Univ.* (Natural Science) iss. 01.

[Foster and Crimer 1980] Foster, K., and Crimer, R. 1980. Nonoscillatory solutions of higher delay equations. *J. Math. Anal. Appl.* 77: 150–164.

[Gelisken and Cinar 2009] Gelisken, A., and Cinar, C. 2009. On the global attractivity of a max-type difference equation, *Discr. Dyn. Nat. Soc.* 2009: Article ID 812674.

[Gelisken et al. 2010] Gelisken, A., Cinar, C. and Yalcinkaya, I. 2010. On a max-type difference equation. *Adv. Difference Eq.* 2010: Article ID 584890.

[Gelisken et al. 2008]   Gelisken, A., Cinar, C. and Yalcinkaya, I. 2008. On the periodicity of a difference equation with maximum. *Discr. Dyn. Nat. Soc.* 2008: Article ID 820629.

[Georgiev and Angelov 2002] Georgiev, L., and Angelov, V. 2002. On the existence and uniqueness of solutions for maximum equations. *Glasnik Matematicki* 37(2): 275–281.

[Golev et al. 2010]   Golev, A., Hristova, S., and Rahnev, A. 2010. An algorithm for approximate solving of differential equations with maxima. *Comput. Math. Appl.* 60:2771–2778.

[Gonzalez and Pinto 2007] Gonzalez, P., and Pinto, M. 2007. Convergent solutions of certain nonlinear differential equations with maxima. *Math. Comput. Modelling* 45(1-2): 1–10.

[Gonzalez and Pinto 2002] Gonzalez, P., and M. Pinto. 2002. Asymptotic equilibrium for certain type of differential equations with maximum. *Proyecciones* 21(1): 9–19.

[Gopalsamy 1992]   Gopalsamy, K. 1992. *Stability and oscillations in delay differential equations of population dynamics*. Dordrecht: Kluwer.

[Gronwall 1919]   Gronwall, T. H. 1919. Note on the derivative with respect to a parameter of the solutions of a system of differential equations. *Ann. of Math.* 20(4):292–296.

[Hadeler 1979]   Hadeler, K. 1979. *Delay equations in biology*. New York: Springer.

[Hale 1977]   Hale, J. 1977. *Theory of Functional Differential Equations*. New York: Second edn. Springer-Verlag.

[Hale and Lune 1993] Hale, J., and Lune, V. S. 1993. *Introduction to Functional Differential Equations*. New York: Springer-Verlag.

[He et al. 2003]   He, Z., Wang, P., and Ge, W. 2003. Periodic
                           boundary value problem for first order impul-
                           sive differential equations with supremum. J.
                           Indian Pure Appl. Math. 34(1): 133–144.

[Henderson and Hristova 2010a] Henderson, J., and Hristova, S. 2010.
                           Eventual practical stability and cone valued
                           Lyapunov functions for differential equations
                           with maxima. Commun. Appl. Anal.: accepted.

[Henderson and Hristova 2010b] Henderson, J., and Hristova, S. 2010.
                           Nonlinear integral inequalities involving max-
                           ima of unknown scalar functions. Math. Com-
                           put. Model. in press.

[Hristova 2010a]   Hristova, S. 2010. Integral stability in terms of
                           two measures for impulsive functional differen-
                           tial equations. Math. Comput. Model. 51: 100–
                           108.

[Hristova 2010b]   Hristova, S. 2010. Stability in terms of two mea-
                           sures for impulsive differential equations with
                           "supremum". Nonlinear Studies 17(4):299–308.

[Hristova 2010c]   Hristova, S. 2010. Stability on a cone in terms
                           of two measures for impulsive differential equa-
                           tions with "supremum". Appl. Math. Lett. 23:
                           508–511.

[Hristova 2010d]   Hristova, S. 2010. Lipschitz stability for im-
                           pulsive differential equations with "supremum".
                           Inter. Elect. J. Pure Appl. Math. 1, no.4 : 345–
                           358.

[Hristova 2009a]   Hristova, S. 2009. Integral stability in terms of
                           two measures for impulsive differential equa-
                           tions. Commun. Appl. Nonlinear Anal. 16(3):
                           37–49.

[Hristova 2009b]   Hristova, S. 2009. Practical stability and cone
                           valued Lyapunov functions for differential equa-
                           tions with "maxima". Intern. J. Pure Appl.
                           Math. 57(3): 313–324.

[Hristova 2009c] Hristova, S. 2009. Razumikhin method and cone valued Lyapunov functions for impulsive differential equations with "supremum". *Intern. J. Dyn. Sys. Diff. Eq.* 2(3-4): 223–236.

[Hristova 2009d] Hristova, S. 2009. Stability on a cone in terms of two measures for impulsive differential equations with "supremum". *AIP Confer. Proc. AMEE09* 1184: 113–120.

[Hristova 2000] Hristova, S. 2000. Boundedness of the solutions of impulsive differential equations with "supremum". *Math. Balkanica, N.S.* 14(1-2): 177–189.

[Hristova 1982] Hristova, S. 1982. Periodic solutions of second-order integro-differential equations with maxima. *Plovdiv. Univ. Nauchn. Trud.* 20(1): 135–148 (In Bulgarian).

[Hristova and Bainov 1993] Hristova, S., and Bainov D. 1993. Monotone-iterative techniques of V. Lakshmikantham for a boundary value problem for systems of impulsive differential equations with "supremum". *J. Math. Anal. Appl.* 172(2): 339–352.

[Hristova and Bainov 1991] Hristova, S., and Bainov D. 1991. Application of the monotone-iterative techniques of V. Lakshmikantham to the solution of the initial value problem for impulsive differential equations with "supremum". *J. Math. Phys. Sci.* 25(1): 69–80.

[Hristova and Gluhcheva 2010] Hristova, S., and Gluhcheva, S. 2010. Lipschitz stability in terms of two measures for differential equations with "maxima", *Intern. Electr. J. Pure Appl. Math.*, 2, (2): 1–12.

[Hristova and Markova 2010] Hristova, S., and Markova N., 2010. Eventual stability for impulsive differential equations with "supremum", *Intern. Electr. J. Pure Appl. Math.*, 2, (2): 21–46.

[Hristova and Roberts 2001] Hristova, S., and Roberts, L. 2001. Monotone-iterative method of V. Lakshmikantham for a periodic boundary value problem for a class of differential equations with "supremum". *Nonlinear Anal., ser. A: Theory, Methods* 44(5): 601–612.

[Hristova and Roberts 2000] Hristova, S., and Roberts L. 2000. Boundedness of the solutions of differential equations with "maxima". *Int. J. Appl. Math.* 4(2): 231–240.

[Hristova and Stefanova 2010a] Hristova, S., and Stefanova K. 2010. Linear integral inequalities involving maxima of the unknown scalar functions. *Funkcialaj Ekvacioj* 53(3): accepted.

[Hristova and Stefanova 2010b] Hristova, S., and Stefanova K. 2010. Two dimensional integral inequalities and applications to partial differential equations with maxima. *IeJPAM* 1(1): 11–28.

[Jankowski 2002]   Jankowski, T. 2002. Differential-algebraic systems with maxima. *J. Math. Anal. Appl.* 274(1): 336–348.

[Jankowski 1997]   Jankowski, T. 1997. Systems of differential equations with maxima. *Dopov. Nats. Akad. Nauk Ukr. Mat. Prirodozn. Tekh. Nauki* 8: 57–60.

[Ivanov et al. 2002]   Ivanov, A., Liz, E., and Trofimchuk, S. 2002. Halanay inequality, Yorke 3/2 stability criterion, and differential equations with maxima. *Tohoku Math. J.* 54(2): 277–295.

[Iricanin and Elsayed 2010] Iricanin, B.D., and Elsayed, E. M. 2010. On the max-type difference equation $x_{n+1} = \max\{A/x_n, x_{n-3}\}$. *Discr. Dyn. Nat. Soc.* 2010, Art. ID 675413.

[Kazakova 1990a]   Kazakova, N. 1990. Application of the finite difference method for finding periodic solutions of systems of differential equations with maxima. *Plovdiv. Univ. Paisii Khilendarski Nauchn. Trud. Mat.* 28(3): 115–122 (In Russian).

[Kazakova 1990b]   Kazakova, N. 1990. Application of the finite difference method for finding periodic solutions of autonomous systems of differential equations with maxima. *Plovdiv. Univ. Paisii Khilendarski Nauchn. Trud. Mat.* 28(3): 105–114 (In Russian).

[Kichmarenko 2009]   Kichmarenko, O. 2009. Averaging of differential equations with Hukuhara derivative with "maxima". *Intern. J. Pure Appl. Math.* 57(3): 447–457.

[Kichmarenko 2006]   Kichmarenko O. 2006. The control problem with vectorial criterion solution on the trajectories with delay. Inform. Syst. Technology, International Seminar, October 2006, *Scientific works collection*, Odessa: 75–82. (In Russian).

[Kiguradze 1964]   Kiguradze, I. 1964. On the oscillation of solutions of the equation $\frac{d^m u}{dt^m} + a(t)|u|^m \mathrm{sign} u = 0$. *Math. Sb.* 65(2): 172–187 (In Russian).

[Kim 2005]   Kim, Y. 2005. On some new integral inequalities for functions in one and two variables. *Acta Mathematica Sinica, English Series* 21(2): 423–434.

[Kolev and Markova 2010]   Kolev, D., and Markova N. 2010. Existence criteria for second order systems with "maxima". *Intern. Electr. J. Pure Appl. Math.* 1 (3): 217–228.

[Kolev et al. 2010a]   Kolev, D., Markova N., and Nenov, S. 2010. Oscillation criteria for n-th order nonlinear differential equations with "maxima". *Int. J. Pure Appl. Math.* 64 (2): 171-182.

[Kolev et al. 2010b]   Kolev, D., Markova N., and Nenov, S. 2010.
                            Non-oscillation of a second order sub-linear dif-
                            ferential equation with "maxima", *Int. Electr.
                            J. Pure Appl. Math.* 2 (2): 131–145.

[Kolmanovski and Nosov 1986] Kolmanovski, V., and Nosov V. 1986.
                            Stability of functional differential equations.
                            London and Orlando: Academic Press.

[Krasnoselskii and Zabreiiko 1975] Krasnoselskii, M., and Zabreiiko P.
                            1975. *Geometrical Methods of Nonlinear Anal-
                            ysis.* English transl. Berlin: Springer-Verlag.
                            1984. Moscow: edn. Nauka.

[Ladde et al. 1987]   Ladde, G., Lakshmikantham, V., Zhang, B.
                            1987. *Oscillation Theory of Differential Equa-
                            tions with Deviating Arguments.* Volume 110 of
                            Pure and Applied Mathematics. New York: Mar-
                            cel Dekker.

[Ladde et al. 1985]   Ladde, G., Lakshmikantham, V., Vatsala, A.
                            1985. *Monotone Iterative Techniques for Non-
                            linear Differential Equations.* Belmonth: Pit-
                            man.

[Lakshmikantham et al. 1990] Lakshmikantham, V., Leela, S., and
                            Martynyuk, A. 1990. *Practical Stability of Non-
                            linear Systems.* Singapore: World Scientific.

[Lakshmikantham and Leela 1969] Lakshmikantham, V., and S. Leela.
                            1969. *Differential and Integral Inequalities.* New
                            York: Volume 1. *Academic Press.*

[Lakshmikantham and Liu 1993] Lakshmikantham, V., and X. Liu.
                            1993. *Stability analysis in terms of two mea-
                            sures.* Singapore: World Scientific.

[Lakshmikantham and Vatsala 1998] Lakshmikantham, V., and A.
                            Vatsala. 1998. *Generalized Quasilinearization
                            for Nonlinear Problems.* Boston, London, Dor-
                            drecht: Kluwer Academic Publ.

[Li et al. 2008]   Li, N., Zhong X., Zhang W., Yu P., and Zhang
                          S. 2008. Oscillation of second order neutral dif-
                          ference equations with maximum. *J. Shandong
                          Univ. of Technology* (Natural Science), iss. 04.

[Li and Zhou 2007]   Li, X., and Zhou, H. 2007. Oscillation and
                          nonoscillation of higher order neutral difference
                          equations with "maxima". *Int. Journal of Math.
                          Analysis*, 1, no. 16, 791–804.

[Liu et al. 2006]   Liu, Y., Yang J., and Xia D. 2006. Oscilla-
                          tion of second order neutral difference equa-
                          tions with "maxima". *J. Jianghan Univ.* (Nat-
                          ural Science), Iss. 04.

[Liz et al. 2003]   Liz, E., Ivanov, A., and Ferreiro, J. 2003. Dis-
                          crete Halanay-type inequalities and application.
                          Nonlinear Anal. 55(6): 669–678.

[Luo 2001]   Luo, J. 2001. Oscillation of neutral difference
                          equations with "maxima." *Soochow J. Math.*
                          27(3): 267–273.

[Luo and Bainov 2001]   Luo, J., and Bainov D. 2001. Oscillatory and
                          asymptotic behavior of second-order neutral
                          difference equations with maxima. *J. Comput.
                          Appl. Math.* 131(1-2): 333–341.

[Lyapunov 1956]   Lyapunov, A. 1956. *Collected Works.* Moscow:
                          Nauka (In Russian).

[Magomedov 1993]   Magomedov, A. 1993. Periodic and almost pe-
                          riodic solutions of differential equations with
                          maxima. *Mat. Fiz. Nelinein. Mekh.* 18(52): 3–6
                          (In Russian).

[Magomedov 1992a]   Magomedov, A. 1992. Existence and uniqueness
                          theorems for solutions of differential equations
                          with maxima that contain a functional parame-
                          ter. *Arch. Math.* 28(3-4): 139–154 (In Russian).

[Magomedov 1992b]   Magomedov, A. 1992. Periodic solutions of dif-
                          ferential equations with maxima and with a

small parameter. *Mat. Fiz. Nelinein. Mekh.* 17(51): 3–8 (In Russian).

[Magomedov 1991]   Magomedov, A. 1991. Stability of solutions of linear periodic systems of differential equations with maxima. *Mat. Fiz. Nelinein. Mekh.* 16(50): 10–12 (In Russian).

[Magomedov 1990]   Magomedov, A. 1990. Some questions concerning periodic solutions for differential equations with maxima and with a small parameter. *Math. Slovaca* 40(3): 321–324 (In Russian).

[Magomedov 1988]   Magomedov, A. 1988. Application of the averaging method for differential equations with maxima. *Akad. Nauk Azerbaidzhan. SSR Dokl.* 44(11): 3–8 (In Russian).

[Magomedov 1983a]   Magomedov, A. 1983. Investigation of solutions of differential equations with maxima for problems with control. *Akad. Nauk Azerbaidzhan. SSR Dokl.* 39(3): 12–18 (In Russian).

[Magomedov 1983b]   Magomedov, A. 1983. Some aspects of stability of solutions of differential equations with maxima. *Akad. Nauk Azerbaidzhan. SSR Dokl.* 39(10): 3–8 (In Russian).

[Magomedov 1981]   Magomedov, A. 1981. Stability of periodic solutions of differential equations with maxima. *Mathematical Cybernetics and Applied Mathematics* (5): 62–66 (In Russian).

[Magomedov 1980a]   Magomedov, A. 1980. Investigation of the solutions of linear differential equations with maxima. *Akad. Nauk Azerbaidzhan. SSR Dokl.* 36(1): 11–14 (In Russian).

[Magomedov 1980b]   Magomedov, A. 1980. On an investigation of linear differential equations with maxima. *Akad. Nauk Azerbaidzhan. SSR Dokl.* 36(9): 12–14 (In Russian).

[Magomedov 1980c]   Magomedov, A. 1980. On the periodic solutions
of a system of differential equations with max-
ima with a small parameter. *Izv. Akad. Nauk
Azerbaidzhan. SSR, ser. Fiz.-Tekhn. Mat. Nauk*
1(3): 22–26 (In Russian).

[Magomedov 1979]   Magomedov, A. 1979. A theorem on the existence and uniqueness of solutions of linear differential equations with "maxima". *Izv. Akad.
Nauk Azerbaidzhan. SSR, ser. Fiz.-Tekhn. Mat.
Nauk* (5): 116–118 (In Russian).

[Magomedov 1977]   Magomedov, A. 1977. Some questions on dif-
ferential equations with "maxima". *Izv. Akad.
Nauk Azerbaidzan. SSR, ser. Fiz.-Tehn. Mat.
Nauk* (1): 104–108 (In Russian).

[Magomedov and Nabiev 1984a] Magomedov, A., and Nabiev, G.
1984. Theorems on nonlocal solvability of the
initial value problem for systems of differential equations with maxima. *Akad. Nauk Azerbaidzhan. SSR Dokl.* 40(5): 14–20 (In Russian).

[Magomedov and Nabiev 1984b] Magomedov, A., and Nabiev, G.
1984. The averaging principle for differential
equations with maxima. *Akad. Nauk Azerbaidzhan. SSR Dokl.* 40(2): 17–20 (In Russian).

[Magomedov and Nabiev 1986] Magomedov, A., and Nabiev, G. 1986.
Some questions of the stability of solutions
of linear differential equations with maxima.
Akad. Nauk Azerbaidzhan. SSR Dokl. 42(2): 3–6
(In Russian).

[Magomedov and Ryabov 1991] Magomedov, A., and Ryabov, Y.
1991. Exponential stability of the solution of
a linear scalar differential equation with maxima. *Mat. Fiz. Nelinein. Mekh.* 15(49): 1–7 (In
Russian).

[Magomedov and Ryabov 1980] Magomedov, A., and Ryabov, J. 1980.
Periodic solutions of linear systems of differential equations with "maxima". *Akad. Nauk*

Azerbaidzhan. SSR Dokl. 36(2): 3–8 (In Russian).

[Magomedov and Ryabov 1975] Magomedov, A., and Ryabov, J. 1975. Periodic solutions of nonlinear differential equations with "maxima". *Izv. Akad. Nauk Azerbaidjan. SSR, ser. Fiz.-Tehn. Mat. Nauk* (2): 76–83 (In Russian).

[Markova and Nenov 2010] Markova, N., and Nenov, S. 2010. Oscillation properties of solutions of n-th order differential equations with "maxima". *Intern. Elect. J. Pure Appl. Math.* 1(2): 113–130.

[Markova and Simeonov 2010] Markova, N., and Simeonov, P.S. 2010. Oscillation of higher order differential inequalities with "maxima", *Inter. Electr. J. Pure Appl. Math.* 2(1): 13–20.

[Migda and Zhang 2006] Migda, M., and Zhang, G. 2006. On unstable neutral difference equations with "maxima". *Math. Slovaca* 56(4): 451–463.

[Milusheva and Bainov 1986a] Milusheva, S., and Bainov, D. 1986. Justification of partial-multiplicative averaging for functional-differential equations with maxima. *Godishnik Vissh. Uchebn. Zaved., Prilozhna Mat.* 22(2): 43–50 (In Russian).

[Milusheva and Bainov 1986b] Milusheva, S., and Bainov, D. 1986. Justification of the averaging method for multipoint boundary value problems for a class of functional-differential equations with "maximums". *Collect. Math.* 37(3): 297–304.

[Milusheva and Bainov 1991] Milusheva, S., and Bainov, D. 1991. Averaging method for neutral type impulsive differential equations with supremums. *Ann. Fac. Sci. Toulouse Math.* 5(12): 391–403.

[Mishev 1989]   Mishev, D. 1989. Oscillation of the solutions of parabolic differential equations of neutral type with "maxima". *Godishnik Vissh. Uchebn. Zaved., Prilozhna Mat.* 25(2): 19–28.

[Mishev 1990]   Mishev, D. 1990. Oscillation of the solutions of hyperbolic differential equations of neutral type with "maxima". *Godishnik Vissh. Uchebn. Zaved. Prilozhna Mat.* 26(2): 9–18.

[Mishev 1986]   Mishev, D. 1986. Oscillatory properties of the solutions of hyperbolic differential equations with "maximum". *Hiroshima Math. J.* 16(1): 77–83.

[Mishev and Musa 2007] Mishev, D., and Musa, S. 2007. Distribution of the zeros of the solutions of hyperbolic differential equations with maxima. *Rocky Mountain J. Math.* 37(4): 1271–1281.

[Mishev et al. 2002]   Mishev, D., Patula, W.T. and Voulov, H.D. 2002. On a reciprocal difference equation with maximum. *Comput. Math. Appl.* 43: 1021–1026.

[Movchan 1960]   Movchan, A. 1960. Stability of processes with respect to two metrics. *Prikl. Mat. Mekh.* 24: 988–1001 (In Russian).

[Muntyan 1987]   Muntyan, V. 1987. On the existence and uniqueness of periodic solutions of integro-differential equations with maxima. *Mat. Issled.* 92 (In Russian).

[Muntyan and Shpakovich 1987] Muntyan, V., and Shpakovich, V. 1987. On the problem of the continuous dependence of the solution of differential equations with "maxima" on the parameter. *Akad. Nauk Ukrain. SSR*: (In Russian).

[Myshkis 1977]   Myshkis, A. 1977. On some problems of the theory of differential equations with deviating argument. *Russian Math. Surveys* 32(2): 181–210 (In Russian).

[Myshkis 1995]   Myshkis, A. 1995. Razumikhin's method in the qualitative theory of processes with delay. *J. Appl. Math. Stoch. Anal.* 8(3): 233–247.

[Nabiev 1985]   Nabiev, G. 1985. Some questions in the theory of stability of solutions of differential equations with maxima. *III. Akad. Nauk Azerbaidzhan. SSR Dokl.* 41(4): 11–14 (In Russian).

[Nabiev 1984a]   Nabiev, G. 1984. Some questions concerning the theory of stability of solutions of differential equations with maxima. *I. Akad. Nauk Azerbaidzhan. SSR Dokl.* 40(8): 14–16 (In Russian).

[Nabiev 1984b]   Nabiev, G. 1984. Some questions in the theory of stability of solutions of differential equations with maxima. *II. Akad. Nauk Azerbaidzhan. SSR Dokl.* 40(9): 16–21 (In Russian).

[Nirenberg 1974]   Nirenberg, L. 1974. *Topics in Nonlinear Functional Analysis.* Courant Institute Lecture Notes. New York: New York University.

[Otrocol and Ioan 2008a] Otrocol, D., and Ioan, R. A. 2008. Functional-differential equations with "maxima" via weakly picard operators theory. *Bull. Math. Soc. Sci. Math. Roumanie* 51(3): 253–261.

[Otrocol and Ioan 2008b] Otrocol, D., and Ioan, R. A. 2008. Functional-differential equations with maxima of mixed type. *Fixed Point Theory* 9(1): 207–220.

[Oyelami and Ale 2010] Oyelami, B., and Ale, S. 2010. On existence of solution, oscillatory and non-oscillatory properties of delay impulsive differential equations containing "maximum". *Acta Appl. Math.* 109(3): 683–701.

[Pachpatte 2002]   Pachpatte, B. 2002. Explicit bounds on certain integral inequalities. *J. Math. Anal. Appl.* 267: 48–61.

[Petrov 1998]   Petrov, V. 1998. Nonoscillatory solutions of neutral differential equations with "maxima". *Commun. Appl. Anal.* 2(1): 129–142.

[Petrov and Proytcheva 1997] Petrov, V., and Proytcheva, V. 1997. Existence of positive solutions of neutral differential equations with "maxima". *J. Tech. Univ. Plovdiv Fundam. Sci. Appl., ser. A, Pure Appl. Math.* 5: 19–31.

[Petukhov 1964]   Petukhov, V. 1964. Questions about qualitative investigations of differential equations with "maxima". *Izw. WUZ, Math.* 40(3): 116–119 (In Russian).

[Pinto and Trofimchuk 2000] Pinto, M., and Trofimchuk, S. 2000. Stability and existence of multiple periodic solutions for a quasilinear differential equation with maxima. *Proc. Roy. Soc. Edinburgh, sect. A* 130(5): 1103–1118.

[Plotnikov and Kichmarenko 2009] Plotnikov, V., and Kichmarenko, O. 2009. A note on the averaging for differential equations with maxima. *Iranian J. Optimization* 1: 132–140.

[Plotnikov and Kichmarenko 2006] Plotnikov, V., and Kichmarenko, O. 2006. The averaging schemes of the controlled movement equations with maximum. *Proc. Odessa Polytech. Uni.* 2(26): 106–114 (In Russian).

[Plotnikov and Kichmarenko 2002] Plotnikov, V., and Kichmarenko, O. 2002. Averaging of differential equations with maximum. *Naukovy Visnyk Chernivetskogo Universitetu: Zbirnyk Naukovyh Prats, Matematyka* 150: 78–82 (In Ukrainian).

[Popov 1966]   Popov, E. 1966. *Automatic regulation and control.* Moscow: (In Russian).

[Qi 2004]   Qi, S. 2004. Infinite boundary value problems for nonlinear impulsive differential equations with "supremum". *Acta Anal. Funct. Appl.* 6(3): 200–208.

[Qi and Chen 2008]   Qi, S., and Chen,Y. 2008. Periodic boundary value problems for first order impulsive differential equations with "supremum". *J. Zhengzhou Univ. Nat. Sci. Ed.* 40(2): 14–19.

[Razumikhin 1988]   Razumikhin, B. 1988. Stability of Hereditary Systems. Moscow: *Nauka* (In Russian).

[Razumikhin 1956]   Razumikhin, B. 1956. On stability of systems with delay. *Prykl. Mat. i Mekh., Appl. Math and Mech.* 20(4): 500–512 (In Russian).

[Rjabov and Magomedov 1978] Rjabov, J. A., and Magomedov, A. 1978. Periodic solutions of linear differential equations with maxima. *Mat. Fiz. Vyp.* 23: 3–9 (1978) In Russian.

[Ronto 1999]   Ronto, A. 1999. On periodic solutions of systems with "maxima". *Dopov. Nats. Akad. Nauk Ukr. Mat. Prirodozn. Tekh. Nauki*, no. 12: 27–31 (In Russian).

[Rus 2001]   Rus, I. 2001. *Generalized Contractions.* Cluj University Press, Cluj, Romania.

[Salle and Lefschetz 1961] Salle, L. J., and Lefschetz, S. 1961. *Stability by Lyapunov's Direct Method and Applications.* New York: Academic Press.

[Samoilenko et al. 1998] Samoilenko, A., Trofimchuk, O., and Bantsur, N. 1998. Periodic and almost periodic solutions of systems of differential equations with maxima. *Dopov. Nats. Akad. Nauk Ukr. Mat. Prirodozn. Tekh. Nauki* (1): 53–57 (In Ukrainian).

[Sarafova 1984]   Sarafova, G. 1984. Periodic solutions of nonlinear systems of differential equations with maxima. *Plovdiv. Univ. Nauchn. Trud.* 22(1): 155–167 (In Russian).

[Sarafova and Bainov 1981] Sarafova, G., and Bainov, D. 1981. Application of A. M. Samoilenko's numerical-analytic method to the investigation of periodic linear differential equations with maxima. *Rev. Roumaine Sci. Tech., ser. Mec. Appl.* 26(4): 595–603 (In Russian).

[Shabadikov and Yuldashev 1989] Shabadikov, K., and Yuldashev, T. 1989. Solvability of a boundary value problem for linear differential equations with maxima. *Studies in integro-differential equations* 22: 115–117 (In Russian).

[Shi and Wang 2010] Shi, P., and Wang, W. 2010. Infinite boundary value problems for second-order nonlinear impulsive differential equations with supremum. *Discr. Dyn. Nat. Soc.*: Article ID 393757.

[Shpakovich and Muntyan 1987] Shpakovich, V., and Muntyan, V. 1987. The averaging method for differential equations with maxima. *Ukrain. Mat. Zh.* 39(5): 662–665 (In Russian).

[Shpakovich and Muntyan 1986] Shpakovich, V., and Muntyan, V. 1986. Periodic solutions of integro-differential equations with "maxima". *Some problems in the theory of asymptotic methods in nonlinear mechanics xiii*, Akad. Nauk Ukrain. SSR, Inst. Mat.: Kiev 186–190 (In Russian).

[Simeonov and Bainov 1985] Simeonov, P., and Bainov, D. 1985. Application of the method of two-sided approximations to the investigation of periodic systems of ordinary differential equations with maxima. *Arch. Math.* 21(2): 65–75 (In Russian).

[Sobeih and Aly 1991] Sobeih, M., and Aly, E. 1991. A study for a unique solution for a vector differential equation with maximum. *An. Stiint. Univ. Al. I. Cuza Iasi Sect. I a Mat.* 37(1): 27–36.

[Soliman and Abdalla 2008] Soliman, A., and Abdalla, M. 2008. Integral stability criteria of nonlinear differential systems. *Math. Comput. Model.* 48: 258–267.

[Stepanov 1997]   Stepanov, E. 1997. On solvability of some boundary value problems for differential equations with "maxima". *Topol. Methods Nonlinear Anal.* 8(2): 315–326.

[Stevic 2010]   Stevic, S. 2010. Global stability of a max-type difference equation. *Appl. Math. Comput.* 216 (1): 354–356.

[Stevic 2009]   Stevic, S. 2009. Global stability of a difference equation with maximum. *Appl. Math. Comput.* 210: 525-529.

[Sun 2008]   Sun, F. 2008. On the asymptotic behavior of a difference equation with maximum. *Discr. Dyn. Nat. Soc.* 2008: Article ID 243291.

[Teryokhin and Kiryushkin 2010] Teryokhin, M., and Kiryushkin, V. 2010. Nonzero solutions to a two-point boundary value periodic problem for differential equations with maxima. *Izv. Vyssh. Uchebn. Zaved. Mat.* 6: 52–63.

[Trench 1975]   Trench, W. 1975. Oscillation properties of perturbed disconjugate equations. *Bull. Amer. Math. Soc.* 52: 147–155.

[Vavilov 1993]   Vavilov, S. 1993. On nontrivial solutions of certain classes of operator equations. *Dokl. Akad. Nauk* 331: 7–10 (In Russian).

[Voulov 2008]   Voulov, H. 2008. On the reciprocal difference equation with maximum and periodic coefficients. *Int. J. Difference Eq.* 3 (1) 153–166.

[Voulov 2003]   Voulov, H. 2003. Periodic solutions to a difference equation with maximum. *Proc. Amer. Math. Soc.* 131: 2155–2160.

[Voulov 1995]   Voulov, H. 1995. Uniform asymptotic stability
                       for a scalar autonomous differential equation
                       with "maxima". *Math. Balkanica* 9: 299–307.

[Voulov 1992a]   Voulov, H. 1992. Uniform and nonuniform sta-
                       bility for ordinary differential equations with
                       "maxima". Int. Conference Differential Equa-
                       tions, Plovdiv (1991). *World Sci. Publ.* NJ:
                       River Edge : 259–263.

[Voulov 1992b]   Voulov, H. 1992. Asymptotic stability of the
                       zero solution of a scalar differential equation
                       with "maxima". Int. Conference Differential
                       Equations, Plovdiv (1991). *World Sci. Publ.* NJ:
                       River Edge : 255–257.

[Voulov 1991]   Voulov, H. 1991. Uniform asymptotic stability
                       of a scalar differential equation with "maxi-
                       mum". *C.R. Acad. Bulgare Sci.* 44(6): 5–7.

[Voulov and Bainov 1995] Voulov, H., and Bainov, D. 1995. Existence
                       and continuous dependence of the solutions
                       of ordinary differential equations with "maxi-
                       mum". *Math. Balkanica, N.S.* 9(2-3): 155–169.

[Voulov and Bainov 1992] Voulov, H., and Bainov, D. 1992. Nonuni-
                       form stability for a nonautonomous differential
                       equation with "maxima". *Rend. Sem. Fac. Sci.
                       Univ. Cagliari* 62(2): 101–113.

[Voulov and Bainov 1991] Voulov, H., and Bainov, D. 1991. On the
                       asymptotic stability of differential equations
                       with "maxima". *Rend. Circ. Mat. Palermo*
                       40(3): 385–420.

[Vrkoc 1959]   Vrkoc, I. 1959. Integral stability.*Czechoslovak
                       Math. J.* 09(1): 71–129 (In Russian).

[Walter 1970]   Walter, W. 1970. *Differential and Integral In-
                       equalities.* New York, Berlin: Springer-Verlag.

[Wu 2003]   Wu D. 2003. Oscillation and nonoscillation of
                       neutral difference equations with "maxima"
                       terms. *J. Ocean Univ. of Qingdao*, Iss. 06.

[Xu and Liz 1996]   Xu, H., and Liz, E. 1996. Boundary value prob-
                           lems for differential equations with maxima.
                           Nonlinear Stud. 3(2): 231–241.

[Yalcinkaya et al. 2007]   Yalcinkaya, I., Iricanin, B. D., and Cinar, C.
                           2007. On a max-type difference equation. *Disc.
                           Dynam, Nat. Soc.* 2007 (1): Article ID 47264.

[Yang et al. 2006]   Yang, X., Liao, X, and Li, C. 2006. On a dif-
                           ference equation wtih maximum, *Appl. Math.
                           Comput.*, 181(1): 1–5.

[Yuldashev 1995]   Yuldashev, T. 1995. An asymptotic formula
                           with respect to small delay for the solution of
                           nonlinear differential equations with maxima.
                           Voprosy Vychisl. i Prikl. Mat. 99: 132–135 (In
                           Russian).

[Yuldashev 1992]   Yuldashev, T. 1992. Periodic solutions of non-
                           linear systems of integro-differential equations
                           with maxima. *Voprosy Vychisl. i Prikl. Mat.* 93:
                           119–129 (In Russian).

[Yuldashev and Kuldashev 1994] Yuldashev, T., and Kuldashev, N.
                           1994. On the stability of systems of differen-
                           tial equations with maxima under constantly
                           acting perturbations. *Voprosy Vychisl. i Prikl.
                           Mat.* 97: 190–197 (In Russian).

[Zhao and Yan 2004]   Zhao, A., and Yan, J. 2004. Oscillation and
                           asymptotic behavior of neutral difference equa-
                           tions with "maxima". *Dynam. Cont. Descr. Im-
                           pul. Syst., ser. Math. Anal.* (2): 300–305.

[Zhang et al. 1995]   Zhang, B., Yu, J., Wang, Z. 1995. Oscillations
                           of higher order neutral differential equations.
                           Rocky Mountain J. Math. 25: 557–568.

[Zhang and Cheng 1999]   Zhang, G., and Cheng, S. 1999. Asymptotic
                           stability of nonoscillatory solutions of nonlin-
                           ear neutral differential equations involving the
                           maximum function. *Intern. J. Appl. Math.* 7:
                           771–779.

[Zhang and Liu 2007] Zhang X., and Liu Y. 2007. Oscillation of neutral difference equations with "maxima", *J. Hainan Normall Univ.* (Natural Science), Iss. 04.

[Zhang and Migda 2005] Zhang, G., and Migda, M. 2005. Unstable neutral differential equations involving the maximum function. *Glasnik Matematicki* 40(2): 249–259.

[Zhang and Petrov 2000] Zhang, B., and Petrov, V. 2000. Existence of oscillatory solutions of neutral differential equations with "maxima". *Ann. Differential Equations* 16(2): 177–183.

[Zhang and Zhang 2000] Zhang, B., and Zhang, G. 2000. Nonoscillations of second order neutral differential equations with "maxima". *Commun. Appl. Anal.* 4(1): 31–38.

[Zhang and Zhang 1999] Zhang, B., and Zhang, G. 1999. Qualitative properties of functional-differential equations with "maxima". *Rocky Mountain J. Math.* 29(1): 357–367.

Index